Field Guide to Grasshoppers, Katydids, and Crickets of the United States

FIELD GUIDE TO

Grasshoppers, Katydids, *and* Crickets *of the* *United States*

JOHN L. CAPINERA,

RALPH D. SCOTT, *and*

THOMAS J. WALKER

Comstock Publishing Associates

A DIVISION OF *Cornell University Press*

ITHACA AND LONDON

First published 2004 by Cornell University Press
First printing, Cornell Paperbacks, 2004

Printed in the United States of America
Color plates printed in China

Library of Congress Cataloging-in-Publication Data

Capinera, John L.
Field guide to grasshoppers, crickets, and katydids of the United States / John L. Capinera, Ralph D. Scott, and Thomas J. Walker.
p. cm.
Includes bibliographical references and index.
ISBN 0-8014-4260-5 (cloth : alk. paper)—ISBN 0-8014-8948-2 (pbk. : alk. paper)
1. Grasshoppers—United States—Identification. 2. Crickets—United States—Identification. 3. Katydids—United States—Identification. I. Scott, Ralph D., 1935– II. Walker, Thomas J., 1931– III. Title.
QL508.A2C25 2004
595.7'26—dc22

2004010727

Cornell University Press strives to use environmentally responsible suppliers and materials to the fullest extent possible in the publishing of its books. Such materials include vegetable-based, low-VOC inks and acid-free papers that are recycled, totally chlorine-free, or partly composed of nonwood fibers. For further information, visit our website at www.cornellpress.cornell.edu.

Cloth printing 10 9 8 7 6 5 4 3 2 1

Paperback printing 10 9 8 7 6 5 4 3 2 1

Contents

Preface

Field Guide to Grasshoppers, Katydids, and Crickets of the United States is a guide to one of the most obvious (abundant, large, colorful, noisy) and important (ecologically and economically significant) insect groups in North America, the order Orthoptera. In the introduction, we describe the biology, behavior, and ecological significance of these insects, give information on how to collect and preserve them, and describe how to use the songs of crickets and katydids in their identification. Pictorial keys expedite identification by distinguishing among the various groups and narrowing the options. Final identification is enhanced by the large number of full-color illustrations. We also provide species profiles: brief, easy-to-understand sections on distribution, identification, ecology, and similar species. Distribution maps are provided for all species considered in the book, about one-third of all Orthoptera species in North America north of Mexico. Black-and-white drawings provide additional details to distinguish among the more-difficult-to-identify species.

This is the first treatment of North American grasshoppers, katydids, and crickets to use color to portray the insects. Because many are cryptically colored (bright colors evident only in flight, or colors that blend in with vegetation) or cryptic in behavior (nocturnal), it will be the first time many amateur naturalists and students have the opportunity to see the amazing and colorful world of Orthoptera. Few of these species have previously been pictured in natural history publications.

This book is designed for amateur naturalists who wish to know the local fauna, for students who seek to identify insects as part of entomology and natural history field courses, and for professional biologists who need to identify invertebrates. Technical language is minimized to enhance readability by nonprofessionals, and diagrams and a glossary are included to ensure complete understanding of terms.

Technical support in map preparation and graphics was kindly provided by Seth Ambler, Carole Girimont, Pat Hope, Jane Medley, and Kathy Milne of the University of Florida Entomology-Nematology Department. Color illustrations were made by Ralph Scott. Kevin O'Neill of Montana State University provided a useful review of the document and made many valuable suggestions. All photographs other than the one by T. G. Forrest (Fig. 10) were taken by or are the property of the authors.

Field Guide to Grasshoppers, Katydids, and Crickets of the United States

How to Use This Field Guide

Usually it is easy to sort any member of the order Orthoptera (grasshoppers and their relatives) into one of the suborders (Caelifera, the grasshoppers; Ensifera, everything else) by the length of the antennae. To a lesser degree, the families and subfamilies are also easily recognizable, and we provide pictorial keys to help you learn their diagnostic features (pp. 43–51). Pictures are the principal aid for identification in any field guide, and you should compare your specimen to the pictures. Once you have an idea of its identity, browse the relevant species accounts and examine the distribution maps. Collectively, the pictures, descriptions, and maps will allow you to confirm or refute your tentative identification. If you are uncertain about your identification, backtrack upward in the identification hierarchy by examining other species in the same genus, other genera in the same family, and so forth, until you are comfortable with the identification. For many species of katydids and crickets, the song is an important way to confirm an identification.

In some cases, you may find it useful to consult other sources of information, such as the ones listed in the section on additional reading. You may find other pictures in recent publications and especially on the Internet. On the website *Singing Insects of North America* (http://buzz.ifas.ufl.edu/), you will find recordings of the songs of katydids and crickets that you can compare with what you heard in the field.

Species Included

Most of the common species found in the United States and Canada are included in this field guide. A few of the rarer species are included as representatives of genera that otherwise would be absent, or to illustrate the limited geographic distribution of some species. More than one-third of the grasshopper, katydid, and cricket species known to occur in Canada and the conterminous United States are treated in some manner.

Similar Species

Features that can help separate similar-appearing species are included when appropriate. Similar species are usually members of the same genus. Some genera (e.g., *Melanoplus* and *Trimerotropis*) contain many species that are similar in appearance; in such cases we make no attempt to distinguish among all similar species.

Body Length

One of the characters included in the species accounts is body length. Information on body size is useful for narrowing down the identifications. Body length, however, is not defined in the same manner in different groups of orthopterans, or by different authors, so you need to know how the measurements are made.

In grasshoppers, length usually is measured from the front of the head to the tip of the wings in long-winged species. In species with wings that do not extend beyond the tip of the abdomen, the length measurement usually refers to the front of the head to the tip of the hind femora (this usually corresponds with the tip of the abdomen in fresh specimens). This approach is used in the species accounts in this book. Some authors, however, have measured length from the head to the tip of the abdomen in dry specimens, so there are discrepancies in the literature if you fail to take into consideration the method of measurement. Therefore, length measurements in grasshoppers should be taken as an indicator of size, but you should not be surprised to find specimens slightly outside the range given.

In katydids and crickets, body length is more complicated. If wings do not extend beyond the tip of the abdomen, length is the maximum measurement from the head to the tip of the abdomen, excluding the ovipositor. If wings do extend beyond the tip of the abdomen, the measurement is made from the head to the tip of the forewings (positioned at rest), except in false katydids (Phaneropterinae), where the measurement includes the hind wings because their tips extend the leaf-like silhouette. The purpose of these conventions is to have body measurements that indicate overall size and are not subject to variation caused by distortion of the abdomen.

Life History

Most orthopterans have one generation annually, and unless noted otherwise, you should assume a single generation per year. Also, the egg is normally the overwintering stage; if this is not the case, it is mentioned in the ecology section.

Distribution Maps

Distribution data can be an excellent aid to identification; maps are included to show the general distribution of each species. The extent of the distribution of some species is poorly known, however, and not all areas within the geographic ranges shown on the maps will be occupied. Grasshoppers, katydids, and crickets have preferred habitats and are often found in association with a particular plant community.

Names

We provide both scientific (binomial) and common (vernacular) names. We have not included the author (describer) of the species, which is commonly part of the scientific name, because although it has significance to scientists, it is of little value in a field guide. Common names are variable (not standardized), and so are less preferred when accurate identification or reporting is required. They are easier to pronounce and may be easier to remember, however. Common names typically reflect the appearance, behavior, or distribution of insects, or occasionally the person who first described the insect. Whenever it is logical, we have followed previous authors and adopted their common names. In a few cases, we created new common names.

Pronunciation of Scientific Names

A section on pronunciation of scientific names (genus names and higher) is included as a means of encouraging you to use scientific rather than common names. Scientists use scientific names because their use is rigidly coded; each is unique. In comparison, the creation and use of common names is not regulated, so they may vary regionally or among authors, which can be confusing.

Introduction to Grasshoppers, Katydids, and Crickets

Relationships

Insects are divided into about 30 major groups called orders. Beetles make up one order, butterflies and moths another, and dragonflies another, for example. In this book, we are concerned with the order Orthoptera. The most commonly encountered members of this order belong to one of its two suborders, Caelifera, and are called *grasshoppers.* Most grasshoppers (and the best known) belong to the family Acrididae and are sometimes referred to as short-horned grasshoppers because of their relatively short antennae. Grasshoppers often are thought of as modest-looking brown or green insects, but many species in this family are brightly colored, and some of the most dull-colored species rival butterflies in beauty when they spread their wings in flight. In addition to the acridid grasshoppers, there are a number of less commonly encountered families in the suborder. These include pygmy grasshoppers or Tetrigidae, monkey grasshoppers or Eumastacidae, desert long-horned grasshoppers or Tanaoceridae, pygmy mole crickets or Tridactylidae, and several other families. In this field guide, we consider the most common of the grasshoppers: the Acrididae, Tetrigidae, and Tridactylidae. Following is a list of North American taxa that are regarded as grasshoppers:

- Suborder Caelifera grasshoppers
  - Superfamily Acridoidea
    - Family Acrididae true grasshoppers and locusts
      - Subfamily Acridinae silent slantfaced grasshoppers
      - Subfamily Cyrtacanthacridinae spurthroated grasshoppers
      - Subfamily Gomphocerinae stridulating slantfaced grasshoppers
      - Subfamily Oedipodinae band-winged grasshoppers
      - Subfamily Romaleinae lubber grasshoppers
  - Superfamily Eumastacoidea
    - Family Eumastacidae monkey grasshoppers
  - Superfamily Pneumoroidea
    - Family Tanaoceridae desert long-horned grasshoppers
  - Superfamily Tetrigoidea
    - Family Tetrigidae pygmy grasshoppers
  - Superfamily Tridactyloidea
    - Family Tridactylidae pygmy mole crickets

The remaining insects treated in this book belong to the other orthopteran suborder, Ensifera, and are more closely related to one another than to grasshoppers of the suborder Caelifera. The *katydids*, family Tettigoniidae, are sometimes referred to as long-horned grasshoppers, but this name blurs their true relationships. Outside North America, katydids tend to be called bush crickets, but this designation obscures their distinctive nature. This is a diverse, strikingly beautiful group that is much more prone to call or "sing" than are the short-horned grasshoppers. Closely related to the tettigoniids is a small group, the hump-winged grigs or Prophalangopsidae. Following is a list of North American families and subfamilies that are regarded as katydids:

Suborder Ensifera katydids, crickets, and relatives
- Superfamily Tettigonioidea katydids and hump-winged grigs
  - Family Tettigoniidae katydids
    - Subfamily Pseudophyllinae true katydids
    - Subfamily Phaneropterinae false katydids
    - Subfamily Copiphorinae coneheaded katydids
    - Subfamily Conocephalinae meadow katydids
    - Subfamily Meconematinae quiet-calling katydids
    - Subfamily Tettigoniinae predaceous katydids
  - Family Prophalangopsidae hump-winged grigs

The *crickets* share many features with katydids but are generally smaller, flatter, and less grasshopper-like. Their familiar songs are a musical delight, in contrast to the buzzy or raspy songs of most katydids and grasshoppers. Following is a list of North American families and subfamilies that are regarded as "true" crickets:

Suborder Ensifera katydids, crickets, and relatives
- Superfamily Grylloidea true crickets, including mole crickets
  - Family Gryllidae crickets
    - Subfamily Gryllinae field crickets
    - Subfamily Nemobiinae ground crickets
    - Subfamily Trigonidiinae sword-tail crickets
    - Subfamily Eneopterinae bush crickets
    - Subfamily Oecanthinae tree crickets
    - Subfamily Mogoplistinae scaly crickets
    - Subfamily Myrmecophilinae ant crickets
    - Subfamily Pentacentrinae anomalous crickets
  - Family Gryllotalpidae mole crickets

Finally, we consider the *gryllacridoids*, which are close relatives to the katydids and crickets but are neither. They include camel crickets and Jerusalem crickets as well as some other groups, such as the wetas of New Zealand. As a group, they have no accepted common name, but because their scientific name is Gryllacridoidea they can be called gryllacridoids. Rarely seen or collected, gryllacridoids are one of the three superfamilies in Ensifera. The other two superfamilies are the katydids (Tettigonioidea) and the crickets (Grylloidea).

Suborder Ensifera katydids, crickets, and relatives
Superfamily Gryllacridoidea gryllacridoids
Family Gryllacrididae raspy crickets
Family Rhaphidophoridae camel crickets
Family Stenopelmatidae Jerusalem crickets
Family Anostostomatidae silk-spinning sand-crickets

Appearance and Body Parts

Grasshoppers, katydids, crickets, and their relatives usually are large insects, which makes it easy to see the different structures and to determine their identity. There are three basic body parts: the head, which bears the sensory structures such as the eyes, antennae, and mouthparts; the thorax, which bears the structures that provide movement—the legs and wings; and the abdomen, which contains the organs of digestion and reproduction.

Learning the body parts can seem daunting because of the foreign-sounding names affixed to many structures. Bear in mind that insect morphology is less complex than that of humans, and many of the names used to describe insect body parts are the same as those used to describe human bones. To accurately identify most orthopterans, it is necessary to examine only the color and shape of a small number of structures, so there is no need to be intimidated by the terminology used to describe them. Identifying orthopterans is similar to learning to identify birds, although it may require some magnification. A 10X hand lens, available from hobby shops, is often sufficient. Sometimes identification is more complex, of course. Among crickets and katydids, some species are most easily identified by their calls. We have attempted to simplify the identification of grasshoppers, katydids, and crickets by minimizing the use of technical terms and by providing whole-insect illustrations, drawings of body parts, depictions of songs, and a glossary.

On the *head*, the principal structures of note are the antennae and the mouthparts (Fig. 1). The *antennae* (singular, antenna) are sensory structures attached to the front of the head between the eyes, and their length and shape are important in identification (Fig. 2). Most antennae consist of a string of small barrel-like segments, although some segments are quite elongate. Other antennae have flattened segments, and many have flattened segments that are larger toward the base of the antenna and smaller toward the tip. Flattened antennae with expanded basal segments are said to be sword-shaped. The antennae of grasshoppers are relatively short, usually less than half the length of the body. The antennae of katydids, crickets, and relatives are long, nearly always exceeding the length of the body. A feature that is important in a few species is the shape of the vertical, flattened, elevated structure on the front of the head, called the frontal ridge. The *mouthparts*, though very important to the insect, have little diagnostic value. The shape of the head is sometimes of diagnostic value. Some species have a head that tapers to a point, whereas in others it is broadly rounded (Fig. 3).

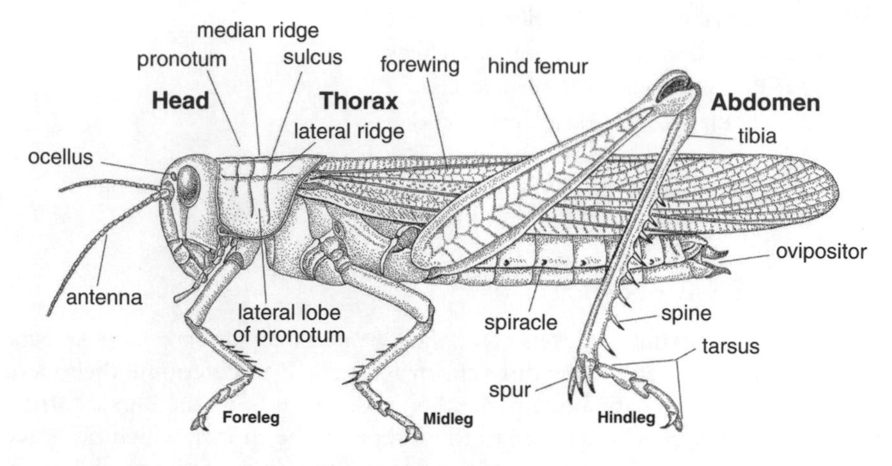

Figure 1. External anatomy of an adult grasshopper (female)

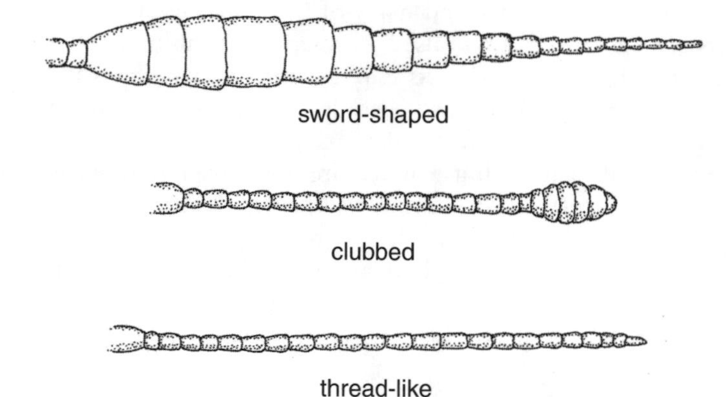

Figure 2. Some antennal shapes

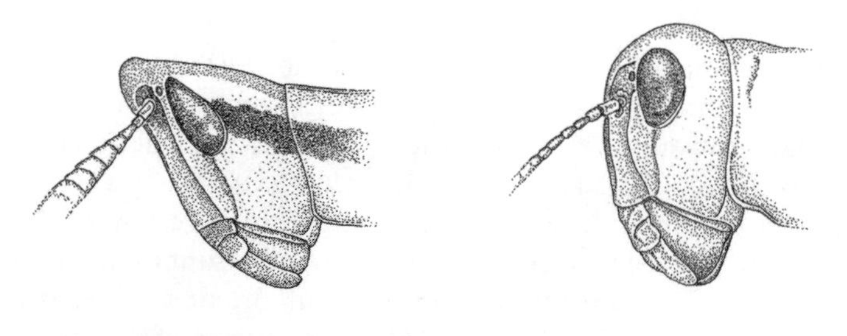

Figure 3. Profiles of two types of grasshopper heads

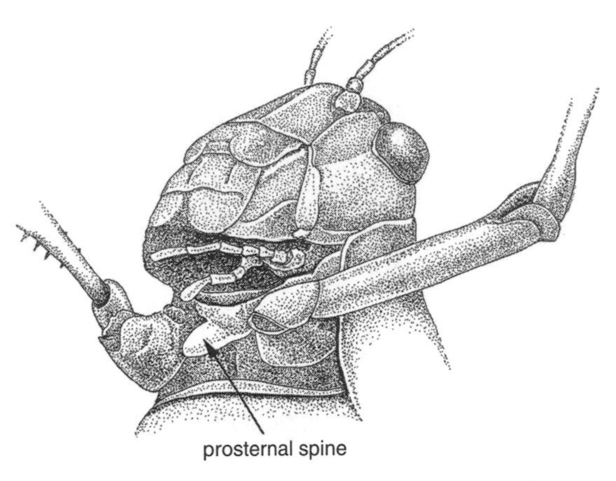

Figure 4. Spur (prosternal spine) of a spurthroated grasshopper

The *thorax* (see Fig. 1) consists of three segments and bears the structures that provide locomotion: the legs and wings. The individual segments of the thorax are hard to see because the upper section of the first thoracic segment, the pronotum, is enlarged and covers the other segments when viewed from above. When viewed from below, the segments are more apparent, and in some insects the presence of a prosternal spine (sometimes called a spur, hence the name "spurthroated" grasshoppers) (Fig. 4) at the base of the front legs is an aid in identification. The pronotum is also important in diagnostics (Figs. 5 to 8); it sometimes bears longitudinal ridges (carinae) on the upper (dorsal) surface. The shape of the medial or central ridge, or the paired lateral ridges, is sometimes diagnostic. In some cases, the ridges are cut by crevices (sulci). The sides of the pronotum (lateral lobes) are often referenced, either for their shape or color. One pair of legs is attached to each of the thoracic segments, with the third pair, or hind legs, enlarged. From the perspective of identification, the important leg components are the large, thickened *femur* (plural, femora), the long thin *tibia* (plural, tibiae), and the multi-segmented *tarsus* (plural, tarsi). Grasshoppers, katydids, and crickets tend to have long legs. The hind legs are especially elongate and enlarged to facilitate leaping and are armed with spines for defense. Despite the widespread belief that grasshoppers and their kin move by hopping, they generally move by walking. The powerful muscles of the hind

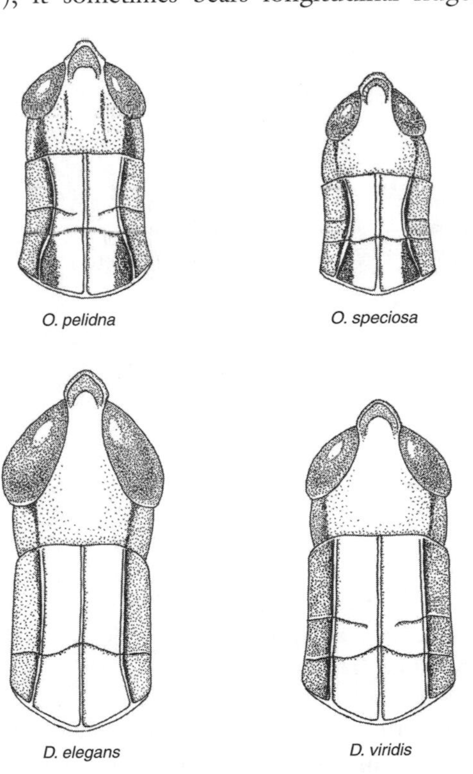

Figure 5. Dorsal view of the pronotum of *Orphulella* species (above) and *Dichromorpha* species (below). The shape of the lateral ridges of the pronotum, and the number of cuts (sulci), are useful for differentiating among the species.

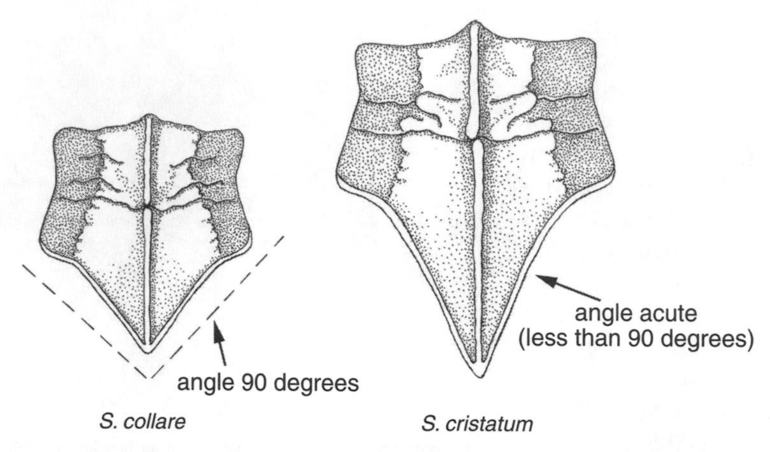

Figure 6. Dorsal view of the pronotum of two similar *Spharagemon* species. The shape of the posterior angle of the pronotum is diagnostic.

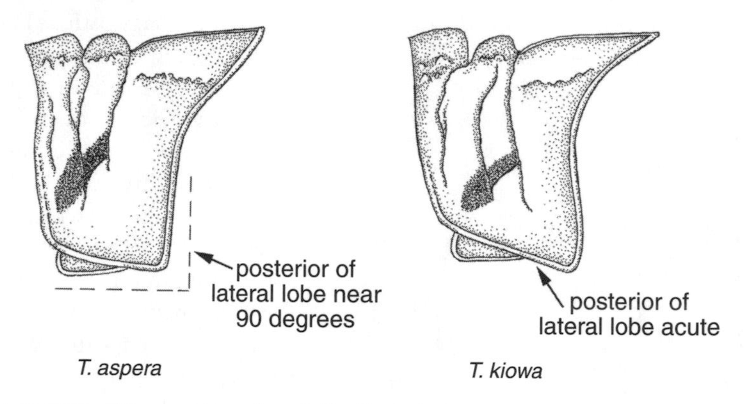

Figure 7. Lateral view of the pronotum of two similar *Trachyrhachys* species. The shape of the lateral lobe is diagnostic.

femora are useful for only a few hops before the insect is exhausted, so their use is generally reserved for an alarm response. Wings may be present or absent, and when present can vary in length. The short-winged forms are flightless whereas the long-winged forms are sometimes strong fliers. The first pair of wings, the *forewings*, in grasshoppers, katydids, and crickets are somewhat thickened and pigmented, and are called tegmina. The *hind wings* are not thickened, and range from unpigmented to brightly colored. The hind wings often are large and fan-shaped, and fold up under the forewings when the insect is not in flight. It is the hind wings that provide lift for flight, the forewings serving as camouflage and for protection of the hind wings.

The *abdomen* (see Fig. 1) is the largest and most posterior component of the body. It bears reproductive structures near the tip (Fig. 9). In some groups, small or large paired appendages called *cerci* (singular cercus) are species-specific in shape among males. Another structure associated with males that has some utility in identification is the *furcula*, which is a forked organ in which only the two tips of the fork are visible, making it appear that there are two structures rather than one. In males, the furcula rests on a broad, flat, dorsal plate near the tip of the abdomen called the *supra-anal plate*. At the tip of the abdomen in males is the *subgenital plate*. It is mostly ventral and is topped dorsally by the supra-anal plate. The shapes of both the supra-anal plate and subgenital plate sometimes have diagnostic value. Beneath the supra-anal plate of males is the aedeagus or penis. We do not use this structure in our descriptions because it is internal, and examination requires difficult manipulation of the specimen and significant magnification. However, the form of the aedeagus seems to be critical in sexual compatibility, and its shape is an excellent indicator of identity in some grasshoppers. In females, the *ovipositor* dominates the tip of the abdomen. In grasshoppers, the ovipositor consists of curved, pointed structures that open upward and down-

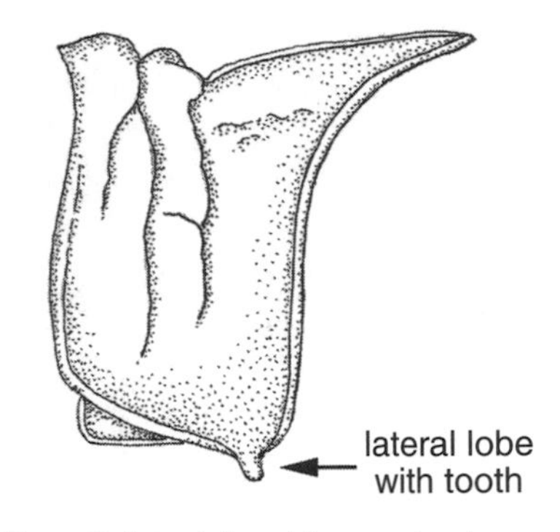

Figure 8. Lateral view of the pronotum in the California band-winged grasshopper, *Trimerotropis californica*, showing a tooth on the lateral lobe.

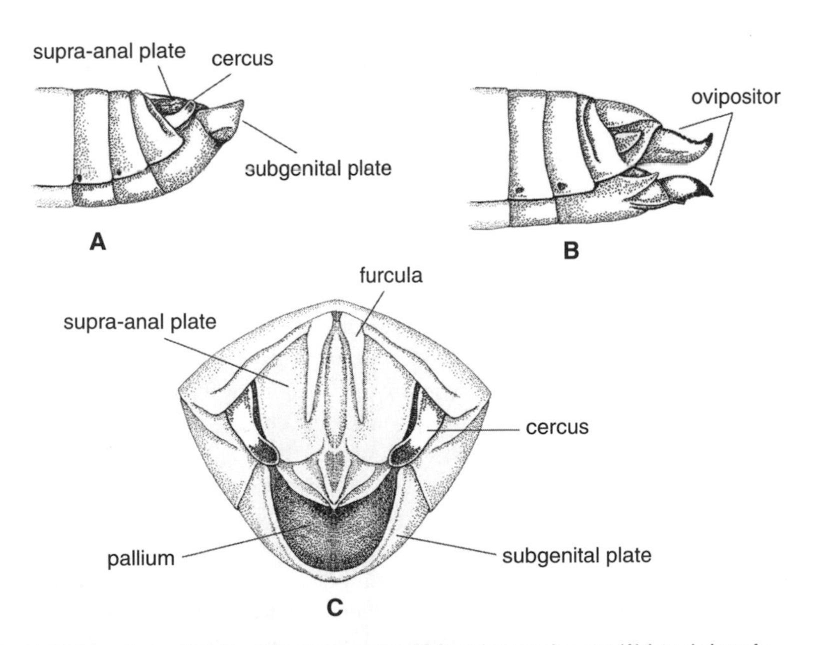

Figure 9. Various views of the tip of the abdomen in a *Melanoplus* grasshopper: (A) lateral view of a male, (B) lateral view of a female, (C) dorsal view of a male.

Figure 10. The female round-winged katydid (right) has just received from the male (left) a spermatophore (bag of sperm) along with a much larger gelatinous mass, visible at the base of her blade-like ovipositor. She will consume this "nuptial meal" while the spermatophore empties its contents into her sperm receptacle. Photo by T. G. Forrest.

ward. In katydids and crickets, the ovipositor is a long, compound structure that may be blade-like (katydids) (Fig. 10) or tube-like (crickets). The ovipositor often is inserted into the soil, sometimes into vegetation, and used to make a hole in preparation for egg laying. The ovipositor has great diagnostic value in distinguishing between the sexes, but limited value in species identification.

Life History

The life history of grasshoppers, katydids, crickets, and relatives is relatively simple, although it varies somewhat among species, and sometimes even among different areas of the country. The principal life forms are the egg, nymph, and adult stages. Orthopterans undergo a change in body form known as *simple* (or gradual) *metamorphosis*, in which the nymph gradually changes to the adult form without a dramatic change in appearance. This is in contrast to insects that undergo complete metamorphosis, in which there is a pupal stage between the immature and adult stages, and a very dramatic change in appearance, as between caterpillars and butterflies.

Adult females produce *eggs* that are deposited in groups. Grasshoppers usually deposit their eggs in the soil (Fig. 11). In most cases, the eggs are in a cluster held together by a frothy secretion that, when dry, forms a rigid covering. The eggs and frothy secretion are collectively known as an egg pod. A pod may contain from 4 to more than 100 eggs, depending on the grasshopper species.

Figure 11. A female grasshopper extends her abdomen deep into the soil during oviposition (egg-laying). She will deposit a cluster of eggs surrounded by a frothy secretion, producing a structure called an egg pod. Females are quite selective in their choice of oviposition site, requiring certain soil and moisture conditions.

Crickets and katydids have various egg-laying habits, with some depositing their eggs in the soil and others depositing them in plant tissues. Some katydids even attach flattened eggs in short rows to twigs or insert them in the edges of leaves (Fig. 12).

Grasshoppers, katydids, and crickets typically pass the most inclement period of the year, which usually is winter, in the egg stage. In all areas, however, some species survive the winter months in the nymphal and adult forms. This is especially true in the southern states.

When a grasshopper egg hatches, the young grasshopper digs its way through the soil to the surface and molts into an active form capable of walking, hopping, and eating. The active stage between hatching and attainment of adulthood is called a *nymph*. The first active form is known as the first instar, and it is followed by additional molts until the nymph has experienced (usually) 5 or 6 instars, and is ready for its final molt to the adult form. Insects molt, or shed their old body covering, because the covering is not elastic and inhibits growth (Fig. 13). Thus, each time the nymph molts it produces a larger body covering, and the nymph gets larger and larger. As nymphs grow, the wings begin to develop, but they are not fully formed until the adult stage. Sexual structures, such as the ovipositor in females and the cerci in males, also develop as the grasshoppers grow. These, too, are not fully formed until the adult stage. This scenario is substantially the same for katydids and crickets, except that the eggs are less frequently laid in soil.

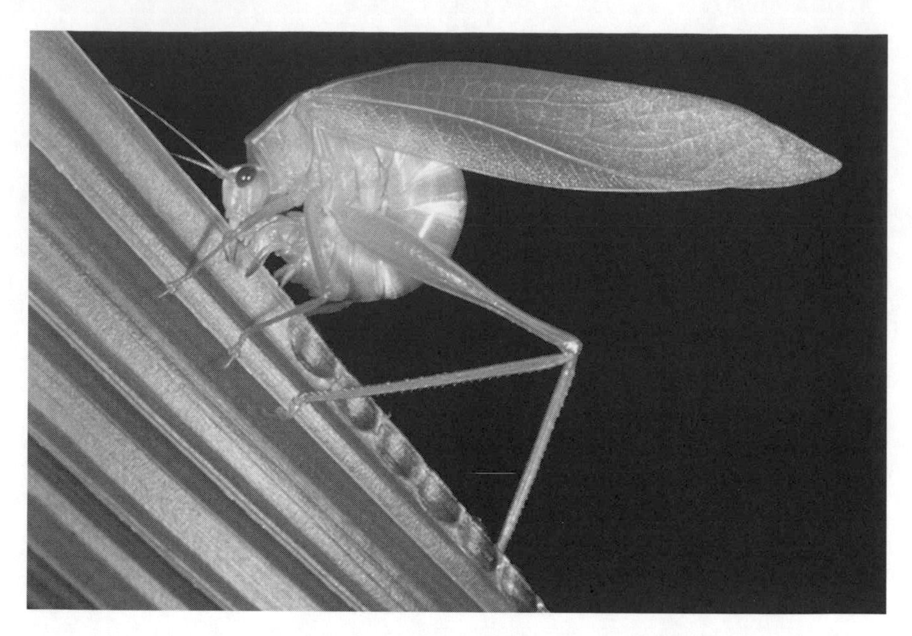

Figure 12. This female narrow-beaked katydid, *Turpilia rostrata*, is inserting her ovipositor between the upper and lower surfaces of a palm leaf. When it is fully inserted, she will add an egg to the row of six that are visible behind her. Certain other katydid females (genus *Scudderia*) deposit their eggs in a similar manner but in more delicate leaves. Photo by Lyle Buss.

The nymphal instars in winged species of grasshoppers can be distinguished by the shape of the wing pads. The wing pads initially are very short and broadly rounded but become slightly elongated in instar 2. Instars 3 and 4 have more elongate, downward pointing wing-pad tips, and display some weak wing veins. In instars 5 and 6, the wing pads are inverted; they point upward or posteriorly instead of downward, and only one pair of wings is visible. In grasshoppers with only five instars, the pattern is much the same, but the form that would be instar 4 is absent from the developmental sequence.

Adults are capable of reproduction (Fig. 14) and have fully formed sexual organs, some of which are visible externally (Figs. 15, 16). The wings have achieved their final state (Fig. 17). Many grasshopper species are macropterous, or long-winged, which means that the wings extend nearly to the tip of the abdomen, or beyond. Others, however, are brachypterous, or short-winged, in the adult stage. Such wings are typically oval, and extend only about one-third the length of the abdomen. A few species are wingless, or nearly so, in the adult stage. It can be difficult to distinguish between immatures and adults when adults can be long-winged, short-winged, or wingless! However, if you look for obvious, fully formed genitalia, and wings extending at least one-third the length of the abdomen, you will identify most adults accurately. Only in a few species are adults likely to look like immatures, and such species are easily recognized from drawings or photographs. Similar variations in wing length occur

Figure 13. The nymph of a Mormon cricket, *Anabrus simplex*, during molting. The nymph attaches itself to a substrate and then hangs, allowing gravity to aid its escape from its old body covering, which must be shed to allow growth.

among cricket and katydid adults, the chief difference being that many species of crickets and some of katydids are dimorphic in wing development, with some individuals being fully flight capable and others having the hind wings reduced to useless vestiges.

In the typical life cycle, eggs hatch in the spring, nymphs develop through the summer, adults mate and produce eggs in the late summer and autumn, and the winter is passed in the egg stage. In northern areas, most species conform to this sequence of events. In southern areas, however, it is not unusual to have nymphs and adults present nearly year-round. Most species are univoltine, which means that only a single generation occurs each year. Some species,

Figure 14. Mating differential grasshoppers, *Melanoplus differentialis*. The smaller male mounts the female from behind and often is transported about while copulating.

Figure 15. A close-up of mating differential grasshoppers. Note that the male has his abdomen twisted below the female's abdomen. This is a remnant of mating behavior found in ancestral species, wherein the female mounts the male.

Figure 16. Most crickets and katydids mate with the female above the male. This male tree cricket (right), with forewings raised, has just given the mounted female a bag of sperm (the small globular spermatophore attached at her rear). She will feed on secretions from glands at the base of the male's wings while the spermatophore empties into her sperm receptacle.

Figure 17. Adults of grasshoppers, katydids, and crickets often have fully formed wings that allow strong, directed flight. Shown here is the High Plains grasshopper, *Dissosteira longipennis*. Species in the genus *Dissosteira* are capable of aerial acrobatics, with most of the lift provided by the hindwings.

however, are multivoltine, producing more than one generation annually. In a small number of cases, more than one year is needed for a species to complete its life cycle; this normally is found only in cold northern climates or at high altitudes.

Habitats and Food

There are few habitats that do not support grasshopper, katydid, or cricket populations. The general rule is that if low-growing plants are present, grasshoppers and katydids can be found feeding on them. Crickets are more secretive but occur even in areas relatively devoid of vegetation, such as the muddy banks of lakes.

Grasshoppers thrive in open sunny environments where vegetation is present but not thick or shaded. The number of species is highest and grasshopper abundance is greatest in the West, particularly the semi-desert southwestern states and the prairie and short-grass prairie of the Great Plains area. In contrast, katydids and crickets thrive in the humid eastern states; however, the predaceous katydids are found principally in the arid West. Crickets are often thought of as ground dwelling but actually are found in all vegetational layers.

One habitat where it is difficult to find grasshoppers is mature, dense forest. In such a habitat, virtually no light reaches the soil, and there is no understory vegetation on which grasshoppers can feed. This does not mean that grasshoppers are totally absent; because small breaks in the canopy are inevitable, grasshoppers will find and inhabit these areas where low-growing vegetation exists. Also, some grasshopper species feed on pine or broadleaf tree foliage, but these are not numerous and are difficult to observe. Katydids and crickets are much more at home in mature forests.

Another habitat where grasshoppers are rare is swamps, both mangrove swamps and freshwater swamps. Although strong fliers, such as the *Schistocerca* species, may sometimes be observed within mangroves, they are infrequent. Mangroves inhabit coastal areas where their roots are immersed by water, usually salt water. Freshwater swamps are often associated with rivers or lakes, which provide continuous or seasonal flooding. Flooding makes swamps an inhospitable environment for most grasshoppers, nearly all of which deposit eggs in soil. The margins of swamps are regularly exploited by grasshoppers, however, with several species adapted to feed on emergent vegetation. Salt marsh vegetation is regularly exploited by grasshoppers, katydids, and crickets, and they can become quite numerous in this habitat.

You might expect cold areas to lack grasshoppers, katydids, and crickets. To some degree this is true, but some species of grasshoppers are well adapted to living in cold, northern climates and on mountaintops above the tree line where vegetation is present for only brief periods in the summer months. In such areas, life cycles lasting more than one year are more common.

The dietary habits of grasshoppers, katydids, and crickets are quite variable. Grasshoppers are almost entirely herbivorous, or plant feeding. Some species graze on a great number of plants, including plants from more than one plant

Figure 18. A Mormon cricket nymph feeding on another Mormon cricket. Mormon crickets and other orthopterans, though usually thought of as plant feeders, can be opportunistic or cannibalistic, feeding on molting, wounded, or dead comrades.

family; such insects are said to be polyphagous. Some species feed on several genera of plants but restrict their feeding to a single plant family; such insects are called oligophagous. Fairly unusual are the species that feed on only a single genus of plants; these are called monophagous. Orthopterans, though generally considered to be plant feeders, often have other feeding habits (Fig. 18). Many are omnivorous, feeding opportunistically on dead as well as living foliage, on other insects (including members of their own species), and on miscellaneous materials, such as fungi, lichens, and animal fecal material. Katydids, and especially crickets, tend to be omnivorous. Some are aggressively carnivorous, readily attacking other insects though still consuming plant material.

Ecological Significance

Orthopterans often appear to be the most abundant aboveground insects. This is especially true in open, sunny, dry habitats, such as prairies and pastures, but it can apply to open woods, salt marshes, and disturbed areas, such as crop fields. Orthopterans exert great ecological impact and may be the dominant herbivores, or plant-feeders, in a community. Plant feeding can result in depletion of plant biomass and crop damage, or shifts in plant community structure due to the food preferences of herbivorous grasshoppers. In extreme cases, herbivory can result in ecosystem damage. This can occur directly, from disruption of habitat by loss of vegetation, or indirectly, through increased rates of erosion resulting from reduction in vegetative cover. Such habitat damage is rare, however, and is largely restricted to areas that are experiencing drought.

The significance of grasshoppers, katydids, and crickets is due not only to their numbers but also to their role in the cycling of nutrients in ecosystems. They consume large amounts of plant tissue, often eating their body weight in plant tissue daily. Their selective feeding behavior can affect the relative abundance of different plant species in an area. More important, these herbivores hasten the degradation of cellulose and other materials by breaking up the plants into smaller pieces that can be attacked by soil flora and fauna. Fecal material, in particular, is easily degraded, resulting in increased solubility of chemical nutrients essential for plant growth. Degradation of fecal material and clipped foliage causes rapid release of nutrients into the soil, favoring new plant growth. Without plant feeders such as grasshoppers, katydids, and crickets, much of the nutrients in an area would be bound up in dead plant tissue, insoluble and unavailable for plant uptake.

Grasshoppers, katydids, and crickets are important food for other invertebates, such as predatory insects and spiders, and parasitic insects. The larvae of many species of blister beetles (Coleoptera: Meloidae), for example, feed only on the eggs of grasshoppers, so the fate of grasshoppers and these blister beetles is intimately linked. Bee flies (Diptera: Bombyliidae), tangle-veined flies (Diptera: Nemestrinidae), and robber flies (Diptera: Asilidae) are other examples of insects that attack grasshoppers. Tachinid flies (Diptera: Tachinidae), one of the largest groups of flies, parasitize many grasshoppers, katydids, and crickets. Some even use the sounds produced by their host to aid in prey location.

Grasshoppers, katydids, and crickets are also ecologically significant because their own bodies have converted plant tissue into "bite-size" units of animal material that serve as food for vertebrate animals. Animal tissue is much more nutritious than plant material, especially for young and rapidly growing animals that need the high levels of protein and lipids found in orthoperans. Grasshoppers, katydids, and crickets are large enough, and abundant enough, that they attract the attention of many types of vertebrate animals, such as reptiles, birds, skunks, raccoons, coyotes, foxes, and mice, which regularly consume them. Most are opportunistic feeders, taking grasshoppers, katydids, and crickets when they are especially available or when other food is in short supply. Most studies of omnivorous mammals find that insects are taken most frequently in the autumn when the insects are most abundant and when preferred summer foods, such as fruits, are in decline. For some insect-feeding birds, such as meadowlarks, flammulated owls, burrowing owls, Franklin gulls, California gulls, and cattle egrets, grasshoppers and katydids are often the principal element of the diet, and their survival and reproductive efficiency may be directly related to the abundance of these insects. Other species, such as the kestrel, phoebe, loggerhead shrike, grasshopper sparrow, house wren, and bluebird, feed extensively on grasshoppers and katydids but readily switch to other insects or small animals if grasshoppers are in short supply.

Despite its name, the "grasshopper sparrow" is one of the species that exemplifies the resourceful nature and opportunistic feeding behavior of birds. In a study conducted in Alaska during a grasshopper outbreak in 1990, grasshoppers were the principal source of food, whereas in the following year after the grasshopper population had collapsed, there was no detectable grasshopper matter among the birds' stomach contents (Miller et al. 1994). Yet the Alaskan grasshopper sparrows did not starve or relocate; they simply switched to feeding on more available types of insects, and their reproduction and survival was but slightly depressed.

Many birds feed heavily on vegetable matter, principally seeds, but feed their young almost exclusively on insects. Grasshoppers and katydids are large enough and abundant enough that hunting specifically for grasshoppers is an energetically efficient activity. Grasshoppers and katydids are 50% to 75% crude protein. Without such nutritious food to consume, many vertebrate animals would suffer from lack of a suitable source of animal protein. In some parts of the world, such as sub-Saharan Africa and Southeast Asia, grasshoppers, katydids, and crickets are a component of the human diet for the same reason.

Biogeography

Nearly all grasshoppers, katydids, and crickets that occur in North America are native species. No grasshoppers are known to have invaded North America successfully in recent times, and only three species of katydid and nine of crickets are known to have reached North America during the past few hundred years. This is in contrast to the approximately 1500 species in other orders of insects that have arrived during this period—mostly as unwelcome stowaways.

The geographic range of orthopteran species varies greatly. Many species have a fairly wide geographic range, often an entire ecological region, such as the Great Plains. A small number of adaptable species are found essentially everywhere in the United States, at least throughout the lower 48 states and sometimes even north to Alaska. There are many species, however, that have become isolated by physiographic features such as mountains and have evolved into separate species. Such a species may have quite limited distribution—in a river drainage, a plateau region, or even a plant community with restricted distribution. Species with restricted distribution seem to be most common in the western and southeastern states, but every region has at least some unique species. A few species have displayed substantial change in range. For example, the differential grasshopper, *Melanoplus differentialis*, has expanded its range from the midwestern states to the arid western edge of the Great Plains as irrigation has made it possible for this species to live in the lush vegetation it prefers. The two-spined spurthroated grasshopper, *Melanoplus bispinosus*, apparently has moved in the opposite direction. Formerly known only from west of the Mississippi River, it has now dispersed to Alabama and Florida.

In this book, we focus on species with fairly broad distribution because these are the species most likely encountered. A few species with limited distribution are included, however, to illustrate the diversity of insect distribution.

What Is a Species?

Unique binomial scientific names (genus and species) are reserved for different species. However, it is not always obvious whether populations differing slightly from one another should be regarded as different species. If the populations are in contact and remain genetically isolated, they qualify as distinct species. If the populations have different geographic ranges but interbreed where populations overlap, they are geographic races rather than species and are sometimes given subspecies names. It is when populations are not in contact, as in geographically isolated populations, that decisions about species status become difficult. Whether the populations would remain genetically isolated if in contact in nature cannot be determined from laboratory crossing experiments, so decisions as to species status must be tentative and are generally based on comparisons with populations that are in contact. Thus, if the disjunct populations differ greatly in appearance or behavior, especially if the differences are associated with reproductive structures or mating behavior, most scientists will consider them separate species. In other cases, disagreement is more likely.

Speciation may be associated with changes in behavior, physiology, and genetics, as well as appearance or morphology. Only morphology is relatively easy to assess and to portray to others. Therefore, systematists (specialists interested in the evolution and classification of organisms) generally depend on morphological characters to distinguish among species. Unfortunately, morphology can vary considerably within a small geographic area, and great variation may occur over the entire range of an insect. Therefore, many scientists have described an orthopteran as a new species, only to discover later that the new "species" was a morphological variant of a known species. Grasshoppers exhibit considerable variation in color and wing length, accounting for many of the erroneous species descriptions. The sexual structures or mating behaviors also are used to distinguish among species, because these characters relate directly to reproductive compatibility and population genetics. However, even here it is difficult to know how much difference in genitalia or behavior constitutes enough variation to result in genetic isolation.

Katydids and crickets present a different problem. They exhibit few morphological variations, leading to the impression of few species. Their calling behavior, however, may be quite distinct. Differences in behavior can lead to reproductively isolated populations and development of species that are exceedingly difficult to distinguish without knowing about their songs.

Exact tabulation of species numbers is probably a fruitless enterprise, for the reasons mentioned above but also because speciation is a dynamic phenomenon. New species are slowly evolving in some areas, and extinction is probably occurring in others. The number of orthopterans in North America north of Mexico is estimated at 1200 species. In this area, the number of acridids is estimated at about 630 species, including about 145 species of band-winged grasshoppers (subfamily Oedipodinae); 70 species of slantfaced grasshoppers (subfamilies Gomphocerinae and Acridinae); and 375 species of spurthroated grasshoppers (subfamily Cyrtacanthacridinae, including Romaleinae and

Melanoplinae of some authors, and others). There are about 30 species of pygmy grasshoppers (family Tetrigidae) and two species of pygmy mole crickets (family Tridactylidae). The number of species of katydids (family Tettigoniidae) is about 255, whereas there are only about 120 species of crickets (family Gryllidae), and 150 species of camel crickets (family Rhaphidophoridae). There are 3 species of hump-winged grigs (family Prophalangopsidae), 7 species of mole crickets (family Gryllotalpidae), at least 14 species of Jerusalem crickets (family Stenopelmatidae), and a single species of raspy cricket (family Gryllacrididae).

Collection, Preservation, and Culture

Why Collect These Insects?

Orthopterans have features that make them especially attractive and rewarding to collect. Travel to far or exotic habitats is not necessarily required to collect many and varied species, and orthopterans are often numerous. Most orthopterans are large in comparison to other insects, which makes them relatively easy to handle and identify. Identification may require no more than picture matching or it may require detailed examination of structures or calling behavior, so there are various levels of challenge. Species vary as to where they can be collected—some species are easy to find whereas others require considerable searching or travel to a specific location. Grasshoppers, katydids, and crickets are inherently interesting and strikingly beautiful if you take the time to look closely. Few people have seen the myriad of hues on the hind wings of band-winged grasshoppers. Fewer yet have examined the inside of the hind legs of these species for the striking array of colors, including bands of yellow, orange, red, blue, and black, or marveled at the intricate leaf-mimic patterns on the forewings of some katydids. Similarly, few people know the sources of the insect songs they hear at night, whereas by collecting crickets and katydids by their calling songs, you will quickly learn to identify the songsters by their songs. Finally, there is considerable physical and intellectual challenge associated with netting these cunning species. The grasshoppers that respond to your net by flying 100 meters away, or 10 meters up into a tree, or by resting on emergent vegetation in a pond frequented by alligators, will quickly earn your respect. Or try to locate a katydid or cricket by its sound, when each time you get close to your target it becomes silent and a neighbor starts to call!

How to Collect Grasshoppers, Katydids, and Crickets

Many orthopterans are collected simply by grabbing the insects from their perches or from the ground. For species that fly readily, however, such as many grasshoppers, an aerial insect net may be needed (Fig. 19). To collect such species, a stealthy approach is recommended, and a long-handled net is helpful. The same net can be used to capture individuals from perches on vegetation, but to avoid damage from briars and stiff branches, a net with a heavy rim and a canvas or muslin bag (a “sweeping net” or “beating net” in entomological

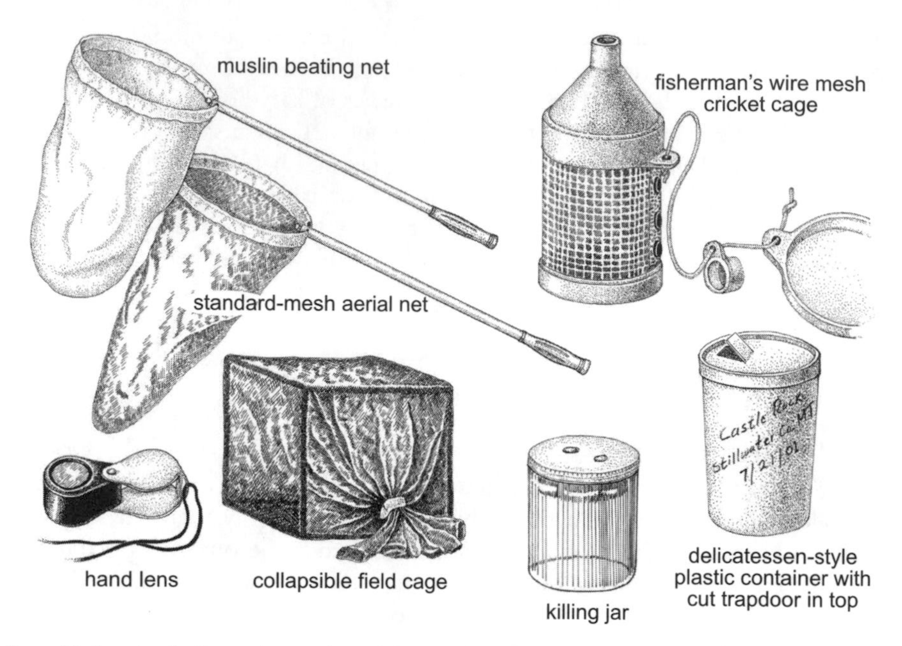

Figure 19. Insect-collecting equipment that will prove useful for collecting grasshoppers, crickets, and katydids.

lingo) may be used instead. Thus, serious collectors have two nets: a lightweight net with a mesh bag for stalking and capturing individual specimens, and a more sturdy net with a heavyweight bag for sweeping vegetation. Many collectors also prefer that the bag of the latter net be rounded rather than tapered to a point; this makes it easier to remove insects and debris. For either net it is important that the bag be long enough that the insects have difficulty escaping and that it can be flipped over the net frame to close off the bag and prevent insect escape.

For ground-dwelling crickets, a jar is a handy device that can be clapped over the insect—but then, before the capture is complete, the jar opening must be closed by sliding a card beneath the cricket. (Tricking the cricket into walking or jumping into the jar is sometimes easier and always more satisfying.) Some crickets can be collected by looking under objects in contact with the soil, and captures can sometimes be increased by providing such objects as pieces of cardboard, wood, or black plastic sheeting for them to hide under. Trickling trails of oatmeal flakes on the ground shortly after sunset is an aid to collecting some nocturnal, ground-dwelling orthopterans. When they encounter a trail, they stop to feed and can be collected by periodically inspecting the trails with a light. Some of the same species can be collected in pitfall traps made by burying No. 10 (1-gallon) tin cans flush with the ground. Each can is furnished with a 15-cm-diameter plastic funnel with the small end cut off to leave a 2.5-cm-diameter opening that makes entry easy but exit difficult. Putting a small amount of oatmeal or cornmeal in the can will increase the catch, and adding a cover resting on small rocks or pieces of wood will help keep out rain.

To attract females, most adult male crickets and katydids produce species-specific calling songs for hours each night. As males broadcast their locations and species identities to potential mates, they are doing the same for the attentive collector. This means that by careful listening, an experienced collector can deduce how many species of crickets and katydids are reproductively active in an area even if none of the songs are familiar. Furthermore, the collector can have a male of each species by capturing the maker of each type of calling song. Capturing singing males can be easy but is often challenging. Most calling is at night, so the songster must be spotted with a light before capture; but bright light will often cause it to stop calling and to fly, drop to the ground, or retreat into a burrow. Therefore, it is best not to shine a light directly at the source of a call until you believe you can quickly spot the caller and net it or grab it. The direction of a caller is easier to establish than its distance. A convenient way to pinpoint the position of a caller is to circle it as you establish imaginary lines leading to it. Where the lines cross should be where you first shine your light.

If the insects captured are to be kept alive to rear, study, photograph, or kill later by way of freezing, they should be transferred to some sort of secure container for transport. For grasshoppers, a portable cage is often used. The cage is usually closed with a gathered cloth, though hinged and sliding doors also work. One popular type of field cage has five sides of aluminum window screen in an aluminum frame, with the sixth side consisting of a cloth closure. Another consists of a large piece of lightweight netting supported by a plastic pipe framework and gathered to form a closure at one end. A smaller, more temporary form of insect retention is a plastic delicatessen-style cup. Three cuts are made in the lid of the cup, forming a small "door" through which individual insects are inserted. The rigidity of the plastic keeps the door closed and the insects contained, although a piece of masking tape makes the closure more secure. The cup container has the advantages of being small and easy to carry, and inexpensive or free, and you can write on it to note your collection site. Some collectors prefer a fisherman's wire mesh cricket cage for temporary storage; this has the advantage of being commercially available in many sporting goods stores. For many species of katydids, individual containers are needed to prevent one captive from killing or biting off the appendages of another. Crickets are relatively small and fit easily into small containers. For individual retention, containers with screw or snap lids work well. Some ventilation must be provided if the insect is large and to be held for more than an hour or so. Small containers can be put in a cooler to prevent the captives from overheating or to immobilize them if you wish to take photographs.

Collecting will be more successful if you know the ecology of the insects. Insects have preferred habitats and a period of the day (or night) when they are most active. Many katydids and crickets are attracted to bright lights at night, and those that are not can be seen and collected by using a light to illuminate foliage at night. A light that can be worn on the head will leave both hands free for collecting. A red filter or a means of significantly dimming the light will allow you to watch crickets and katydids without triggering their aversion to light. Weather influences activity, and your ability to capture insects. For strong fliers, it may be easier to collect them when it is cool, as in early morning periods, although they are active and most easily seen when it is warm.

One of the advantages of collecting grasshoppers, katydids, and crickets is that it is always possible to find suitable habitats because nearly all habitats support some interesting species. You can stop along the side of nearly any road that is bounded by grass, weeds, and low-growing or open vegetation, and find large numbers of grasshoppers and katydids. To get a wide assortment, of course, you will have to vary the habitat and move around to different locations. You can collect throughout the year, but the most productive periods are generally April to November. Many landowners are cooperative and freely allow insect collecting, but inquire before collecting on private property. National and state parks prohibit removal of anything from the property, so it is necessary to inquire in advance and to request a permit if you want to collect in these habitats.

How to Culture Grasshoppers, Katydids, and Crickets

You may wish to raise your captured insects to the adult stage before preserving them, or you may prefer to keep them alive to study their behavior or enjoy their calling. Indeed, crickets are often kept as pets in Asian countries to permit their melodious tones to be heard clearly and dependably. Most orthopterans can be kept alive readily, and, if you pay attention to their temperature, food, and moisture requirements, most nymphs can be reared to the adult stage. In some cases, you will be surprised at how much the nymphs and corresponding adults differ in form and color.

Grasshoppers and katydids are best maintained in screen cages, and it is important that the mesh is metal rather than plastic so the captives do not chew through and escape. The cage can be any shape, but it needs an opening that allows you to add food and water. Often, the easiest way to secure the opening is to use gathered cloth as the closure; you can tie or clamp off the opening to keep your pets inside—except for those that chew through and escape! (A hinged wood, metal, or plexiglass door is more complicated but will contain all species.) A cage that is partially mesh, such as a canning jar with wire mesh substituted for the inner part of the two-piece lid, works well for many species, including most crickets. Ideally, you should feed your insects what they were eating before they were collected. If in doubt, try a mixture of fresh grass and broadleaf plants, and pulverized dry cat food. Romaine lettuce, but not iceberg lettuce, is a good substitute food for natural vegetation. You may want to insert the cut end of the plant into water to keep the plant turgid, or change it at two-day intervals. Some species benefit from access to dry grains, such as unsweetened dry breakfast cereals or a mixture of wheat bran, soy flour, and wheat flour. Many species will do better if free water is available from a wet sponge or a bottle of water with a narrow neck and a cotton wick.

Providing a means for insects to perch is quite important, and wire mesh window screen is good for this. If you do not have window screen on the side of the container, try inserting a cylinder of screen or paper into your cage. When insects molt, they hang down from their perch as they expand to a larger size and harden. Denied the ability to hang from a perch as they molt, your captive insects may not be able to fully escape from their old body covering, or their wings and legs may be badly deformed.

Grasshoppers, but not katydids or most crickets, like to adjust their body temperature, and the optimal temperature is surprisingly high. Put a desk lamp or other light source close to one side of the cage. The grasshoppers will move around on the screen until they find their optimal temperature, and perch there.

How to Photograph Grasshoppers, Katydids, and Crickets

A good way to enjoy orthopterans, and an attractive alternative to killing them, is photography. Photographs have the advantage of preserving insect color better than other methods of preservation, and also may record behavior or important elements of the habitat, such as the food plants. It is difficult, however, to capture all the important elements in adequate detail through photography. In band-winged grasshoppers, for example, the color of the hind wings is important in identification, but it is quite difficult to photograph a grasshopper in flight. Thus, photographs serve as a complement to actual specimens.

Many cameras are suited to photographing grasshoppers, katydids, and crickets. Only a few modifications are necessary, because these insects are large enough to be photographed easily. Traditionally, photographers use a single-lens reflex (SLR) camera with a lens that allows close approach to the insect, such as a 55-mm macro lens. The minimum working distance of such camera-and-lens systems is often as little as one centimeter, but they also can focus on distant subjects, so SLR photography is a very flexible approach. Use of extension tubes with the 55-mm lens or a close-up macro lens can enhance such photography, allowing you to obtain even bigger images of small subjects, but this is not a necessity.

A useful piece of equipment for close-up photography is a flash attachment for the camera, or a pair of flashes. Use of flashes provides additional lighting and better light balance. The extra light allows you to use a faster shutter speed and a smaller lens opening, which reduces the possibility of blurry pictures due to movement of the insect or camera and increases the depth of field that is in focus.

Use of a lens that allows close-up photography, especially when supplemented with flashes, allows manipulation of the insect under controlled conditions. An insect can be captured and chilled in a refrigerator prior to photographing it; a chilled insect is less likely to jump or fly away, and yet it still behaves fairly normally. Thus, you can pose the insect on whatever plant you wish, or manipulate the background, and still have time to take several photographs before the insect warms up and walks or jumps away.

An alternative to a lens that allows you to get close to the subject is a telephoto macro lens. A telephoto lens effectively brings the subject close to the photographer rather than the photographer approaching the subject closely. It has the advantage of allowing you to photograph the insect with minimal disturbance and in its natural habitat. It is difficult to get the insect to "pose" in its natural habitat, however, and difficult to get optimal lighting under natural conditions.

Digital cameras are beginning to replace cameras that use film. A major advantage of using a digital camera is that you can quickly see the quality of your shots. If none meet your standards, you can take more—assuming your subject is still available. No matter how many shots you take to get that perfect picture, you need not buy film or wait to have it developed. A lesser advantage of digital photography is that you can send your photos as e-mail attachments or post them on the World Wide Web without first scanning them. A major disadvantage of using a digital camera is the expense of a high-resolution camera. To take high-quality close-ups with a film camera of equal capability is far less expensive. Another possible disadvantage is that editing digital images requires a computer, computer skills, and appropriate software. Most digital cameras do not provide resolution that is acceptable for production of high-quality printed matter, so if you are contemplating publication, consult with a publisher before you depend on digital photographs for your illustrations.

How to Preserve Grasshoppers, Katydids, and Crickets

Orthopterans are normally preserved by killing, pinning, and drying. They can be killed by freezing or with chemicals. The easiest, safest, and possibly most humane technique is to leave insects in a freezer for several hours. Or they can be killed by exposing them to a small amount of toxic fumigant, such as ethyl acetate. Toxicants should be used in a specially prepared killing jar. Commercially prepared killing jars usually have a porous dispenser mounted on the jar lid. Homemade killing jars have a layer of plaster of paris (also a porous material) in the bottom of the jar. Once ethyl acetate is poured on the porous substrate and is absorbed, the jar will produce toxic fumes for several days if kept closed. Insects killed with ethyl acetate are easier to mount than those killed by freezing. The hind legs of frozen orthopterans often remain stiff after thawing.

For most purposes, it is not useful to kill nymphs because they lack the characters needed for identification. Also, because of their soft bodies they do not preserve well in a dry state and are best placed in alcohol to prevent excessive distortion.

To mount an adult grasshopper, katydid, or cricket on a pin, first insert the pin into the dorsal surface until the point protrudes from the ventral surface. The preferred insertion point for the pin is the posterior area of the prothorax, and to the right of the midline (Fig. 20). Push the insect up on the pin, leaving enough shaft above to enable you to pick up the dead insect without touching its body. Collection data should be recorded on one or more labels below the insect's body. Labels can be made by writing or printing data on stiff paper and

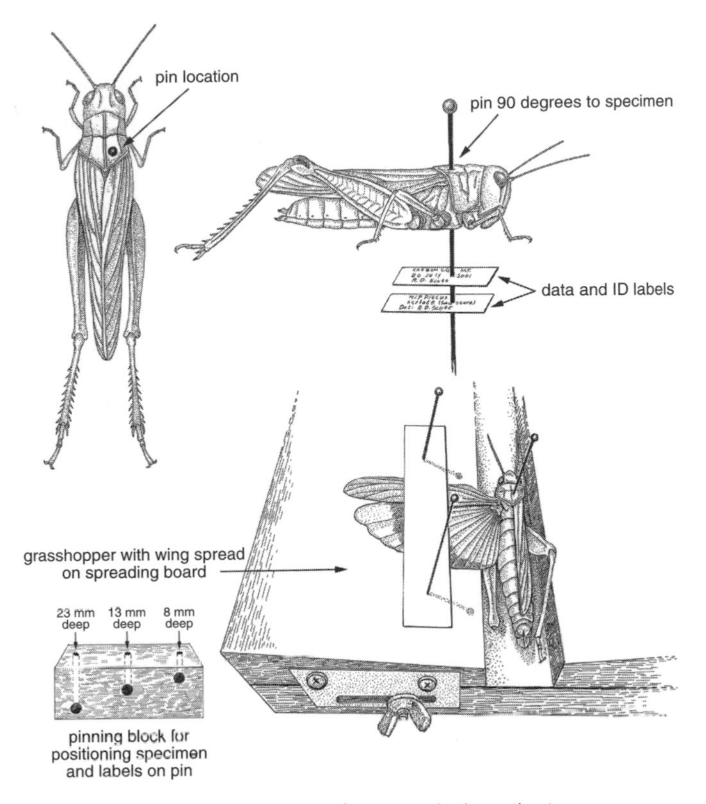

Figure 20. How to pin grasshoppers and other orthopterans

cutting the label to a small rectangle. A pinning block is often used to align the insect body and label(s) to standard heights. The label should indicate the date of collection, place of collection, and collector's name. Ecological data, such as habitat or host plant, may also be included. Pins vary in size and quality. It is best to use insect pins, which are rustproof and are longer and sharper than standard pins, allowing attachment of labels and easy mounting. Insect pins, as well as other collection and preservation equipment, are available from biological supply houses.

To fully appreciate the beauty of the band-winged grasshopper species, and to assist in identification, you should spread at least one forewing and hind wing. Species other than band-winged grasshoppers are rarely spread, even though the spread wing may sometimes be an aid in identification. The usual procedure is to spread the left forewing perpendicular to the grasshopper body. Similarly, the leading edge of the hind wing is spread perpendicular; this results in full extension of the remainder of the hind wing. To properly spread a grasshopper's wings, some support is needed to keep the wings elevated and flat. Such support can be provided by a rectangle of Styrofoam or other suitable pinning material to which has been glued a strip of similar material to provide a pinning surface about 1.5 to 2 cm higher than the first. The lower pinning surface provides support for the grasshopper's body on its pin, and the elevated portion supports the wings. Strips of paper and pins are used to hold the wings in place. Whether or not the wings are spread, grasshoppers must be dried as an aid to preservation. Drying can be accomplished by placing the pinned insect, often with its wings spread, in an oven at low temperature (less than 200 °F) until the subject is dry and stiff. Once dried, the wings, antennae, and legs cannot be moved without breaking, so it is important to align the body parts before drying.

Large-bodied insects often contain so much fat and fluid that it is difficult to obtain good drying without serious discoloration. To facilitate rapid drying and minimize discoloration, make a slit along the underside of the abdomen, and then reach fore and aft with forceps or tweezers to remove the internal organs. Some orthopterists dust the inside of the body cavity with a small amount of "stuffing powder" (3 parts talcum powder + 1 part boric acid powder) to further aid drying. They also insert a little loose cotton or rolled tissue into the cavity to reduce distortion as the body dries.

The colors of insects tend to fade after the insect dies. Keeping your insects out of sunlight will slow the fading process. Severe discoloration may be due to the accumulation of body oils at the surface of the body. The oil can be extracted, preventing some of the color change, by placing the dried insect on its pin in a bath of acetone. Usually a few hours is adequate; prolonged extraction causes the insect body to bleach to a light color.

Insect collections need a suitable storage facility. Storage requires nothing more than a tight box with pinning material in the bottom. However, it is important that the box be very tight; otherwise, ants, carpet beetles, and cockroaches will gain access and devour the pinned insects. To help prevent damage to specimens, you can place moth balls or moth crystals in the box with the specimens. This will kill any insects that gain access, and protect your preserved insects for years.

Sound Production

Adult grasshoppers, katydids, and crickets frequently have specialized structures for producing sounds that they use to communicate with other members of the same species. Males are the noisy sex and nearly all the noises that males make are broadcast sounds, termed *calling songs*. When a male produces a calling song, he proclaims his species, location, sexual readiness, and quality as a mate. Conspecific females may use the song to travel to the male, be courted, and mate. When a female hears more than one male, she may use the songs to discriminate among competing males. In some katydids, females answer the male's song with a brief, nondescript sound (one or several *ticks*) that allows the male to locate and travel to the female. Calling songs may also benefit the males that produce them by keeping competing males at a distance. However, calling songs

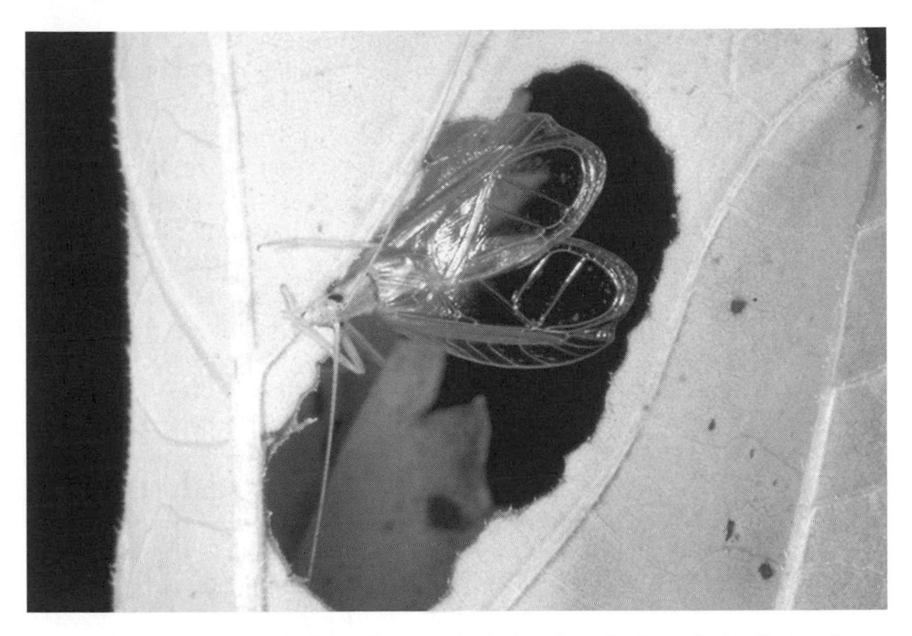

Figure 21. Tree crickets often make their calling songs louder by using natural acoustical baffles to reduce the interference between sound waves emanating from the front and rear of the upraised wings. This tree cricket is using a hole in a leaf for that purpose.

have costs as well. For example, producing a calling song may allow a parasite, predator, or competitor to home in on the caller and to kill or harm it.

Calling songs are directed at a potential but unknown audience of conspecifics rather than at a particular individual. In many species, the same apparatus that makes the calling songs is used to produce sounds ("songs") directed at known targets. For example, many male crickets produce *courtship songs* when contacted by a female, and male field crickets produce *fight songs* when they are in contact with another male. Finally, some katydids, including some females, produce *protest songs* when teased or threatened. These songs generally differ substantially from the calling song and have physical properties appropriate to the context.

Means of Sound Production and Hearing

The principal means of sound production in grasshoppers, katydids, and crickets is *stridulation*. This is accomplished by rubbing together parts of the body that are modified to enhance the sound. Grasshoppers use a variety of body parts, but the most prevalent modes of stridulation are the hind femur rubbing against the forewing and the forewing rubbing against the hind wing. They often accompany their stridulation with visual signals, usually some form of movement or color flashing involving the hind legs.

Crickets and katydids stridulate by rubbing their forewings together (Fig. 21). At the base of each forewing is a specialized vein with a series of regularly spaced, hard, downward-projecting ridges (*teeth*). In most crickets, the functional *stridulatory file* is on the right wing, and in most katydids, it is on the left. The file on the other wing rarely functions and, in the case of most katydids, is poorly developed. Stridulatory files cannot be seen when the wings are at rest, but the *stridulatory vein*, which bears the functional stridulatory file on its underside, is clearly visible. The stridulatory vein and file sometimes have features that are helpful in identifying species. On the inner, lower edge of the forewing that lacks the functional file is the functional *scraper*: a hard, sharp, upward-projecting structure that is positioned to engage the file when the wings are opened and closed during stridulation. Katydids elevate their forewings very little when they stridulate, but crickets raise theirs to about 45 or 90 degrees (e.g., field crickets and tree crickets, respectively). The stridulatory structures of most katydids are highly asymmetrical, whereas those of crickets are not. When females of katydids or crickets stridulate, as when they answer males or protest a disturbance, they use structures on the forewings that are less elaborate than those of the male but nonetheless effective.

Another means of sound production is *crepitation*, or wing snapping, which grasshoppers accomplish as they are flying. It is a common means of sound production only in the band-winged grasshoppers. Crepitation is often accompanied by wing flashing, which is a dramatic visual display in species that have colorful hind wings. The sounds of crepitation are apparently made when the membranes between large veins in the hind wings are suddenly popped taut. Yet another means of sound production is *drumming*, in which the hind legs are shaken vigorously against the substrate on which the insect is standing.

Whereas in most species only males produce sound, both sexes of sound-producing species have organs that allow them to hear airborne sounds. These organs have *tympana*, eardrum-like structures that serve as the interface between the sounds and the insect's nervous system. In grasshoppers, the tympana are fully exposed and located laterally on the first abdominal segment. In katydids and crickets, the tympana are less conspicuous; they are located on the front legs near the base of the tibiae and are sometimes hidden in cavities behind slits.

Identifying Crickets and Katydids by Their Songs

Sound production occurs throughout the order Orthoptera, but it is best developed and most important among the katydids and crickets. Males of nearly every species of katydid and cricket make loud and persistent calling songs that attract sexually ready females. Species that live together have different calling songs, which enable females to go to males of their own kind and to avoid going to males of other species. These species-specific songs are of great value to those who study katydids and crickets. Species can be identified without being seen or collected, facilitating studies of seasonal, ecological, and geographical distributions. Even more important, studies of calling songs have revealed the existence of species that are so similar morphologically as to have been missed in careful studies of museum specimens. In most cases, species that are first recognized on the basis of calling song differences turn out to have clear-cut morphological differences as well. This is because it is vastly easier to find differences among groups of specimens separated into species on the basis of calling songs than it is to decide which morphological variations within a group of specimens are differences among species and which are merely differences within a species.

To use songs for identifying and studying katydids and crickets, you will need to know more about them.

Describing and Depicting Songs

The sounds made by crickets and katydids are best learned by hearing the sounds, but because this field guide has no audio component, we must rely on words and pictures. In some cases, our words describe a subjective impression ("a melodious trill"), or attempt an onomatopoeic rendition ("chlonk"), or use words that are themselves onomatopoeic ("buzz"). In other cases, we state the physical properties of the song. This allows us to be specific about features of the songs that are important to identification. On the down side, the terminology may be unfamiliar and some important physical features cannot be gauged adequately by ear. In this section, we explain the terminology, and in a later section we describe how those who wish to determine the physical features of cricket and katydid songs can do so.

The two common means of depicting sounds are *waveforms* and *spectrograms*. Waveforms are the simpler (Fig. 22). Silence is a horizontal trace, and any sound causes deviations up and down from that line. The more intense (louder) the sound, the greater are the deviations. In spectrograms (see Fig. 22), time again moves along the horizontal axis, but silence leaves no trace. Sounds produce dark marks that show intensity by their darkness, and audiofrequency (pitch) by their vertical position.

If a cricket or katydid song is depicted with enough detail in the time domain (horizontal scale), the cycling of the stridulatory apparatus is revealed. For example, the cricket song depicted in Fig. 22A continues indefinitely and consists of a uniform sequence of *pulses*, each of which is made during the closing stroke of a cycle of forewing movement. During the 1 second of "continuous trilling" that is depicted, there are 45 pulses; thus the cricket is stroking its stridulatory file 45 times per second. In the cricket song depicted in Fig. 22B, the calling song consists of brief groups of pulses (chirps) produced at a rate of about 2 per second. Each chirp has four pulses, which correspond to four wing closures. The waveform and spectrogram reveal that the first pulse is much weaker than the remaining three, and that the third and fourth pulses last longer than the first and slightly longer than the second. The katydid song depicted in Fig. 22C is more difficult to interpret, but wingstroke cycles can still be discerned, especially in the waveform. An interesting feature of this song is that wingstroke cycles occur in pairs. Like calling crickets, this katydid produces a burst of sound during forewing closure but none on the return movement. In producing its song, the katydid is closing its wings about 156 times per second and is alternating short and long times between closures. Because the bursts

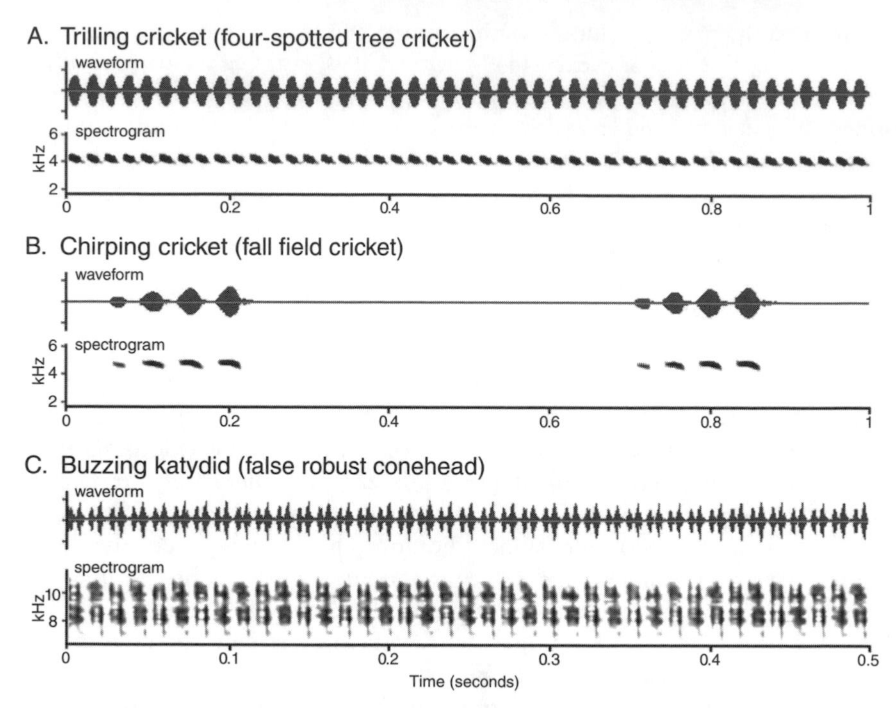

Figure 22. Waveforms and spectrograms of cricket and katydid songs. (A) Four-spotted tree cricket, *Oecanthus quadripunctatus*, producing a 45-pulse-per-second trill with a carrier frequency of about 4.2 kHz. (B) Fall field cricket, *Gryllus pennsylvanicus*, producing four-pulse chirps at a rate of about 2 per s. Pulses have a carrier frequency of 4.7 kHz and are produced at an average rate of 23 per s within each chirp. (C) False robust conehead, *Neoconocephalus bivocatus*, producing 156 paired bursts of sound per s. The broadband carrier frequency has modes of 8.1 and 9.9 kHz.

come in pairs, this song is more "buzzy" than similar songs in which the bursts are evenly spaced.

The tones of cricket and katydid calling songs differ: cricket songs are musical to human ears whereas katydid songs are whiny, buzzy, or raspy. This is because their *carrier frequencies* (subjectively interpreted as pitch) are different. Cricket songs have relatively pure carrier frequencies between 2000 and 9000 cycles per second (i.e., 2 to 9 kiloHertz [kHz]), whereas katydid songs have noisier (less pure) carrier frequencies that are usually higher (8 to 20 kHz). These differences are illustrated in the sound spectrographs, where frequency (in kHz) can be read on the vertical scale and pure tones have minimal vertical dimensions (see Fig. 22). The trilling and chirping crickets have nearly pure frequencies averaging 4.2 and 4.7 kHz, respectively; in both examples, the frequency drops slightly during each pulse. The buzzing katydid produces a mixture of frequencies with two modes, one at about 8.1 kHz and the other at about 9.9 kHz.

In calling crickets, the relationship between sound and wing movements is nearly always as simple as in the two examples given here. The closing portion of each cycle of forewing movement produces a single burst of a nearly pure tone (here termed a pulse). The fundamental oscillations of the pulse correspond to *toothstrikes* of the scraper on the file. In longer pulses, more file teeth are used; in stronger pulses, more force is applied.

In calling katydids, toothstrike rate and carrier frequency do not usually coincide, and the relationship between cycles of wing movements and sounds is complex. Individual toothstrikes may produce discrete sounds, and as many as four markedly different file-stroking techniques may be used in sequence during a single calling song. In such cases, special techniques are needed to establish the relationship between wing movements and parts of the calling song. Repeated patterns are usually evident, however, and each repetition of the pattern is likely to correspond to a complete cycle of wing movement.

Because of their tonal purity and simple amplitude modulation patterns, cricket songs are easily interpreted from spectrograms, whereas the complex tonal mixtures of katydid songs make waveforms the more useful means of showing distinguishing features.

Temperature Effects

Like other insects, crickets and katydids generally have body temperatures that match their immediate environment. The chemistry of nerve and muscle slows at lower temperatures and accelerates at higher ones. Consequently, the rate at which crickets and katydids move their forewings during calling depends on temperature: cool crickets and cool katydids call more slowly than warm ones.

The speed of calling is important in identifying species by their songs because closely related species may have songs that differ importantly only in that regard. A pair of species that differs in this way will always produce different songs when at the same temperature, but if the slower species is warmed or the faster one is cooled (relative to the other), the songs may then have identical wingstroke rates. Because outdoor temperatures are variable, one might anticipate that female crickets or katydids that relied on species-specific wingstroke rates to find their mates would become confused. No problem arises because females also take on the temperature of their surroundings, and the rates to which they respond vary with temperature in the same manner as the rates that males produce.

Naturalists who try to use song speed to distinguish species sometimes have problems. If the fast and slow species are calling together, correctly judging which is faster and which is slower is usually easy. If only one is singing, identification is difficult. However, if the song is recorded *and* the temperature of the caller is noted, identification becomes easy again once the wingstroke rate is determined (see below). The wingstroke rates given in this book are for a temperature of 77 °F. Rates at other temperatures can be adjusted to 77 °F by applying the formula

$$R_{77} = 36R_x/(x - 41)$$

where R_{77} = rate at 77 °F, and R_x = rate at x°F. For example, a cricket that produced 40 pulses per second at 70 °F would be expected to produce 50 at 77 °F [$36 \times 40/(70 - 41)$].

This formula assumes that wingstroke rate is a linear function of temperature and that the line of best fit for rate versus temperature extrapolates to a zero rate at 41 °F. The first assumption is usually true or very nearly so; the second is a useful approximation. The closer x is to 77 °F, the smaller is the error introduced by the formula.

Because wingstroke rates of calling crickets and katydids are linear functions of temperature, simple formulas can estimate the temperatures of callers from their wingstroke rates. However, wingstroke rates are generally much too fast to determine by ear, so using the calls of most species to estimate temperature in the field is impractical. Only species that produce song elements (e.g., chirps) at regular, countable rates make usable thermometers. Of special note in this regard is the snowy tree cricket, which has been dubbed the "thermometer cricket."

Ultrasonic Calls and Detectors

Human ears hear a limited range of frequencies, and sounds of equal energy are perceived as louder if they are in the low end of this range, the part most used in human speech. Sensitivity decreases as frequencies increase. For young, healthy ears the highest audible frequencies are 16 to 20 kHz. No calling song of a North American cricket or katydid is completely ultrasonic: all include frequencies less than 20 kHz. However, most of the energy in some katydid calling songs is above 20 kHz, and many have nearly all of the energy above 10 kHz. The effect is that even with the best ears, some katydid songs are difficult to hear and especially so when mixed with calls of crickets and katydids that have lower carrier frequencies.

Added to the inherent limitations of human ears is the fact that as we age, our ability to hear high frequencies declines. In some individuals (for example, those who don't protect their ears from sounds that overload them), the decline is more rapid than in others, but no one is immune. The effect is that persons who have no hearing handicap in normal conversation may nonetheless be deaf to most katydid songs and even to higher frequency cricket songs.

A variety of electronic devices are available that change high-frequency sounds to lower frequency sounds that nearly everyone can hear. Most of these are sold as ultrasonic detectors and find their principal use in studying the high- and very-high-frequency sounds used by flying bats to detect prey or obstacles. Of special interest to those who have trouble hearing sounds in the upper and middle parts of the human sonic range are devices made for field studies of bird and insect songs. A search of the Internet will reveal sources, models, and prices of "bat detectors" and "songfinders."

Recording and Analyzing Songs

The physical properties of cricket and katydid songs are best determined from recordings of the songs. Recording and analyzing insect songs once required costly, highly specialized equipment. Now anyone with a personal computer equipped with a sound card can, with modest effort and little or no additional expense, render insect songs as digital files and determine their physical characteristics. The paragraphs below summarize how to proceed.

CAGE ADULT MALES. It is possible to record calling songs from free males in the field. With special equipment, you can even record males that are calling in treetops or from other hard-to-approach places. However, because it is important to know what male is producing the call you are recording, and because close proximity eliminates the need for special equipment, recording captured males is the way to begin. Most crickets and katydids thrive in captivity on pulverized dry cat food and a source of moisture (fresh romaine lettuce or a piece of apple), and, in time, most males will call frequently and normally. Useful recordings can be made from any cage that imprisons the male while giving sounds a way to get out, but quality recordings require that the cage have no parallel hard surfaces. Some crickets quickly chew through cloth mesh and escape; metal screening will confine all species. (See the section on how to culture grasshoppers, katydids, and crickets.)

RECORD THE CALLING SONGS. You can use a recorder that is independent of a computer or you can use the computer's sound card to record directly into a digitized sound file. Any independent audio recorder that will record music with high fidelity will serve to record calling songs. To use a computer to record directly, plug the microphone into the microphone port and make the recording with whatever software you have for the purpose—for example, Sound Recorder, a program that comes with Microsoft Windows. If you have a choice of file format for saving the recorded song, choose *wav*, because most audio programs accept *wav* files. No matter how you record a calling song, be sure to note what male produced the song and its temperature at the time of recording. A recording made of an unknown male or at an unknown temperature is of little use. You must devise a means of keeping essential information permanently associated with the recording and the recording associated with the specimen. For example, you can give each recording a unique code that is spoken at the end of the "take," and use that code to label the specimen and to record the circumstances of the recording in a notebook or database.

DIGITIZE THE RECORDINGS. If you make the recordings with the computer, the songs are digitized (saved as 0's and 1's) as you record them. If you make the recordings with an analog tape recorder, you can convert them into digital files by feeding the output of the recorder into the port on the computer meant for acoustical inputs from devices other than microphones.

ANALYZE THE SONGS. If your only concern is counting the number of pulses in a chirp or determining the wingstroke rate, you may be able to use the program you used for digitizing to slow a playback of the song sufficiently to accomplish your ends. For example, Sound Recorder's Effects|Decrease Speed command will slow a song to one-half, one-fourth, one-eighth, etc. normal speed. At one-sixteenth normal speed, it is possible to count pulses that were

Figure 23. Orthopterans sometimes become numerous and form aggregations consisting of winged individuals, called *swarms*, or of flightless individuals, called *bands*. Shown here is a band of Mormon crickets, *Anabrus simplex*, crossing a roadway. Once bands form, they march daily in the same direction. At night, they form small clusters in protected locations. The mechanism of coordination among members of the bands is unknown.

originally emitted at 50 pulses per second and to use a stopwatch to determine the slowed rate. (The normal rate will be 16 times the slowed rate.) For analyses beyond what can be done by ear, you need a program that displays digitized sounds as waveforms or spectrograms. Several basic programs are available for free on the Internet.

Grasshoppers, Katydids, and Crickets as Pests

Grasshoppers and katydids usually consume considerable amounts of foliage during their nymphal development, and also as adults. Most crickets are less inclined to damage living plants, but on occasion become numerous enough to be a nuisance or to cause plant injury. Grasshoppers are by far the most likely to be damaging. They may injure forage, field, vegetable, fruit, and ornamental crops by defoliating them. Damage is especially common in states west of the Mississippi River, where the arid climate allows more frequent development of grasshopper plagues. Even in eastern states, however, grasshoppers sometimes reach alarmingly high and damaging densities. In the West, shieldback katydids also reach high densities on occasion. One shieldback species known as the Mormon cricket is notorious for becoming abundant and damaging crops (Fig. 23). Tree and bush crickets cause another type of injury, sometimes killing berry canes and the twigs of trees as a result of egg laying.

The Origin of Plagues

Abnormally high densities of grasshoppers and other insects are called outbreaks by entomologists, and plagues by the general public. Irrespective of the terminology applied, the phenomenon occurs throughout the world, and in grasshoppers its origin is invariably related to weather and food. To a lesser degree, outbreaks occur among some katydid and cricket populations, though the mechanisms are less well understood than in grasshoppers.

Grasshoppers require warm and sunny conditions for optimal growth and reproduction. Warmth alone seems to be inadequate. Grasshopper activity diminishes during cloudy weather. Thus, drought stimulates grasshopper populations to increase, apparently because there is less rainfall and cloudy weather to interfere with grasshopper activity. A single season of such weather is not adequate to stimulate massive population increase; two to three years of drought usually precede grasshopper plagues. Warm winter temperatures also seem to be a factor because less mortality of overwintering nymphs and adults occurs.

Food is a prerequisite for grasshopper success. Optimal weather alone, in the absence of adequate food supply, is insufficient for rapid population growth. For outbreaks or plagues to occur, both requisites must be satisfied. Thus, some precipitation must be present at the appropriate time to stimulate plant growth, but an overabundance of rainfall results in too much cloud cover. Optimal weather conditions and food supply rarely coincide, so outbreaks are not frequent.

The balance of weather and food vary according to location. In cooler northern locations, food is usually abundant and grasshoppers are more often limited by absence of sufficient warm weather, so abnormally hot, dry weather can produce a population increase. In contrast, in warmer southern locations, grasshoppers are more limited by the availability of adequate food, so increases in rainfall and associated vegetative growth stimulate grasshopper populations.

Grasshoppers, katydids, and crickets have numerous natural enemies, especially other insects, such as parasitic flies or predatory beetles. Under most conditions, these natural enemies are abundant and effective, preventing orthopteran outbreaks. However, if special circumstances allow an orthopteran population to exceed some threshold, the usual natural enemies no longer keep pace, and an outbreak ensues that continues until starvation, disease, or other mortality factors return the population to a non-outbreak level (Fig. 24).

Controlling Outbreaks

INSECTICIDES. When orthopteran populations reach outbreak levels, those with commercial interests at risk, such as farmers and livestock producers, spend large sums to buy and apply chemical insecticides in hopes of avoiding greater losses. Because insecticide applications can result in collateral environmental damage, the government and concerned citizens support the development of pest suppression strategies that reduce or eliminate the use of insecticides. What are the alternatives to using insecticides to control orthopteran outbreaks, and how effective are they?

Figure 24. A dead grasshopper clinging to the upper portions of a plant stem, a sign of "summit disease." Such grasshoppers are infected with a fungal disease, *Entomophaga grylli*. During periods of grasshopper abundance, it is normal to see dead grasshoppers like this. They may be widespread, but in arid environments they often are found adjacent to ponds and other water sources.

CULTURAL PRACTICES. Prevention is the best control of an outbreak. This has little application for orthopteran outbreaks that originate in relatively natural habitats and disperse great distances. Grasshoppers may take flight and Mormon crickets may walk many miles to reach crops and gardens. However, for pest species that have a local origin, changes in land management practices will help. Mechanical mowing or livestock grazing of the source areas will prevent outbreaks unless grass pastures are damaged by overgrazing and broadleaf weeds invade.

PHYSICAL BARRIERS. Valuable plants can be covered with netting, floating row cover, or similar material to deny access to grasshoppers, katydids, and crickets.

This is suitable for small gardens, and is even applied commercially for ornamental plant production, wherein shade houses are sealed tightly to prevent insects from entering. The potential for this approach on a larger scale is limited by its cost. For flightless species, physical barriers in the form of a ditch with steep sides, or a short metal or plastic wall, can be an effective impediment. If such a wall is contemplated, however, consider that grasshoppers can ascend vertical surfaces with amazing agility, so the top of barriers should be bent out at a 45-degree angle, forcing the insects to fall back.

BIOLOGICAL SUPPRESSION. People often ask about the potential for biological suppression rather than using chemical insecticides. It is true that wild birds and domestic fowl (especially turkeys) readily consume vast quantities of insects, but this solution is not appropriate for most circumstances because the birds are not sufficiently abundant. There are insect disease agents under investigation, with some that are already sold commercially, but so far few have been shown to provide adequate suppression. For some people, neem products are attractive. Neem products are derivatives of the neem tree that, when applied to plants, act as a feeding deterrent, reducing damage. Also, if applied to nymphs, neem can act as a growth regulator, disrupting the normal growth and development, and sometimes resulting in death or sterilization of grasshoppers, katydids, and crickets. Although neem products are chemicals, many people take comfort in knowing that they are derived from plants, and therefore somewhat natural. Like many natural controls, effectiveness in not consistent. Biological suppression remains a promising area for research, but thus far the successes have been minor.

CLASSICAL BIOLOGICAL CONTROL. For nonnative species that became pests when they reached North America because their natural enemies were left behind, a potentially effective and self-perpetuating control strategy is the introduction of natural enemies from the pests' homeland. Termed classical biological control, this strategy has the danger that the introduced natural enemies may become pests themselves. Therefore, regulations require a thorough evaluation of potential harm before permission is granted to release any such control agent.

Three species of two-clawed mole crickets (genus *Scapteriscus*) are the only nonnative orthopterans that have become major pests in North America, and introduction of their enemies has greatly decreased the damage they do. Between 1899 and 1925, tawny, southern, and short-winged mole crickets were accidentally introduced from South America at several ports in the southeastern United States, probably in ships' ballast. They gradually spread until one or more of the species were pests of turf and pastures throughout the Southeast. Studies at the University of Florida that began in 1978 showed that in their South American homeland they were kept at non-pest levels by a complex of natural enemies. Several of these species were evaluated for release in the United States, and by 1988 three had been approved and released: a nematode that enters a host mole cricket, kills it by introducing lethal bacteria, and produces progeny that feed off the carcass; a parasitoid fly that finds its hosts by homing in on the calling song; and a digger wasp that chases mole crickets from their burrows to make them hosts for its lethal larvae. All attack only two-clawed mole crickets. All have become established in Florida, where they have reduced mole crickets to non-pest levels in most contexts.

TOLERANCE. If total elimination of insect damage is demanded, few control measures other than applications of insecticides will be judged successful. In the case of the classical biological control of mole crickets, they can be kept at low levels with no ongoing expenditures, but they are not eliminated. Unfortunately for turf managers of golf courses, one mole cricket burrow on a golf green is more than can be tolerated. To generalize from this example, the reduction or elimination of applications of insecticides may be thwarted by homeowners or consumers who demand perfect lawns or blemish-free produce. Greater tolerance of minor damage would greatly reduce the need for chemical insecticides.

Pictorial Keys

Following is a pictorial guide consisting of insect images and simple diagrams. They are designed to help you identify your unknown insect. Of course, the unknown must be an insect in the order Orthoptera: a grasshopper, katydid, or cricket. These insects are quite distinctive, though, and if the hind femora are expanded you can be fairly sure that you are using the correct field guide.

To use the pictorial guide, start at the beginning with the key to the major groups. In all cases, you must select between two options, as indicated by arrows leading to the options. Look carefully at your insect before deciding which option is most correct, then follow the path, making choices as you go. After determining the correct group (grasshopper, katydid, or cricket), proceed to the appropriate key and continue making choices to determine the family or subfamily. Once you have narrowed down the options to one of these smaller groups, find the group using the table of contents or index, then use the pictures, diagrams, maps, and descriptions in the species accounts section to identify your insect. Don't be discouraged if you initially find yourself in the wrong group, or your unknown insect resembles two or more insect descriptions. You also may find that your insect fails to match any species described, suggesting that your insect either is immature, or is one of the rarer species not covered in the field guide. Orthopterans are among the easiest insects to identify; it just takes time and effort to gain expertise and confidence.

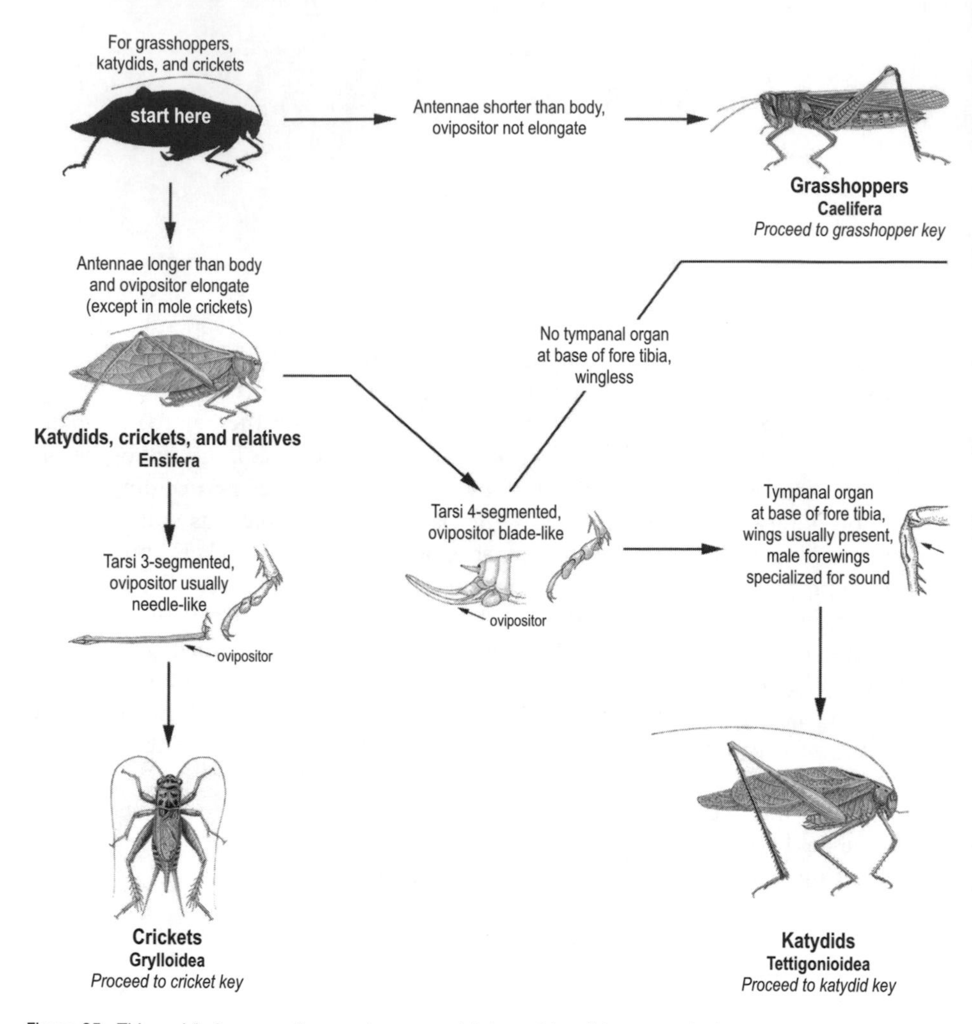

Figure 25. This and facing page. For grasshoppers, crickets, and katydids, turn to the keys on the following pages; for gryllacridoids, see plate 48 and the text: camel crickets, p. 215; Carolina leaf-roller, p. 217; Jerusalem crickets, p. 216.

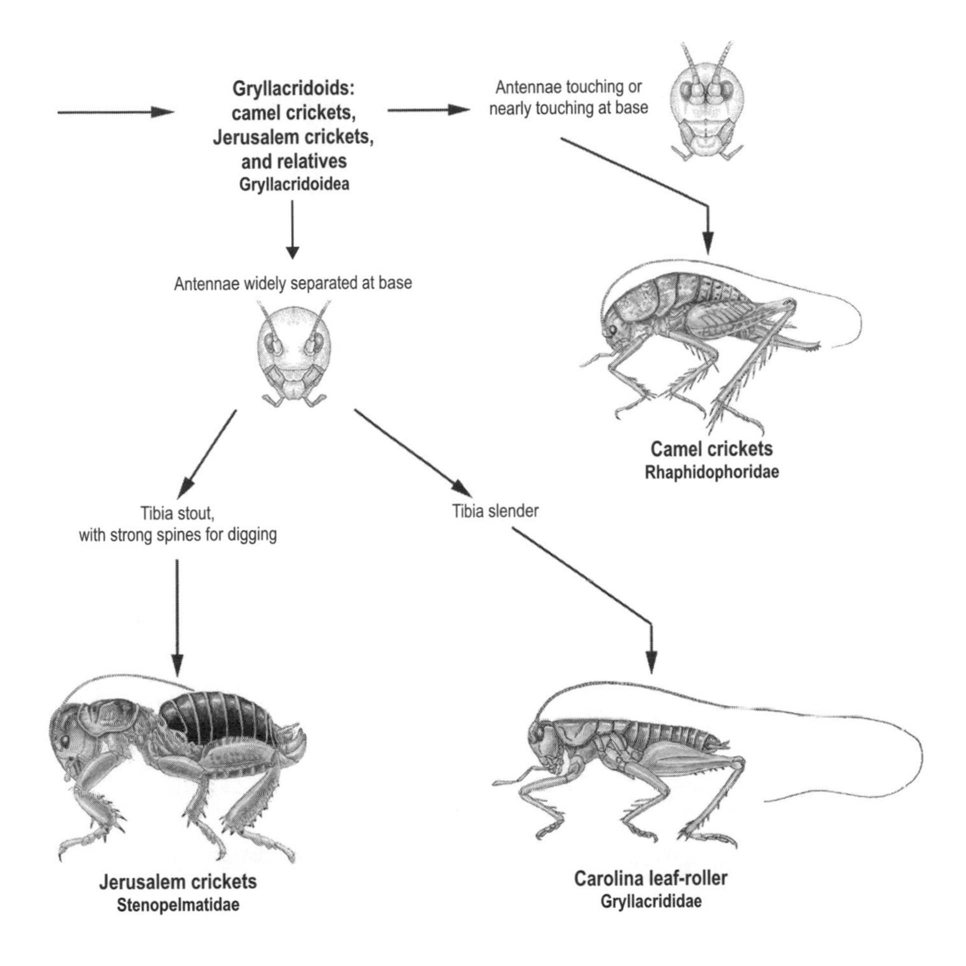
Gryllacridoids:
camel crickets,
Jerusalem crickets,
and relatives
Gryllacridoidea
Antennae touching or
nearly touching at base
Antennae widely separated at base
Camel crickets
Rhaphidophoridae
Tibia stout,
with strong spines for digging
Tibia slender
Jerusalem crickets
Stenopelmatidae
Carolina leaf-roller
Gryllacrididae

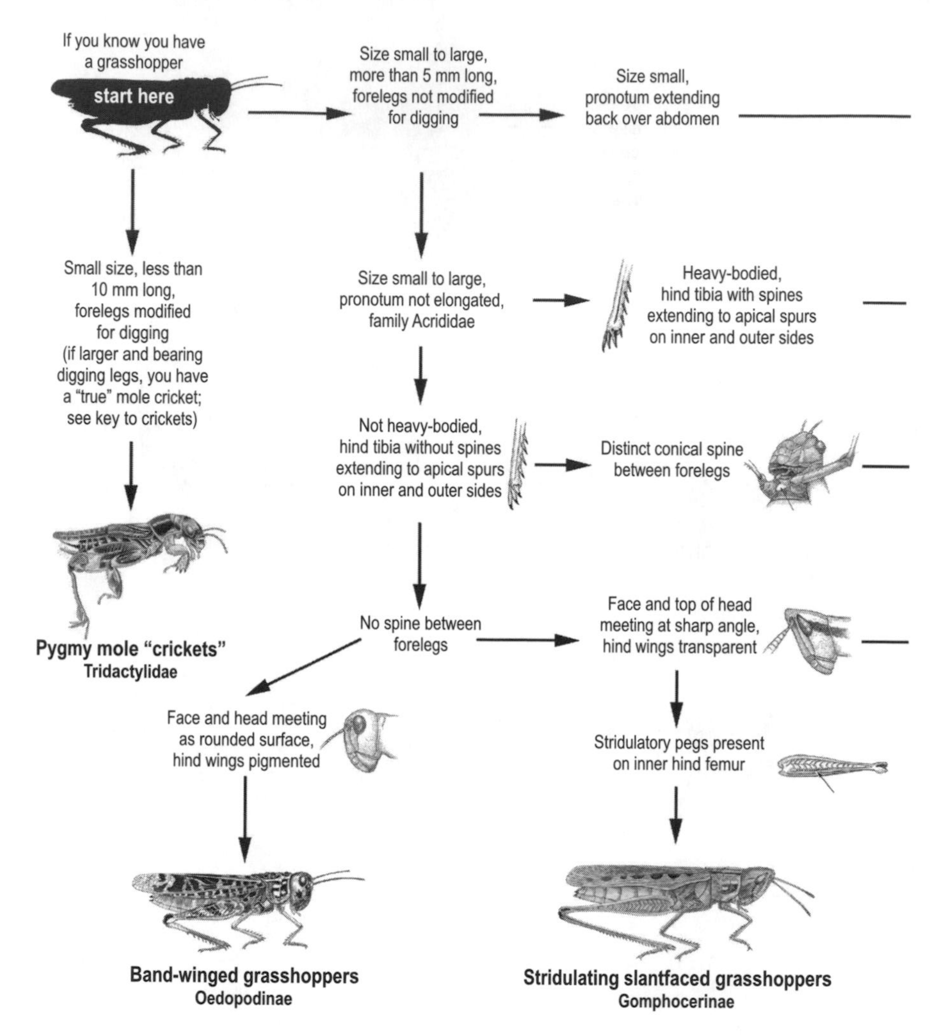

Figure 26. This and facing page. See the text and plates to identify further: band-winged grasshoppers, p. 77, pls. 8–21; lubber grasshoppers, p. 148, pl. 32; pygmy grasshoppers, p. 150, pl. 33; pygmy mole "crickets," p. 152, pl. 33; silent slantfaced grasshoppers, p. 76, pl. 7; spurthroated grasshoppers, p. 107, pls. 22–31; stridulating slantfaced grasshoppers, p. 55, pls. 1–7.

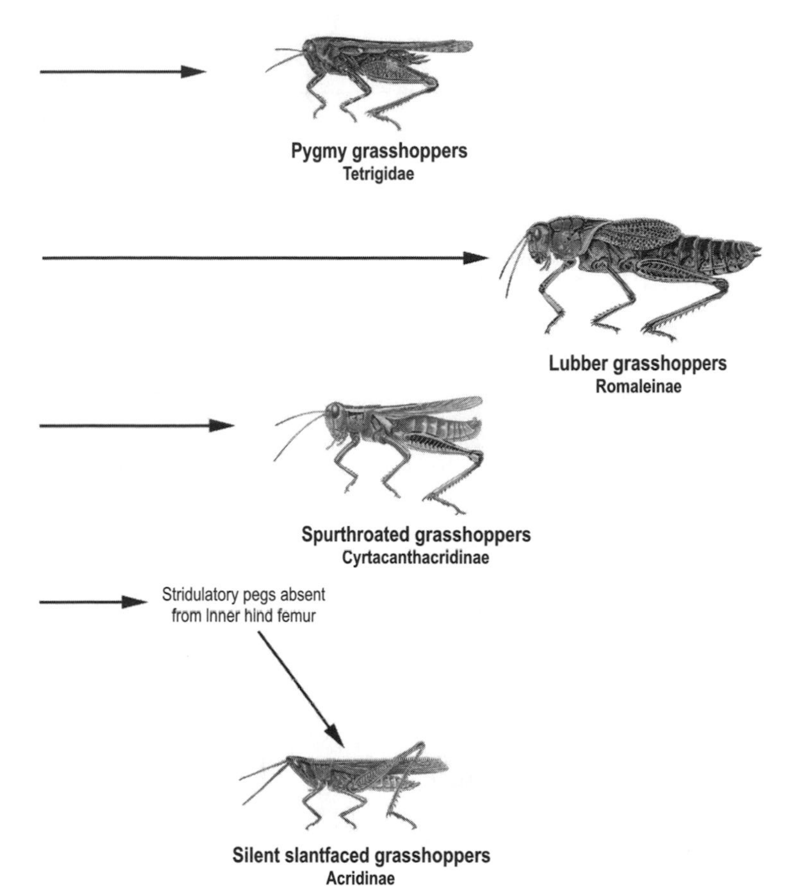
Pygmy grasshoppers
Tetrigidae
Lubber grasshoppers
Romaleinae
Spurthroated grasshoppers
Cyrtacanthacridinae
Stridulatory pegs absent
from inner hind femur
Silent slantfaced grasshoppers
Acridinae

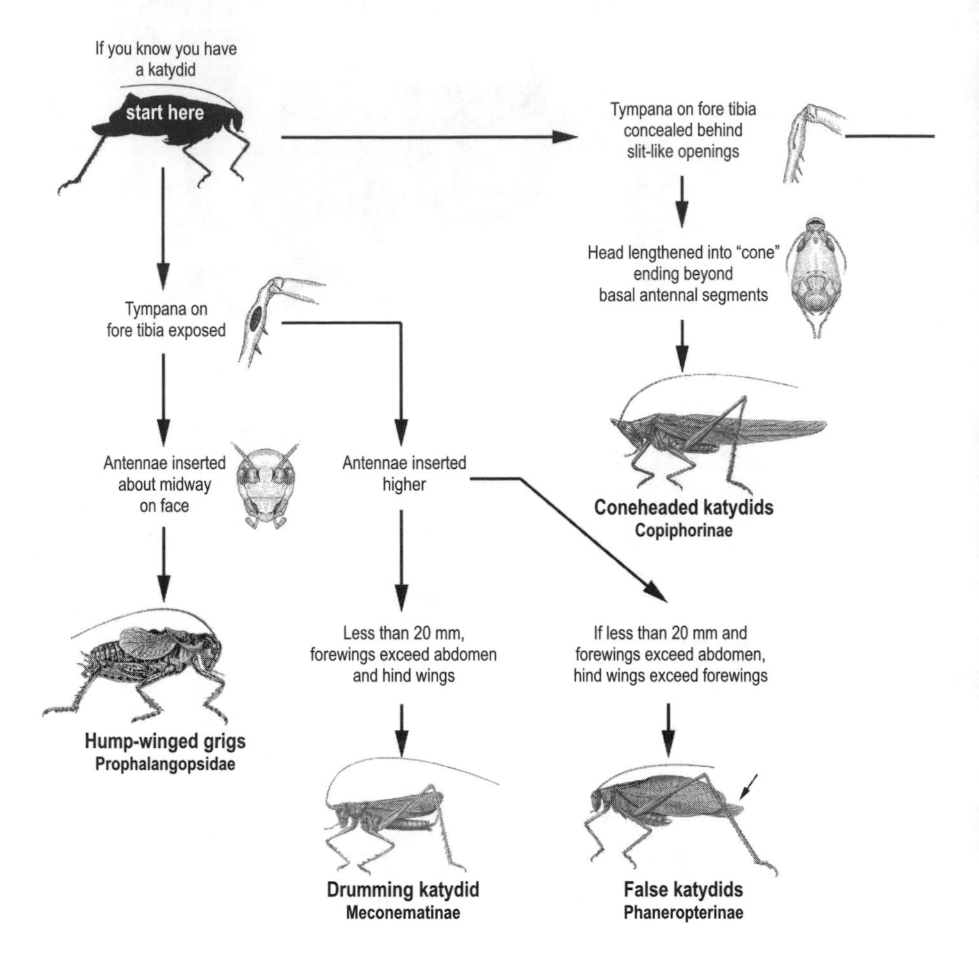

Figure 27. This and facing page. See the text and plates to identify further: coneheaded katydids, p. 169, pls. 37–38; drumming katydid, p. 183, pl. 40; false katydids, p. 157, pls. 34–36; hump-winged grigs, p. 191, pl. 41; meadow katydids, p. 177, pls. 39–40; predaceous katydids, p. 184, pls. 41–42; true katydids, p. 155, pl. 34.

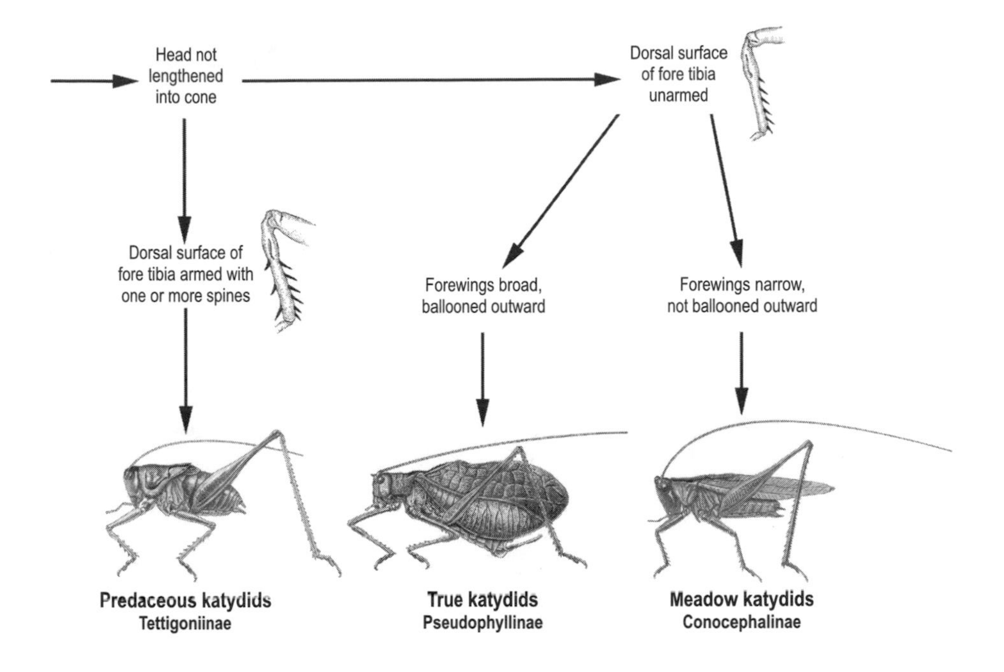
Head not lengthened into cone
Dorsal surface of fore tibia unarmed
Dorsal surface of fore tibia armed with one or more spines
Forewings broad, ballooned outward
Forewings narrow, not ballooned outward
Predaceous katydids
Tettigoniinae
True katydids
Pseudophyllinae
Meadow katydids
Conocephalinae

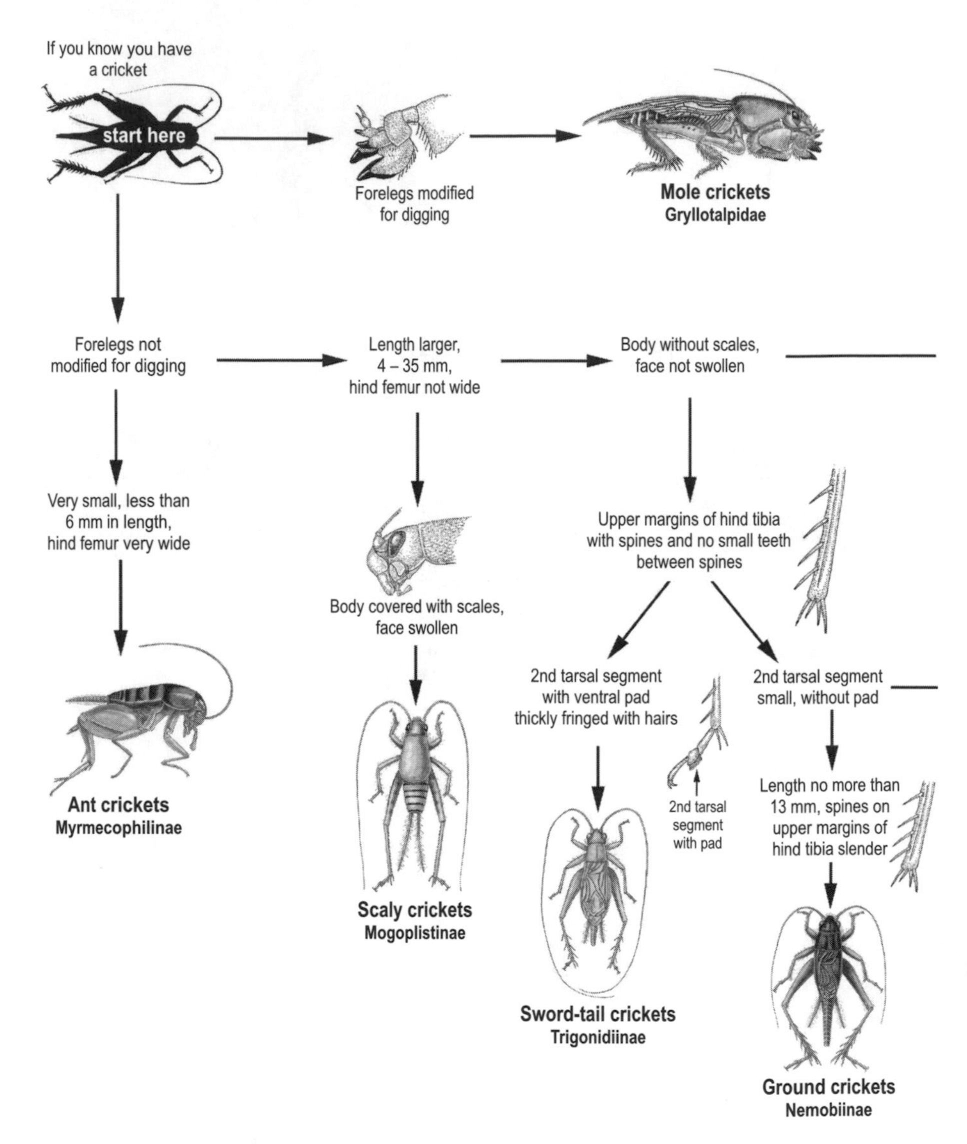

Figure 28. This and facing page. See the text and plates to identify further: ant crickets, p. 211, pl. 47; bush crickets, p. 203, pl. 45; field crickets, p. 193, pl. 43; ground crickets, p. 199, pl. 44; mole crickets, p. 212, pl. 47; scaly crickets, p. 210, pl. 45; sword-tail crickets, p. 201, pl. 44; tree crickets, p. 205, pl. 46.

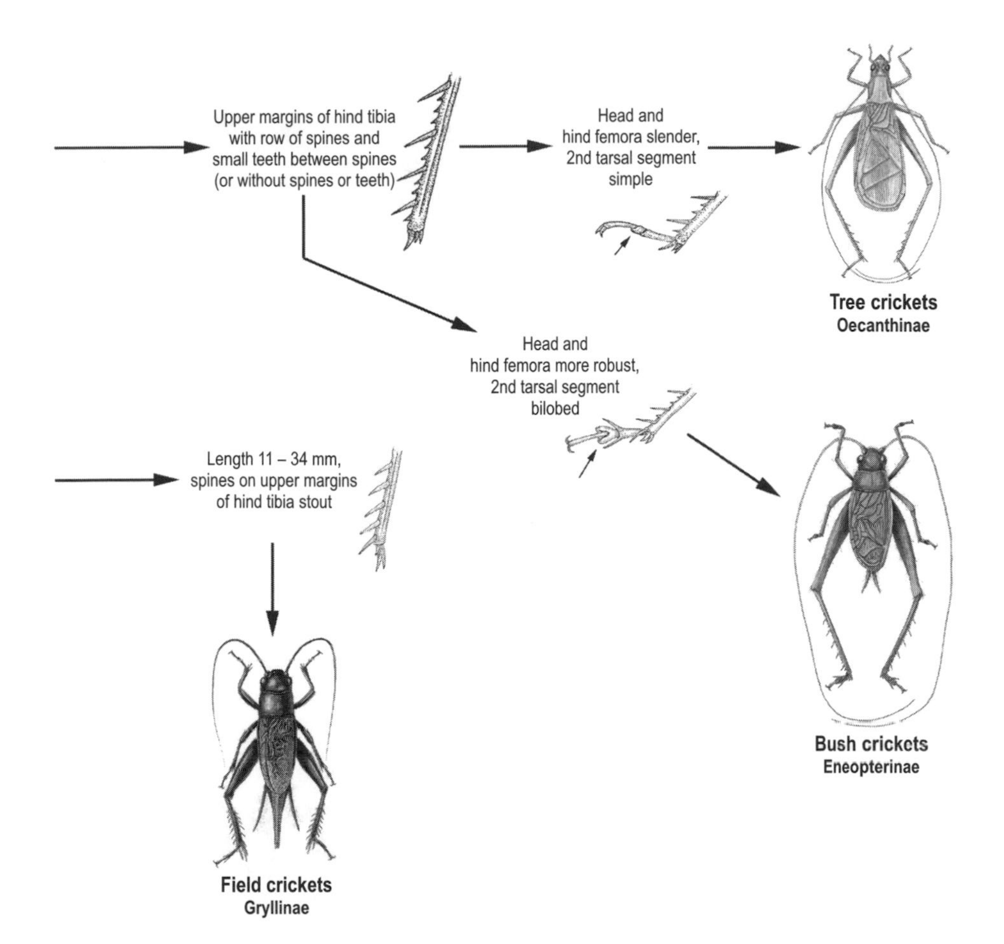
Upper margins of hind tibia with row of spines and small teeth between spines (or without spines or teeth)
Head and hind femora slender, 2nd tarsal segment simple
Tree crickets
Oecanthinae
Head and hind femora more robust, 2nd tarsal segment bilobed
Length 11 – 34 mm, spines on upper margins of hind tibia stout
Bush crickets
Eneopterinae
Field crickets
Gryllinae

SPECIES ACCOUNTS

Grasshoppers

Suborder Caelifera

Most grasshoppers are easily recognized as belonging to one of the three families or five subfamilies treated in this field guide. This will quickly narrow down the potential choices and speed up your identification. Keep in mind that although most grasshoppers are easily sorted, a few are quite deceptive. For example, look at the pictures of the cattail toothpick grasshopper, *Leptysma marginicollis* (Plate 30), or the glassy-winged toothpick grasshopper, *Stenacris vitreipennis* (Plate 30). Superficially, these species do *not* look like spurthroated grasshoppers (subfamily Cyrtacanthacridinae); they look like stridulating slantfaced grasshoppers (subfamily Gomphocerinae). However, even casual examination of the underside of the grasshoppers will reveal a prosternal spine protruding from the base of the front legs, which identifies them as belonging to the spurthroated group. Similarly, try to find the band on the hindwing of the band-winged grasshopper *Camnula pellucida* (Plate 9). Good luck! It is called the clear-winged grasshopper for a very good reason; it lacks the typical pigmentation on the hind wings of band-winged species (subfamily Oedipodinae). Although it resembles a spurthroated grasshopper in wing color, this species' lack of a prosternal spine is a sure sign that it is not a spurthroat.

So, if you have an unknown grasshopper, take a few minutes to review the family and subfamily characteristics of the grasshoppers given in the pictorial key and look for each character. General appearance can be deceiving.

Stridulating Slantfaced Grasshoppers

Family Acrididae, subfamily Gomphocerinae

Grasshoppers in this subfamily tend to have slender bodies and long, slender legs. Their heads are elongate and often cone-shaped, usually having a highly slanted face (see Fig. 3). They usually lack the ventral spine between the front legs (the prosternal spine) that is found in the spurthroated grasshoppers (subfamily Cyrtacanthacridinae) and lubber grasshoppers (subfamily Romaleinae). Gomphocerine grasshoppers tend to be green or brown in color; sometimes distinctly brown or green forms occur within the same species. The hind wings usually are not colorful.

Gomphocerines have relatively short wings, rendering them incapable of sustained flight. When disturbed, these grasshoppers leap and use their wings, but their wings do little more than increase the distance jumped. They do not make sounds during flight (called crepitation), as occurs in the band-winged grasshoppers (subfamily Oedipodinae). This does not mean these grasshoppers are silent—they can make noise by rubbing the inner surface of the hind femur on the edges of the forewing. They create this sound (called stridulation) while resting, however, not while flying. Because the males of this subfamily usually have a row of stridulatory pegs on the inner surface of the hind femora, they are also known as toothlegged grasshoppers.

The habitat of gomphocerines tends to be tall grasses in open fields. The form and color of many species allows them to blend in with stems and blades of grass, making them difficult to detect until they move. Most species feed predominantly on grasses.

Although the common species are discussed here, more complete treatment of the subfamily is found in *The North American Grasshoppers,* volume 1 (1981) by Daniel Otte.

LONG-HEADED TOOTHPICK GRASSHOPPER

Achurum carinatum (Plate 1)

Distribution: Found throughout the southeastern states from South Carolina to Mississippi.

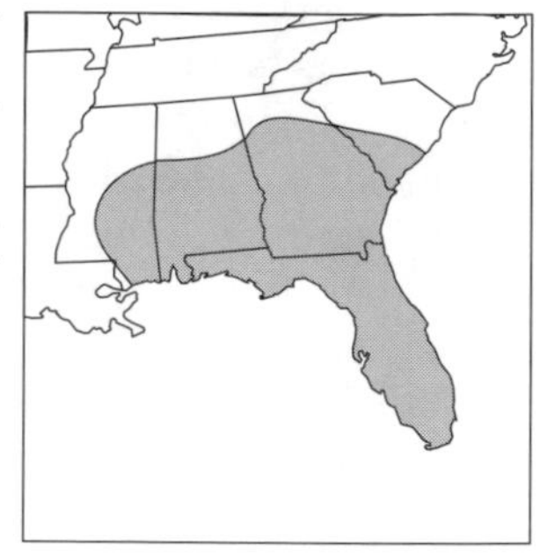

long-headed toothpick grasshopper

Identification: As suggested by its name, this slender grasshopper has an extremely long, thin body form. It is pale brown or grayish brown in color, often with the forewings and legs partly green. Thus, it easily blends in with grasses and pine needles and is difficult to detect. Forewings of this flightless species are small, no longer than the length of the head. Face is extremely slanted, the antennae large and sword-shaped. Some individuals bear numerous black dots. Males are 24–36 mm long, females 33–50 mm.

Ecology: Commonly found in grass of open woodlands, particularly pine woodlands, and areas with tall grasses, such as old fields and pond margins. Both nymphs and adults can be found through most of the year in Florida, though nymphs predominate in winter, and eggs likely hatch in late summer.

Similar Species: There are several "toothpick" grasshoppers that are similar in appearance. The short wings and absence of a spine between the front legs serve to differentiate the long-headed toothpick grasshopper from some other thin-bodied species, such as the cattail toothpick grasshopper, *Leptysma marginicolis* (Plate 30), and the glassy-winged toothpick grasshopper, *Stenacris vitreipennis* (Plate 30). In western states, the Wyoming toothpick grasshopper, *Paropomala wyomingensis* (Plate

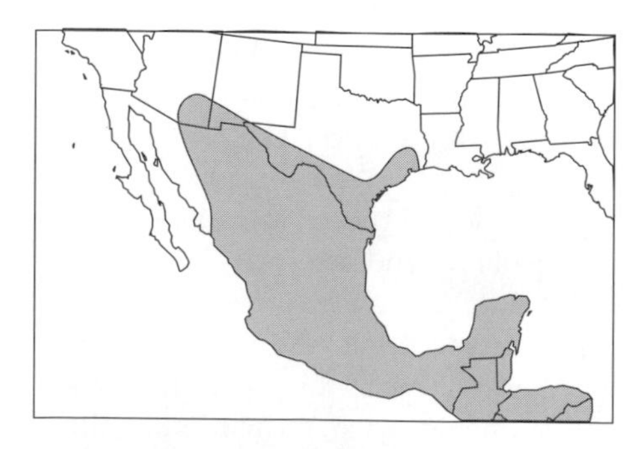

Sumichrast's toothpick grasshopper

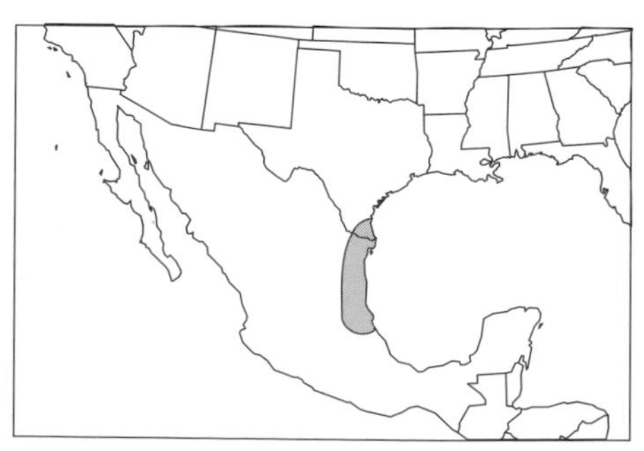

stubby toothpick grasshopper

3), is similar in appearance and lacks a spine between the front legs, but this species is not as thin and the wings are longer. A rare species from southern Arizona and Mexico, *Porocorypha snowi*, is spineless and its body is extremely long and thin. It greatly resembles the long-headed toothpick grasshopper but bears a pale stripe running from the antennal bases to the hind femora.

Two additional species of *Achurum* occur in the United States, both in western states. Sumichrast's toothpick grasshopper, *A. sumichrasti*, is easily separated from *A. carinatum* because it bears long wings. The stubby toothpick grasshopper, *Achurum minimipenne*, is more similar in appearance but the wings are longer than the head, whereas in the long-headed toothpick grasshopper they are shorter than the head.

GREEN FOOL GRASSHOPPER

Acrolophitus hirtipes (Plate 6)

Distribution: Found in the western Great Plains from Canada's Prairie Provinces south to Texas and northeastern Mexico.

Identification: The pointed head and high arching pronotum serve to distinguish this grasshopper from others. In northern regions of its range, it is entirely uniform green, but in southern areas the thorax and forewings are mottled. Unlike most other stridulating slantfaced grasshoppers, it has hind wings that are greenish yellow basally and marked with a black band centrally. Antennae are red. Males are 25–42 mm long, females 32–51 mm.

green fool grasshopper

Ecology: An inhabitant of rolling shortgrass prairie, this species is unusual among striped slantfaced grasshoppers in preferring broadleaf plants, particularly those in the Boraginaceae family. Males are known to stridulate. Adults are found from June to August.

Similar species: Though superficially similar to the green fool grasshopper, the great crested grasshopper, *Tropidolophus formosus* (Plate 21), is distinguished by the lack of a pointed head and the presence of teeth on the pronotal crest. Also, the hind wings of the great crested grasshopper are orange, with an incomplete brown band.

CLUB-HORNED GRASSHOPPER

Aeropedellus clavatus (Plate 1)

Distribution: Found widely in the northern states and southern Canada, though it is western or west-central in general distribution. Occurs on the High Plains in many areas; its southern occurrence along the Rocky Mountains into Colorado and New Mexico is limited primarily to foothill and mountain areas.

Identification: As suggested by its name, a diagnostic character of this species is the enlarged tips of the antennae. This species is green or brown in color. Ridges on the dorsum of the pronotum are strongly constricted in front of the midpoint. In males, the wings are about the same length as the

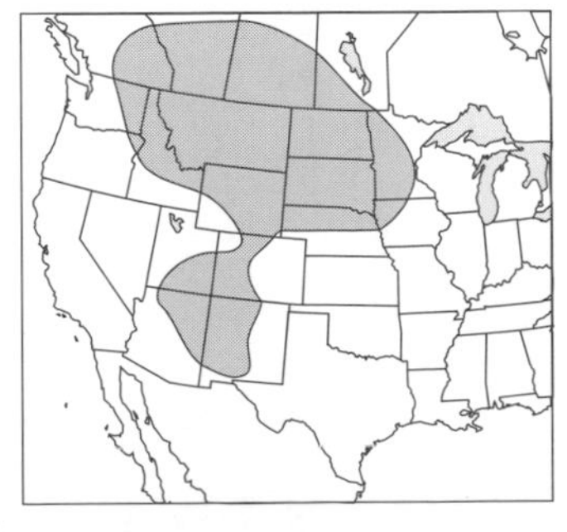

club-horned grasshopper

abdomen, whereas in females the wings are shorter, about half the length of the abdomen. Abdomen is marked on the sides with a dark spot on each segment. Hind tibiae are yellow to brown. Males are 16–20 mm long, females 18–25 mm.

Ecology: Considered to be an alpine and boreal species, but in many northern areas it also occurs on grassy plains. Feeds on grasses and sedges, and occasionally becomes abundant enough to cause damage to pastures. Males are active, constantly running along the ground or flying. Females are more sedentary. Adults are found from July to September.

WHITE-WHISKERED GRASSHOPPER

Ageneotettix deorum (Plate 1)

Distribution: Found widely in western North America west of the Mississippi River and also occurs in the Great Lakes area.

white-whiskered grasshopper

Identification: This small to medium-sized species is brown above and yellow below. Antennae are threadlike and whitish. White stripes extending back from the head along the upper surface of the pronotum are constricted near the middle of the pronotum. Forewings vary in length but usually are about the length of the abdomen and are marked with brown speckles centrally. Hind femora bear a dark triangular spot centrally on the upper surface. The junction of the hind femora and tibiae is black. Hind tibiae are orange or red. Males are 11–28 mm long, females 15–28 mm.

Ecology: Inhabits areas with little vegetation. Feeds on grasses and sedges, sometimes causing damage to rangeland grasses. Adults are found from July to October.

BROWN WINTER GRASSHOPPER

Amblytropidia mysteca (Plate 1)

Distribution: Found widely in the southern United States from North Carolina to Arizona, and south through Mexico and Central America.

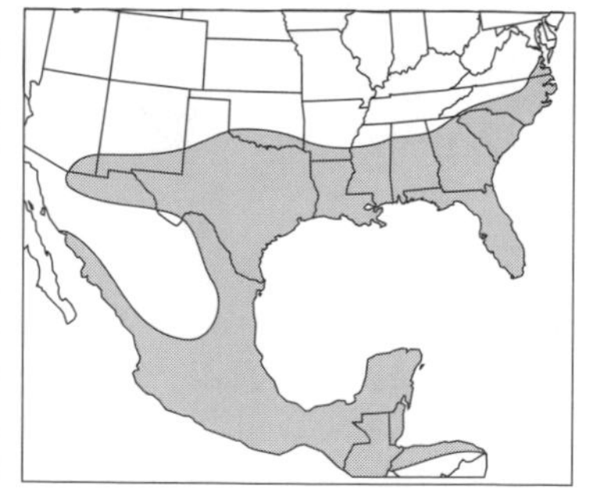

brown winter grasshopper

Identification: This yellowish-brown to brownish-black grasshopper is robust in form. Although the face is strongly slanted, the top of the head is broadly rounded. Antennae are relatively short. The top of the head, thorax, and sometimes a portion of the forewings are yellowish brown or gold. The junction of the upper surface and sides is marked by a distinct change in color. Forewings are dark brown or blackish distally. Hind femora are light brown or gold, sometimes with a longitudinal black line. Hind tibiae are light brown basally and brownish black distally. Males are 19–27 mm long, females 27–38 mm.

Ecology: Inhabits short and moderately high grasses, usually in open woodlands. Along the Gulf Coast, this species can be found during the winter months in both nymphal and adult stages but also is present through most of the summer. The egg stage occurs during the summer months. When disturbed, brown winter grasshoppers fly short distances, dive into vegetation, and burrow out of sight among the foliage and debris; this easily recognizable behavior aids in field identification.

STRIPED SLANTFACED GRASSHOPPER

Amphitornus coloradus (Plate 1)

Distribution: Found throughout the western United States and Canada, west of the Mississippi River.

Identification: This is a brownish-yellow or green grasshopper with two dark stripes running from the top of the head, broadening and fading on the forewings. The sides of the head and pronotum are brown. The femora bear two dark bands on the outer face, and the junction of the femur and tibia is black. Hind tibiae are bluish. Males are about 20 mm long, females about 25 mm.

Ecology: This grass-feeding species is found in areas with short or sparse grass. At times, this species is sufficiently abundant to damage grasses, though usually as part of a species complex. Adults are found from July to September.

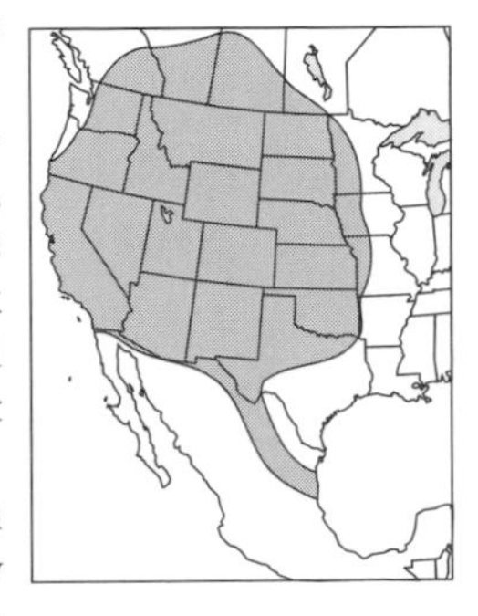

striped slantfaced grasshopper

BIG-HEADED GRASSHOPPER

Aulocara elliotti (Plate 1)

Distribution: Found widely in western North America, ranging from the central Great Plains to the Pacific Ocean. It also occurs in southern areas of Canada's Prairie Provinces, and in north-central Mexico.

Identification: This yellowish-brown species is marked by an oversized head, a weak whitish cross or X on the dorsal surface of the pronotum, and blue hind tibiae. Lateral lobes of the pronotum often are marked with a blackish patch. Forewings are brownish or grayish with dark speckles. Hind femora bear two weak black bands. Wings extend beyond the tip of the abdomen. Males are 16–25 mm long, females 22–35 mm.

big-headed grasshopper

Ecology: Inhabits shortgrass prairie and desert grassland habitats, preferring areas of sparse vegetation. Feeds on grasses. Males may stridulate during courtship. At times, this species is sufficiently abundant to damage grasslands. Adults are found from June to September.

Similar species: The big-headed grasshopper is very similar in appearance to the white-crossed grasshopper, *Aulocara femoratum* (Plate 1). In the white-crossed grasshopper, however, the wings do not reach the tip of the abdomen. Also, the lateral lobes of the pronotum and the hind femora of the white-crossed grasshopper are more distinctly marked with black.

WHITE-CROSSED GRASSHOPPER

Aulocara femoratum (Plate 1)

Distribution: Found from the western edge of the Great Plains nearly to the Pacific Coast, including southernmost Canada and northern Mexico, but infrequent west of the Rocky Mountains.

Identification: This yellowish-brown species is marked by an oversized head, a white cross or X on the dorsal surface of the pronotum, and blue hind tibiae. Lateral lobes of the pronotum are marked with a large black patch, and the hind femora each bear two black bands. Wings are shorter than the abdomen. Males are 16–23 mm long, females 22–34 mm.

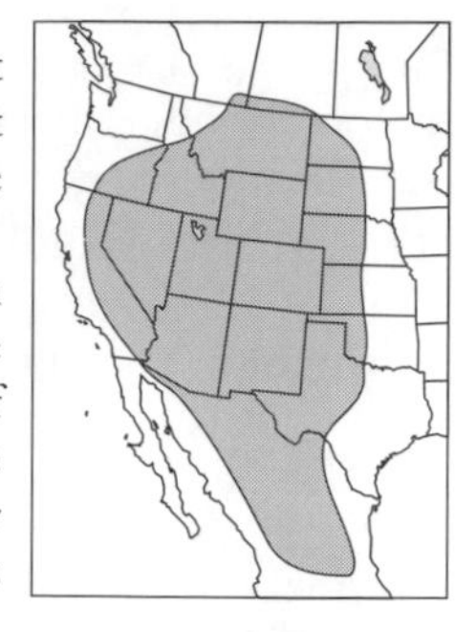

white-crossed grasshopper

Ecology: Inhabits shortgrass prairie and desert grassland habitats, preferring areas of sparse vegetation. Feeds on grasses. Males stridulate during courtship. Adults can be found from June to October.

Similar species: The white-crossed grasshopper is very similar in appearance to the big-headed grasshopper, *Aulocara elliotti* (Plate 1). In the big-headed grasshopper, however, the wings extend beyond the tip of the abdomen, and the lateral lobes of the pronotum and the hind femora are less distinctly marked with black.

GRACEFUL RANGE GRASSHOPPER

Boopedon gracile (Plate 6)

Distribution: Limited to the southern Great Plains, primarily Kansas south through Texas and into northeastern Mexico.

Identification: This is a moderately large, heavy-bodied grasshopper with an unusually large, rounded head. It is brown or yellow brown but marked with cream or ivory, and black. Hind tibiae are red, black, cream, or a combination of these colors. Lateral lobes of the pronotum are partially black, and three broad bands cross the hind femora. In males, the wings usually reach the end of the abdomen, but in females the wings usually are shortened. Males are 24–37 mm long, females 30–48 mm.

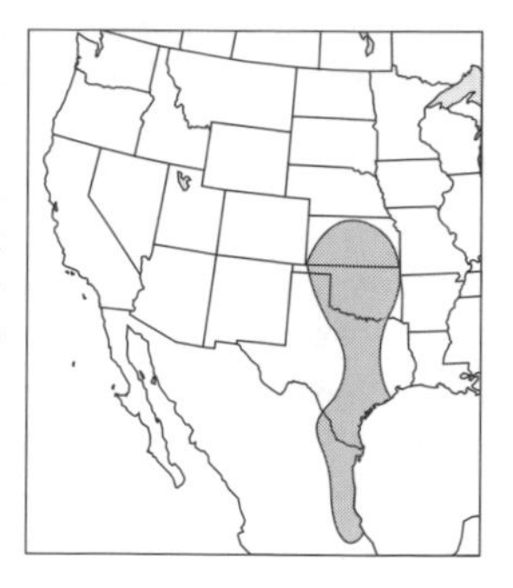

graceful range grasshopper

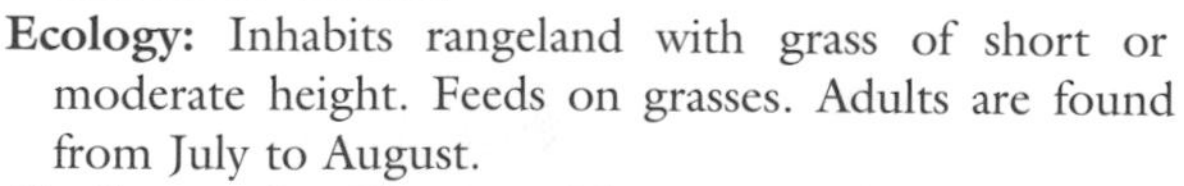

Ecology: Inhabits rangeland with grass of short or moderate height. Feeds on grasses. Adults are found from July to August.

Similar species: The graceful range grasshopper is easily confused with other *Boopedon* species, particularly the ebony grasshopper, *B. nubilum* (Plate 6). The black color of ebony grasshopper males is the most apparent distinguishing character in these two species. The other common *Boopedon* found in the United States is the pale range grasshopper, *B. auriventris.* This latter species bears even shorter wings, shorter than the length of the head and pronotum, and lacks distinct black bands on the outer face of the hind femora.

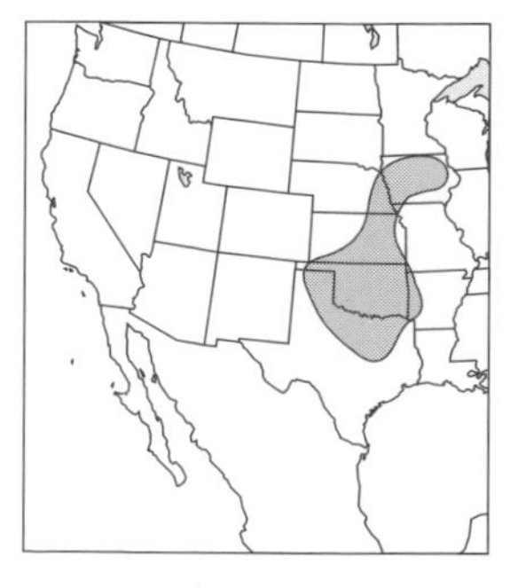

pale range grasshopper

EBONY GRASSHOPPER

Boopedon nubilum (Plate 6)

Distribution: Found widely in the Great Plains region, from Montana and North Dakota to Arizona and Texas, and south into northern Mexico.

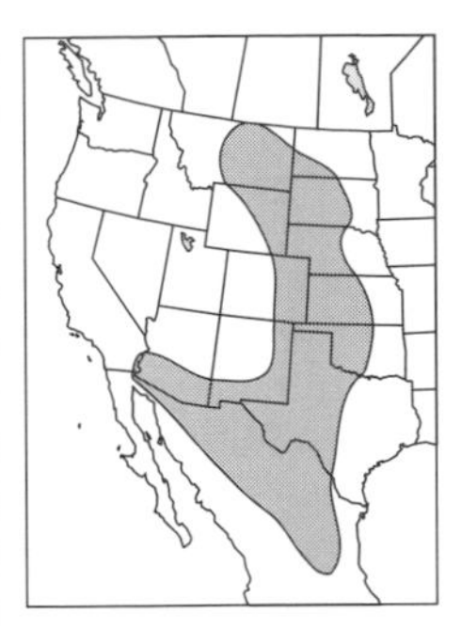
ebony grasshopper

Identification: A moderately large, heavy-bodied grasshopper. Like the graceful range grasshopper, *Boopedon gracile*, its head is large and round. Hind tibiae are red, black, cream, or a combination of these colors. The male, as suggested by the common name, is entirely black, and so this species is sometimes called the black males grasshopper. The female is brown or yellow brown but marked with cream or ivory and black. Occasionally it, too, is entirely black. In males, the wings do not reach the end of the abdomen, and are about the length of the head and pronotum combined. The wings of females are even shorter than those of males. Males are 24–34 mm long, females 33–52 mm.

Ecology: Inhabits shortgrass prairie where it feeds on grasses. In such habitats, it can become quite abundant. Males are active and readily observed; females are more cryptic. Adults are found from July to September.

Similar species: The ebony grasshopper is easily confused with other *Boopedon* species, particularly the graceful range grasshopper, *B. gracile* (Plate 6). The black color of ebony grasshopper males is the most apparent distinguishing character in these two species. Also, in females the lateral lobes of the pronotum and the hind femora are not as distinctly marked with black as in the graceful range grasshopper. The only other common *Boopedon* found in the United States is the pale range grasshopper, *B. auriventris.* This latter species bears even shorter wings, shorter than the length of the head and pronotum, and lacks a black form.

CREOSOTE BUSH GRASSHOPPER

Bootettix argentatus Bruner (Plate 2)

Distribution: Found only in the southwestern United States and northern Mexico.

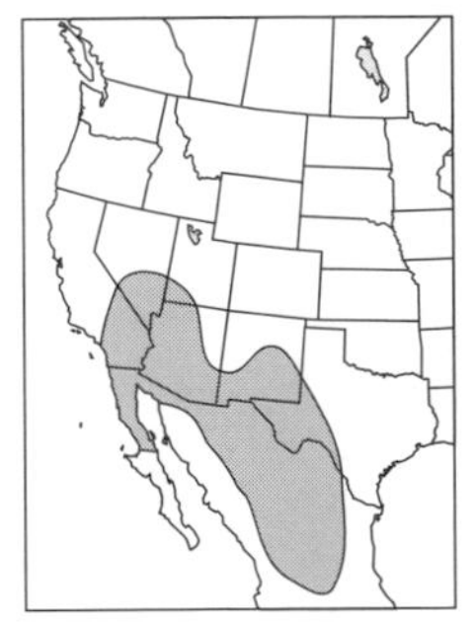
creosote bush grasshopper

Identification: This green grasshopper is marked with silvery white on the body and hind femora, and with white and black on the lateral lobes of the pronotum. Hind tibiae are green. Head is pointed. Forewings are spotted and long, extending beyond the tip of the abdomen. It is identified as much by its host plant as by appearance, as it lives exclusively in creosote bushes. Males are about 18 mm long, females 25 mm.

Ecology: The most important feature of the creosote bush grasshopper is its strong association with the creosote bush, which is found in southwestern deserts. It is well camouflaged while resting in these bushes and leaves only reluctantly, though it does feed on some other plants. Males stridulate during the evening. Adults are most abundant from September to December but can be found throughout the year.

SPRINKLED BROAD-WINGED GRASSHOPPER

Chloealtis conspersa (Plate 2)

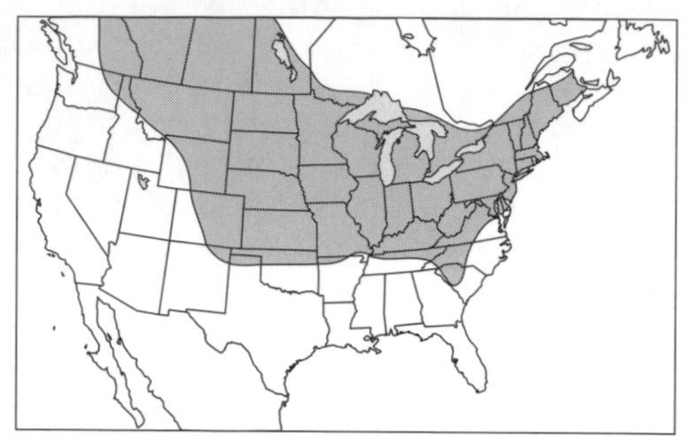

sprinkled broad-winged grasshopper

Distribution: Found widely in the northern United States and southern Canada, extending as far south as South Carolina and northern New Mexico. It is absent from the southwestern and Pacific Coast states.

Identification: This fairly small grasshopper is brown, though in males the lateral lobes of the pronotum are shiny black. Face is quite slanted. Forewings of males are bluntly rounded and expanded at the tips, but do not attain the tip of the abdomen. In females, the forewings are even more abbreviated, measuring only about the length of the head and prothorax combined. Basal abdominal segments are black, but the black color dissipates toward the tip of the abdomen. Hind tibiae are orange or red. Males are 15–20 mm long, females 20–28 mm.

Ecology: Favors wooded areas and bushy pastures, though preferring relatively dry areas. Feeds on grasses. This species is relatively unusual in its selection of oviposition site, normally depositing its eggs within decaying wood. Adults are found from July to September.

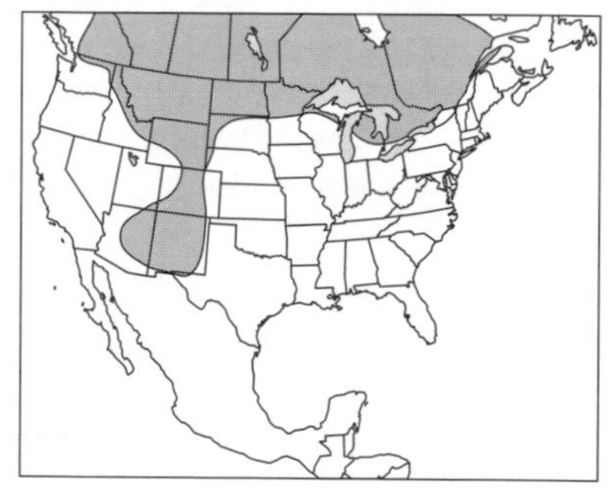

Thomas's broad-winged grasshopper

Similar species: The *Chloealtis* species tend to have forewings that are markedly expanded near the tip (wider than the basal region) and stubby; this combination of characters, which gives the wings an unusually broad appearance, is not common in grasshoppers. This feature is more evident in the long-winged forms than the short-winged forms. Males of the sprinkled broad-winged grasshopper are distinguished from the similar Thomas's broad-winged grasshopper, *Chloealtis abdominalis*, by the presence of black lateral

welcome broad-winged grasshopper

California broad-winged grasshopper

graceful broad-winged grasshopper

pronotal lobes. In Thomas's broad-winged grasshopper, the lateral lobes are shaded black only on the dorsal portions, not entirely as in the sprinkled grasshopper. Other rare *Chloealtis* species occur in California and Oregon: the welcome broad-winged grasshopper, *C. aspasma*; the California broad-winged grasshopper, *C. dianae*; and the graceful broad-winged grasshopper, *C. gracilis*.

MARSH MEADOW GRASSHOPPER

Chorthippus curtipennis (Plate 2)

Distribution: Widely distributed, ranging from far northern Canada and Alaska to southern California, Arizona, and New Mexico in the West, and North Carolina in the East. It is absent from most southeastern states and west to Texas. In southern regions, it is found mostly at high altitudes.

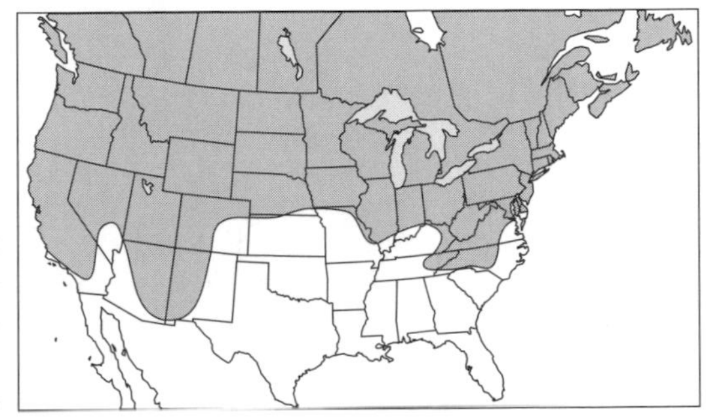

marsh meadow grasshopper

Identification: This species is light brown dorsally and greenish or brownish laterally. Often a black stripe extends back from the eye across the lateral lobes of the pronotum. Forewings are grayish or yellowish, broadly rounded at the tip, and variable in length. Abdominal segments are yellowish, with black patches laterally. In most cases, the females are short-winged, and the males bear wings that extend to about the tip of the abdomen. Male wing length may vary, however, from about half the length of the abdomen to much longer than the abdomen. Antennae are long, yellow basally but often black at the tip. Tips of the femora are black. Hind tibiae are yellow or orange. Males are 12–20 mm long, females about 20–35 mm.

Ecology: Inhabits moist areas with tall grasses. Feeds only on grasses. Both sexes stridulate. In some areas, eggs may persist for more than one winter before hatching. Adults are found from July to October.

CRENULATED GRASSHOPPER

Cordillacris crenulata (Plate 2)

Distribution: Found in the western Great Plains from Montana and North Dakota south to New Mexico, western Texas, and northern Mexico. It also is found in the Great Basin area of the Southwest, including Utah, Arizona, Nevada, and southern California.

Identification: This small, brownish grasshopper is marked with a wavy or crenulated dark stripe on the forewings. A dark stripe also extends from the top of the head, through the eye, and across the lateral lobe of the pronotum. Hind femora bear a dark stripe, and the hind tibiae are cream in color. Males are 13–19 mm long, females 16–26 mm.

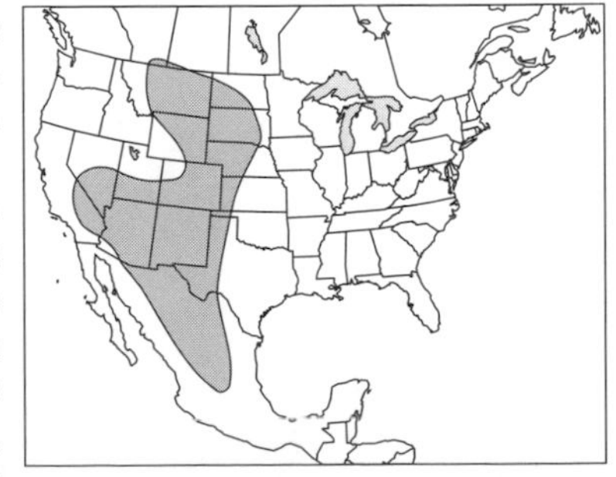

crenulated grasshopper

Ecology: Widely inhabits arid grasslands where it feeds on grasses. Though widespread, it rarely is abundant. Adults are found from May to September.

WESTERN SPOTTED-WING GRASSHOPPER

Cordillacris occipitalis (Plate 2)

Distribution: Found throughout the western states and provinces, occurring as far east as Manitoba and North Dakota in the north, and central Oklahoma and western Texas in the south.

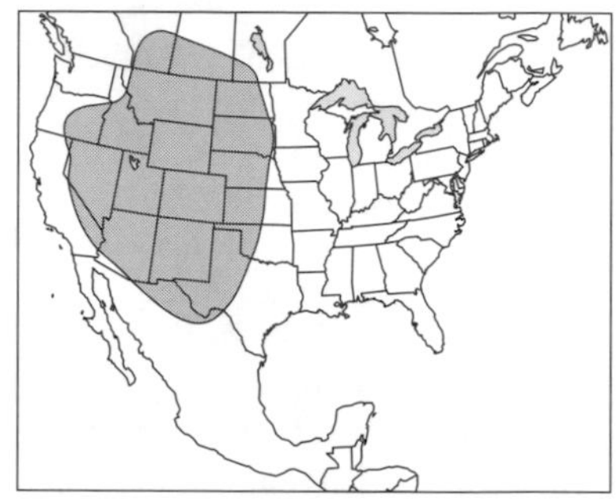

western spotted-wing grasshopper

Identification: This small grasshopper has few distinguishing characteristics. It is yellowish brown in overall color and bears a dark stripe extending from the eye across the lateral lobe of the pronotum to the base of the forewing. The base of the forewing also has a thin white stripe, and a row of brown spots extends the length of the forewing. Hind femora are marked with a dark stripe. Hind tibiae are light orange. Males are 15–24 mm long, females 17–29 mm.

Ecology: Inhabits areas of sparse vegetation where it often becomes numerous and damaging. Adults are found from June to September.

Similar Species: Though superficially similar to the crenulated grasshopper, *Cordillacris crenulata* (Plate 2), the lack of a crenulated pattern on the front wings and larger overall size serve to distinguish the western spotted-wing grasshopper.

ELEGANT GRASSHOPPER

Dichromorpha elegans (Plate 4)

Distribution: Largely coastal in distribution. Most records are from within 160 km (100 miles) of the seacoast. It occurs in the eastern and southern United States, along both the Atlantic and Gulf Coasts, from New Jersey to Texas.

elegant grasshopper

Identification: This grass-green or brownish-green grasshopper has a slanted face but a broadly rounded head. The slightly enlarged head usually is marked by a narrow black line extending from the posterior of the eye, across the prothorax, and onto the forewings. Forewings are variable in length, ranging from about one half the length of the abdomen to the tip of the abdomen. Males are much smaller and more slender than females. Hind tibiae are brownish. Males are 17–23 mm long, females 19–30 mm.

Ecology: Inhabits moist areas such as lakesides, freshwater swamps, and coastal salt marshes. Does not inhabit the dryer areas frequented by the short-winged green grasshopper. Adults are found from July to September.

Similar Species: This species is easily confused with the short-winged green grasshopper, *Dichromorpha viridis* (Plate 4). Close examination of the dorsal surface of the pronotum will differentiate the two species (see Fig. 5). The elegant grasshopper has a single, narrow, line-like transverse cut or crevice (sulcus) that bisects the median and lateral pronotal ridges. The short-winged green grasshopper has another crevice that bisects the lateral pronotal ridges but not the middle ridge. The elegant grasshopper has a larger head than the short-winged green grasshopper.

SHORT-WINGED GREEN GRASSHOPPER

Dichromorpha viridis (Plate 4)

Distribution: Found throughout the eastern United States and west to the Great Plains. It also occurs in Mexico.

short-winged green grasshopper

Identification: This species normally is green in color, though sometimes brown, or both. The upper surfaces and sides of male grasshoppers sometimes are contrasting colors; the common forms are a green upper surface and pale or dark brown sides, or light brown upper surface and dark brown sides. Females are uniformly colored but may be either green or brown. The forewings are, as the name suggests, usually abbreviated in length, but occasional long-winged individuals occur. In males, the forewings are about three-fourths the length of the abdomen, whereas in females the forewings extend just half the length of the abdomen. Males are 14–22 mm long, females 23–30 mm.

Ecology: This is a common species in grassy areas, including edges of ponds and woods, low areas of pastures, and along roadsides. It also feeds readily in improved pastures and on lawn grasses, which accounts for its wide distribution and abundance. In some grassy habitats, it is the most abundant grasshopper. Males are reported to stridulate. Adults generally are found in July–September, but in Florida they can be observed from May to December.

Similar Species: This species is very similar in form to the elegant grasshopper, *Dichromorpha elegans* (Plate 4). The short-winged green grasshopper can be distinguished by the presence of two crevices or cuts in the lateral pronotal ridges (see Fig. 5). The elegant grasshopper bears only a single cut, and it also has a larger head than the short-winged green grasshopper.

TEXAS SHORT-WING SLANTFACED GRASSHOPPER

Eritettix abortivus (Plate 3)

Distribution: Found in Texas, eastern New Mexico, and northeastern Mexico.

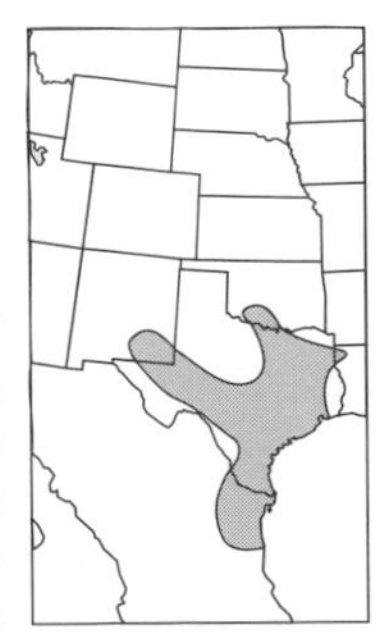

Texas short-wing slantfaced grasshopper

Identification: This small grasshopper greatly resembles other *Eritettix* species, possessing a slanted face and broadly rounded head. Dark stripes extend from the head along the upper surface of the pronotum. Lateral ridges on the upper surface of the pronotum are strongly constricted near the midpoint and white in color. Body color is quite variable. Forewings are marked with a row of spots near the midpoint and do not attain the tip of the abdomen; in males the wing length is about equal to the head and pronotum whereas in females it is shorter. Males are 11–15 mm long, females 19–22 mm.

Ecology: Like the other *Eritettix* species, this species inhabits grassy areas, though in most cases the grass is thin and short. In addition to inhabiting extensive grassland, it is found within small grassy breaks of thicker vegetation. Adults overwinter, and are found from October to April. Eggs occur in spring and summer.

Similar species: Wing length serves to distinguish the Texas short-wing slantfaced grasshopper from the long-winged velvet-striped grasshopper, *Eritettix simplex* (Plate 3). There is no range overlap with the obscure slantfaced grasshopper, *Eritettix obscurus* (Plate 3).

VELVET-STRIPED GRASSHOPPER

Eritettix simplex (Plate 3)

Distribution: Found widely in the Rocky Mountains and Great Plains regions from southern Canada to northern Mexico. It is also present on the eastern slope of the Appalachian Mountains from Connecticut to Georgia. It is absent from Florida and most of Texas, where it is replaced by the obscure slantfaced grasshopper, *Eritettix obscurus*, and the Texas short-wing slantfaced grasshopper, *Eritettix abortivus*, respectively.

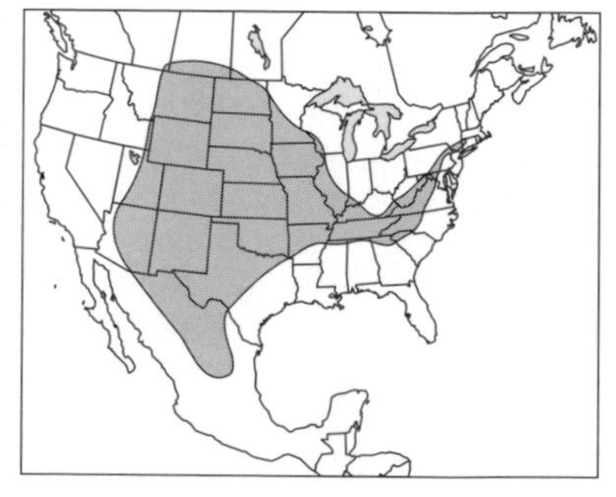

velvet-striped grasshopper

Identification: This species is fairly small in size, with the face slanted back and the top of the head broadly rounded. Antennae are usually expanded at the tip. Dark stripes extend from the head along the dorsal surface of the pronotum. Lateral ridges on the upper surface of the pronotum are strongly constricted near the midpoint and white in color; the white color is a strong contrast to the dark pronotal stripes. Usually grayish or grayish brown in general color, but sometimes green. Forewings are marked with a short white stripe basally and a row of spots near the midpoint; they attain the tip of the abdomen. Hind tibiae bear a dark streak along the upper edge and are brownish. Males are 16–20 mm long, females 21–26 mm.

Ecology: Inhabits grassy areas, both extensive grasslands and small grassy openings in dense vegetation. Feeds on grasses. Males stridulate noisily. Nymphs overwinter, and adults are found from April to October.

Similar species: The velvet-striped grasshopper is distinguished from the other *Eritettix* species by the length of the forewings. It is the only *Eritettix* species with long wings.

OBSCURE SLANTFACED GRASSHOPPER

Eritettix obscurus (Plate 3)

Distribution: Found only in Florida, where it occurs widely in the peninsular region.

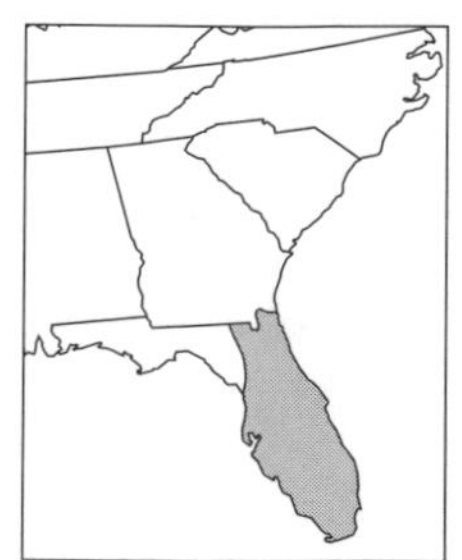

obscure slantfaced grasshopper

Identification: This small, brownish, flightless grasshopper is usually distinguishable by the broad whitish or yellowish stripe that extends from the top of the head to the tip of the abdomen. This character is sometimes absent, however, making recognition more difficult. Forewings are always shortened, covering about one-half to three-fourths the length of the abdomen. Antennae are slightly flattened and sword-shaped. Face is markedly slanted. Upper surface of the pronotum bears a small ridge along each side, and the ridges are constricted near the middle of the pronotum. These lateral ridges may be black or white in color. Males are 13–17 mm long, females 21–27 mm.

Ecology: Preferred habitat is dry oak forests, where it occurs among wiregrass and low-growing oak. It is never abundant. Adults and nymphs are found throughout the year in Florida.

Similar species: The geographic range of this species serves to distinguish it from other *Eritettex* species. Within Florida, no other species is readily confused with the obscure slantfaced grasshopper, especially when the dorsal stripe is present.

RUFOUS GRASSHOPPER

Heliaula rufa (Plate 2)

Distribution: Found in the southern Rocky Mountains and the western margin of the Great Plains, and north to Wyoming and Nebraska.

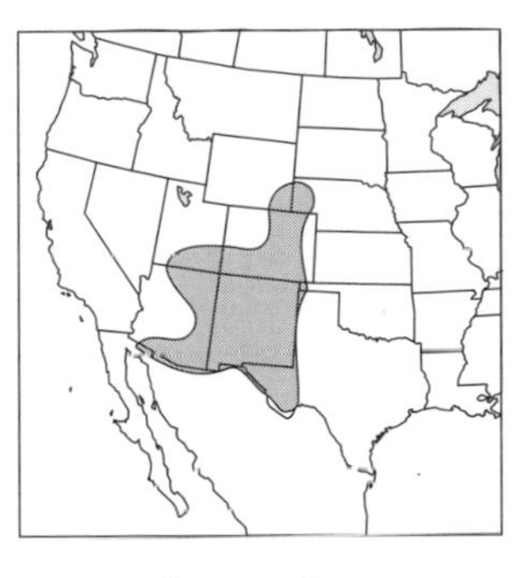

rufous grasshopper

Identification: This fairly small grasshopper is generally uniformly colored, though the color varies from cream to reddish. Brown speckling is sometimes found on the forewings. Hind wings are transparent to pale blue. In some individuals, a distinctive marking is a dark area at the dorsal hind margin of the pronotum, which results in the appearance of a dark crescent on the back of the grasshopper. Head is large and often gray. Face is only weakly slanted. Antennae are slender. Dark bands may be present on the upper and outer face of the hind femora, but they are lacking from the inner face. Hind tibiae are orange or pink. Males are 14–20 mm long, females 20–27 mm.

Ecology: Inhabits deserts, and prairies with sparse vegetation, sometimes on rocky outcrops or hillsides. Generally not numerous. Adults are found from June to October.

DESERT CLICKER GRASSHOPPER

Ligurotettix coquilletti (Plate 2)

Distribution: Found in the Sonoran Desert of Nevada, southern California, southern Arizona, and northwest Mexico.

desert clicker grasshopper

Identification: A uniformly grayish or grayish-brown insect, but sometimes the thoracic area is mostly black on the sides. Wings extend well beyond the tip of the abdomen, and the front margin of the male's forewing bears unusually large cells. Hind wing is largely transparent, but sometimes smoky at the tip. Antennae are threadlike. Upper and inner surfaces of the hind femora bear three black bands, and may be weakly banded on the outer face. Males are 16–23 mm long, females 21–30 mm.

Ecology: Dwells in shrubs. Favors creosote bushes, but also found in *Atriplex*, *Franseria*, mesquite, and ironwood. This species is elusive, and hard to see and to capture. Males stridulate throughout the day and until about midnight. Males are territorial, so it is unusual to find more than one per bush.

Similar species: The Pecos clicker grasshopper, *Ligurotettix planum*, is found in the Chihuahuan Desert of southern New Mexico, western Texas, and north-central Mexico. Like the desert clicker, it favors bushes; however, the Pecos clicker is never found in creosote bushes. The Pecos clicker is slightly larger, and the large cells on the leading edge of the male's forewing are not as large as in desert clicker.

Pecos clicker grasshopper

TWO-STRIPED MERMIRIA GRASSHOPPER

Mermiria bivittata (Plate 7)

Distribution: Most common in the Great Plains region, although it is found nearly everywhere in the United States except the northwestern and northeastern states. It also occurs in southern Canada and northern Mexico.

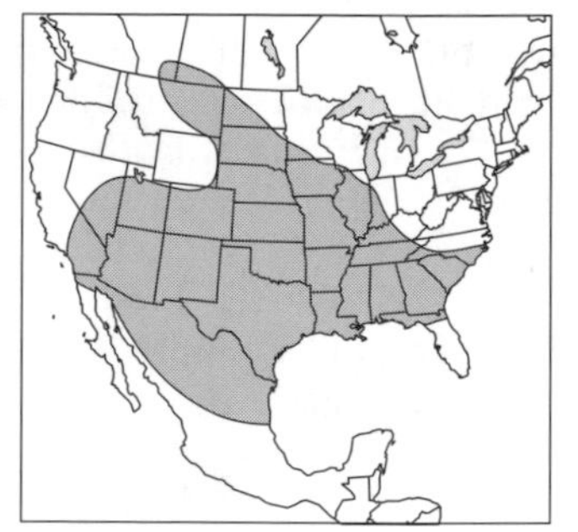

two-striped mermiria grasshopper

Identification: This narrow-bodied species is quite large. Though variable in color, it is marked with a dark stripe originating behind the eye and running across the pronotum. The stripe extends weakly onto the forewings, where a narrow white streak also may be found basally. Generally, the body is brownish or greenish above and yellow below. Face is strongly slanted. Antennae are sword-shaped. Hind tibiae are reddish. Males are 28–38 mm long, females 39–56 mm.

Ecology: Inhabits areas of tall grass, including coastal salt marsh habitats. Feeds exclusively on grasses. Adults are found from June to October.

Similar species: The two-striped mermiria grasshopper is separated from the eastern mermiria grasshopper, *Mermiria intertexta* (Plate 7), by the lack of a median dark stripe along the dorsal portion of the body. It is distinguished from the lively mermiria grasshopper, *Mermiria picta* (Plate 7), by the absence of distinct lateral ridges on the dorsal surface of the pronotum.

EASTERN MERMIRIA GRASSHOPPER

Mermiria intertexta (Plate 7)

Distribution: Found in the coastal region of the eastern United States from New Jersey to Florida.

eastern mermiria grasshopper

Identification: This species is long and narrow in general appearance, and quite large. Face is strongly slanted. Antennae are sword-shaped. General color is yellowish or greenish, but a reddish or dark-brown stripe often is present dorsally, especially in males, from the tip of the head to the posterior margin of the pronotum. Another distinct dark brown or black stripe extends from the posterior margin of the eye and onto the base of the forewings, where it merges into the brown forewings. The stripe in the basal region of the forewings contains a narrow white streak. The long, thin hind tibiae are reddish in color. Males are 32–38 mm long, females 33–58 mm.

Ecology: Preferred habitat is tall grasses and wet areas. Flies freely when disturbed, but is not a strong flier. Feeds exclusively on grasses. Adults are found from July to September in most areas, but from May to November in Florida.

Similar Species: The narrow white streak at the base of the forewing is normally lacking in the similar lively mermiria grasshopper, *Mermiria picta* (Plate 7). Also serving to distinguish the eastern mermiria grasshopper from the lively mermiria grasshopper, but much more difficult to see, is the absence of lateral ridges on the dorsal surface of the pronotum. The two-striped mermiria grasshopper, *Mermiria bivittata* (Plate 7), is separated from the eastern mermiria grasshopper by the lack of a median dark stripe along the dorsal portion of the body.

LIVELY MERMIRIA GRASSHOPPER

Mermiria picta (Plate 7)

Distribution: Found throughout the eastern and central United States, north to Virginia and South Dakota, and west to Arizona. It also is known from Mexico.

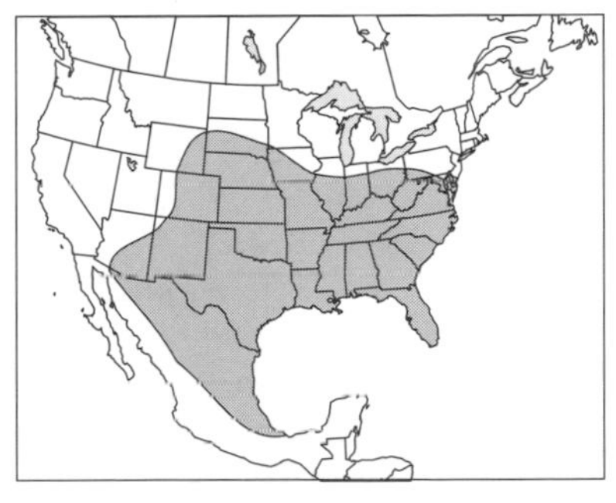

lively mermiria grasshopper

Identification: Quite large in size, long and narrow in general appearance, and greenish or brownish in general color. Face is strongly slanted. Antennae are sword-shaped. A reddish or dark brown stripe often is present dorsally, especially in males, from the tip of the head to the posterior margin of the pronotum. Another distinct dark brown or black stripe extends from the posterior margin of the eye onto the base of the forewings, and merges into the brown forewings. Hind tibiae are reddish. Males are 28–41 mm long, females 33–58 mm.

Ecology: Occurs widely in habitats containing tall grasses, including wooded environments. Feeds exclusively on grasses. Adults are found from July to September.

Similar Species: This large, thin species greatly resembles the eastern mermiria grasshopper, *Mermiria intertexta* (Plate 7); however, it does not have the narrow white streak at the base of the forewing. Also serving to distinguish the lively mermiria grasshopper from the eastern mermiria grasshopper, but much more difficult to see, is the presence of lateral ridges on the dorsal surface of the pronotum. The presence of lateral ridges also distinguishes the lively mermiria grasshopper from the two-striped mermiria grasshopper, *Mermiria bivittata* (Plate 7).

OBSCURE GRASSHOPPER

Opeia obscura (Plate 3)

Distribution: Found widely in the western United States, extending to the southern areas of the Canadian Prairie Provinces and central Mexico.

obscure grasshopper

Identification: A common grasshopper of moderate size, this species is modestly colored with pale brown, yellow, or green. The face is quite slanted, but the top of the head is broadly rounded. Antennae are sword-shaped. A broad brown stripe extends from the back of the eye across the lateral lobes of the pronotum and onto the forewings. A narrow brown stripe is sometimes found on the top of the pronotum. Forewings of females bear a dark stripe bordered beneath by a narrow white stripe. Wings are long, extending to the tip of the abdomen. Males are about 16 mm long, females 25 mm.

Ecology: Inhabits grassland, both thin and thick stands. Feeds exclusively on grass, but accepts numerous species. Can attain high levels of abundance, damaging western rangelands. Adults are found from July to October in most regions, but earlier in southern areas.

SPOTTED-WING GRASSHOPPER

Orphulella pelidna (Plate 5)

Distribution: A very widespread and frequently encountered grasshopper. It occurs throughout the United States, southern Canada, and northern Mexico.

spotted-wing grasshopper

Identification: This slender species is variable and indistinct in appearance. Under most conditions, the spotted-wing grasshopper is brown or green, bearing both black and white accents. Large black triangular marks are found dorsally along the posterior margin of the pronotum. Lateral ridges on the upper surface of the pronotum are markedly compressed near the mid-point of the pronotum. A series of small dark rectangular spots is present along the forewings, and is the basis for the name of this species, but numerous speckles are also generally present. A broad dark stripe usually extends from the back of the eye to the base of the forewing. In some coastal locations, larger forms appear, which may be completely green, or bear only some of the aforementioned stripes and marks. On dark soils or in burned forests, blackish forms may occur. Forewings normally extend well beyond the tip of the abdomen, but individuals with shorter wings are sometimes observed. Hind tibiae are usually brown, but sometimes bluish. Antennal segments are not markedly flattened, appearing thread-like. Males are 18–25 mm long, females 18–28 mm.

Ecology: Found in all habitats except those that are heavily shaded. Despite its adaptable nature, it rarely is numerous in any habitat, or at any location. When disturbed, it flies swiftly, but for fairly short distances, before diving to the soil or vegetation to seek shelter. Upon landing it often runs a short distance to hide. Feeds on grasses. Adults are found from July to October; in Florida they can be found throughout the year.

Similar species: Although this species superficially resembles the handsome grasshopper, *Syrbula admirabilis* (Plate 5), it usually can be separated based on its overall smaller size. The strongly compressed lateral ridges also are diagnostic, although in the handsome grasshopper they are slightly compressed. Also, the forewings are usually spotted and speckled, characteristics usually absent from the handsome grasshopper. The spotted-wing grasshopper also is easily confused with the pasture grasshopper, *Orphulella speciosa* (Plate 5). They are distinguished by the number of cuts in the lateral ridges on the dorsal surface of the pronotum. In the pasture grasshopper there is but one cut in each lateral ridge, whereas in the spotted-wing grasshopper there are two cuts (see Fig. 5).

PASTURE GRASSHOPPER

Orphulella speciosa (Plate 5)

Distribution: Found widely east of the Rocky Mountains, though it is absent from some southeastern states.

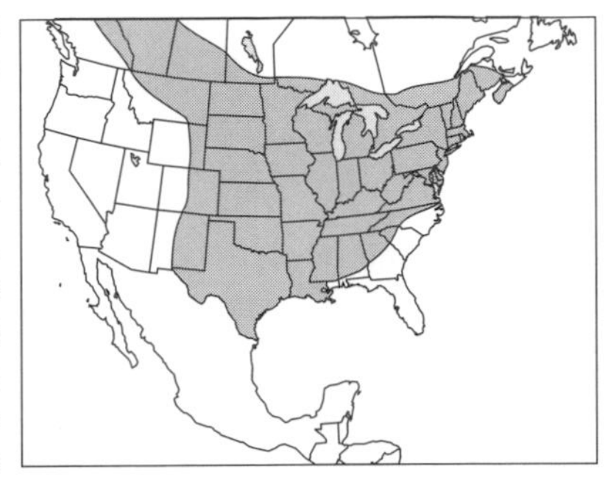

pasture grasshopper

Identification: This small- to medium-sized grasshopper has few distinguishing characteristics. The face is not strongly slanted, the head rather broadly rounded. The color is variable: brown or green is common but a combination of brown and green is also frequent. A series of elongate spots occurs near the center of the forewings, which normally extend to the tip of the abdomen or beyond. Large black triangular marks are often found dorsally along the posterior margin of the pronotum. Lateral ridges on the dorsal surface of the pronotum are markedly compressed near the mid-point of the pronotum. A broad dark stripe usually extends from the back of the eye and across the lateral lobes of the pronotum. Males are 14–21 mm long, females 18–27 mm.

Ecology: Inhabits areas with grass that is short or medium in height. Prefers dry areas over wet areas. Adults are found from August to October in most areas, but from May to December in Texas.

Similar species: The pasture grasshopper is easily confused with the spotted-wing grasshopper, *Orphulella pelidna* (Plate 5). In the pasture grasshopper there is but one cut in each lateral ridge on the upper surface of the pronotum, whereas in the spotted-wing grasshopper there are two cuts (see Fig. 5).

WYOMING TOOTHPICK GRASSHOPPER

Paropomala wyomingensis (Plate 3)

Distribution: Found in the Great Plains region from Wyoming and South Dakota south to central Mexico, and west to Nevada and southern California.

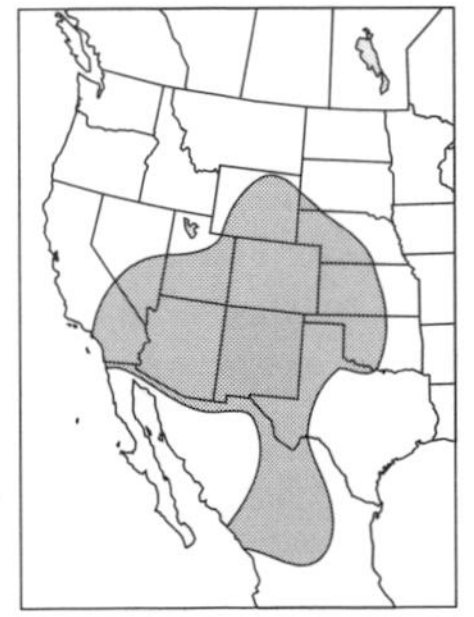

Wyoming toothpick grasshopper

Identification: Long and slender-bodied, like the other so-called toothpick grasshoppers. Face is strongly slanted. Antennae are sword-shaped. Color is variable but tends to be uniform within an individual. Colors may be light green, gray, light brown, pink, or yellowish. A whitish stripe extends from the lower border of the eyes to the bases of the middle legs. Forewings are pointed and shortened, only reaching the fifth abdominal segment. Abdomen is long and cylindrical; the subgenital plate of the male is markedly pointed. Males are 19–31 mm long, females 25–40 mm.

Ecology: Feeds on grasses and is typically found among tall grasses. In shortgrass prairie regions, it occurs in low areas where moisture accumulates and taller grasses are found. Adults are found from June to October.

Similar species: The short wings and absence of a spine between the front legs serve to differentiate the Wyoming toothpick grasshopper from other common thin-bodied species, such as the cattail toothpick grasshopper, *Leptysma marginicolis*

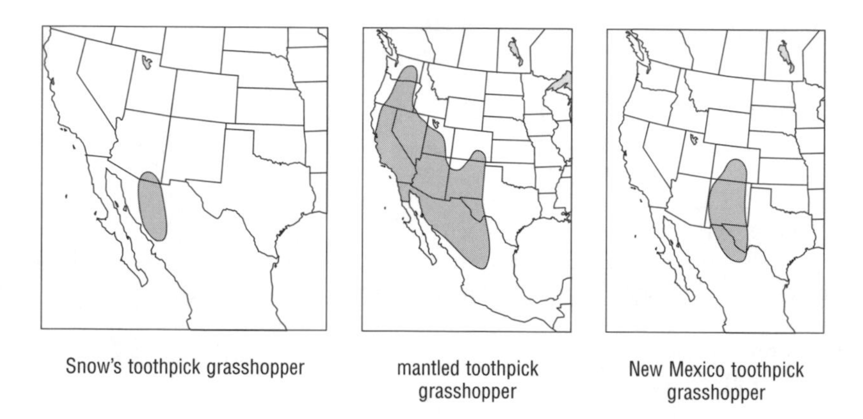

Snow's toothpick grasshopper

mantled toothpick grasshopper

New Mexico toothpick grasshopper

(Plate 30), and the glassy-winged toothpick grasshopper, *Stenacris vitreipennis* (Plate 30). In eastern states, the long-headed toothpick grasshopper, *Achurum carinatum* (Plate 1), is similar in appearance and also lacks a spine between the front legs, but this species is thinner and the wings are shorter. In border areas and in Mexico, there are other short-winged, spineless species, the stubby toothpick grasshopper, *Achurum minimipenne*, and Snow's toothpick grasshopper, *Prorocorypha snowi*, which are distinguished by the relatively short wings (extending to the third abdominal segment or less).

Two other species in the genus occur in western states: the mantled toothpick grasshopper, *Paropomala pallida*, and the New Mexico toothpick grasshopper, *P. virgata*. They are similar in appearance to the Wyoming toothpick but have longer wings. In the Wyoming toothpick the tips of the wings do not reach the tip of the hind femora, whereas in the mantled and the New Mexico toothpicks the tips of the wings extend beyond the hind femora. The latter two species are distinguished by the relative length of the abdomen; in the New Mexico toothpick the tip of the abdomen does not extend beyond the tip of the hind femora, whereas in the mantled toothpick the abdomen extends beyond the end of the hind femora.

FOUR-SPOTTED GRASSHOPPER

Phlibostroma quadrimaculatum (Plate 3)

Distribution: Found in the western portion of the Great Plains from Canada's Prairie Provinces to central Mexico.

Identification: This is a handsome, well-marked species bearing 3 to 5 large dark spots on the forewings. The spots may be discrete or in a crenulate pattern. Overall, the four-spotted grasshopper is brown and green, but

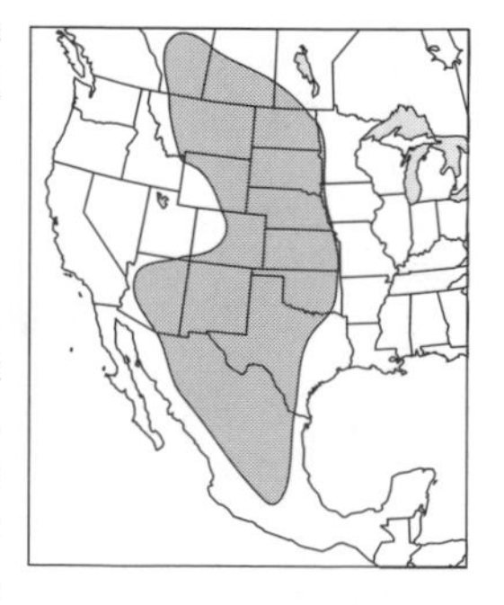

four-spotted grasshopper

bears black, ivory, and white markings. The pronotum, when viewed from above, often is marked with two black triangles. Wing length is variable but most often is slightly shorter than the abdomen. Hind femora are banded. The junction of the hind femora and tibiae is blackened. Hind tibiae are orange. Males are 13–23 mm long, females 21–32 mm.

Ecology: Inhabits regions of shortgrass prairie where it feeds predominantly on grass. On occasion, it is abundant enough to damage rangeland. Adults are found from July to September in most regions.

SHORT-WINGED TOOTHPICK GRASSHOPPER

Pseudopomala brachyptera (Plate 7)

Distribution: A northern species, occurring from Maine and New Jersey west to British Columbia and northern Nevada. In the central Great Plains, distribution extends south to Oklahoma.

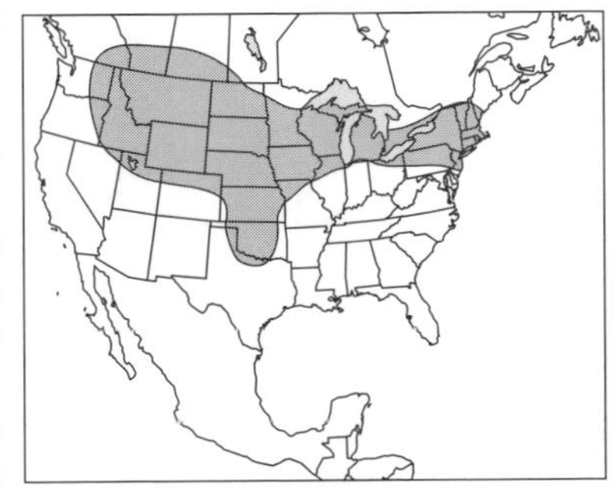

short-winged toothpick grasshopper

Identification: This elongate slantfaced species is not as thin-bodied as the other so-called toothpick grasshoppers. Nevertheless, it is more elongate than most grasshoppers. Often it is uniformly brown or grayish in color, but sometimes bears light stripes on the head, thorax, and forewings. Antennae are sword-shaped. Lateral ridges of the pronotum are distinct. Length of the pointed forewings is variable but is usually considerably shorter than the tip of the abdomen. Hind tibiae are brown. Males are 23–27 mm long, females 27–30 mm.

Ecology: Preferred habitat is open areas of tall bunchgrass, though it can be found in other locations including wooded areas and along the edges of salt marshes. It stridulates freely and is a grass feeder. Adults are found from June to October.

BROWN-SPOTTED RANGE GRASSHOPPER

Psoloessa delicatula (Plate 3)

Distribution: Found throughout the western United States, southwestern Canada, and northwestern Mexico.

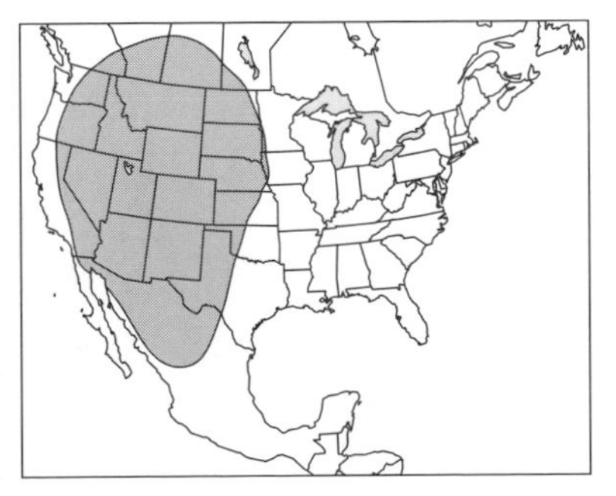

brown-spotted range grasshopper

Identification: Variable in color, ranging from brown to grayish or greenish. It bears dark stripes behind the eyes and across the lateral lobes of the pronotum. Lateral ridges on the pronotum are strongly constricted centrally. Wings are long, and the forewings bear a row of 4 to 6 rectangular black spots centrally. Upper surfaces of the hind femora bear dark spots; the central spot, at least, is triangular in shape. Hind tibiae are orange. Males are 18–20 mm long, females 16–27 mm.

Ecology: Inhabits shortgrass prairie and feeds on grasses. It is common in many areas but rarely builds to high levels of abundance. Adults are found from July to September.

Similar species: The other *Psoloessa* species found north of Mexico is the Texas spotted range grasshopper, *P. texana*, which is more southwestern in distribution. The hind tibiae of the Texas spotted range grasshopper are brown or gray rather than orange.

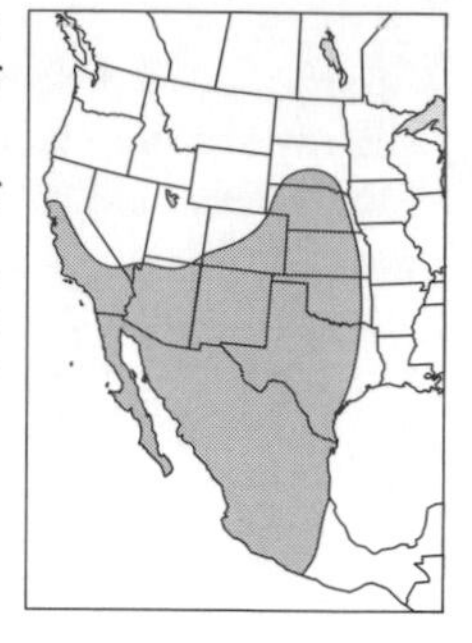
Texas spotted range grasshopper

GRACEFUL SEDGE GRASSHOPPER

Stethophyma gracile (Plate 6)

Distribution: Found in the northern United States and southern Canada.

Identification: Yellowish green in color and relatively free of contrasting markings, although a dark stripe is usually found on the head behind each eye. Long-winged, and medium in size. Forewings are wider near the tips. Junction of the hind femora and tibiae is marked with black, as are the spines on the hind tibiae. Hind tibiae are yellow. Males are 25–29 mm long, females 30–38 mm.

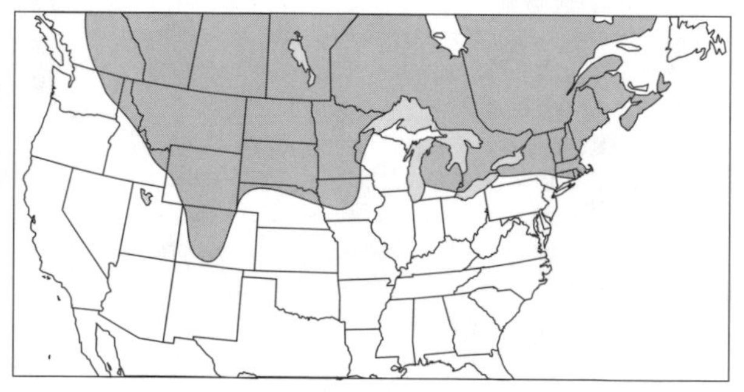
graceful sedge grasshopper

Ecology: Inhabits wet areas and areas of thick grass, where it feeds on grasses and sedges. Males stridulate loudly. Adults are found from July to September.

Similar species: Other *Stethophyma* species are less common, and readily distinguished. The striped sedge grasshopper, *S. lineata*, bears white markings on the forewings. Otte's sedge grasshopper, *S. celata*, bears broad dark markings on the lateral lobes of the pronotum.

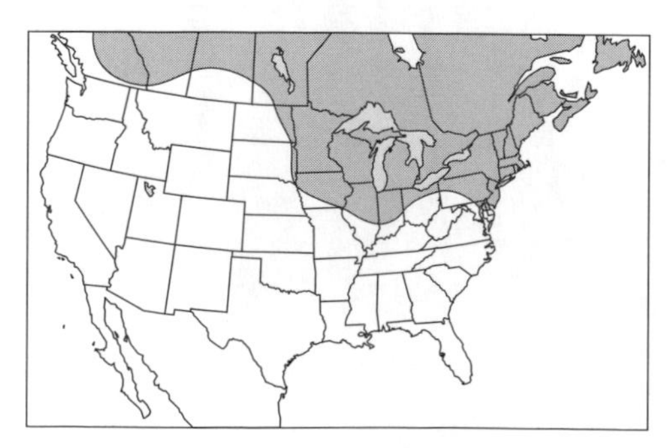
striped sedge grasshopper

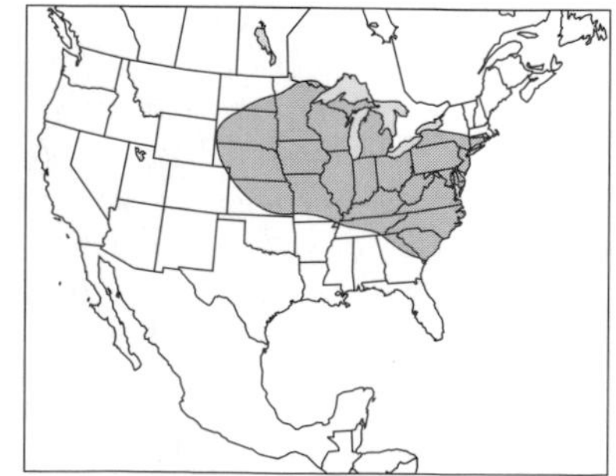
Otte's sedge grasshopper

HANDSOME GRASSHOPPER

Syrbula admirabilis (Plate 5)

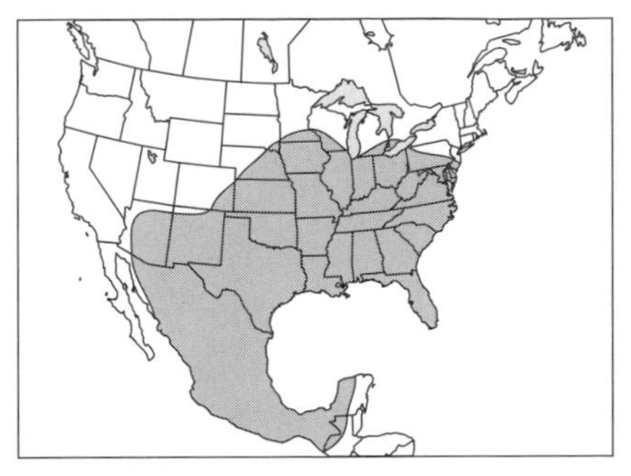

handsome grasshopper

Distribution: Found throughout the eastern states west to the Rocky Mountains and south into Mexico.

Identification: This slender species deserves its name; it is a strikingly attractive insect. Face is quite slanted; hind legs are especially long and slender. Antennae, though generally thread-like, are slightly expanded at the tip in males. Hind tibiae are brownish or grayish. Pronotum bears small ridges laterally, which are marked by white stripes. Lateral ridges are constricted slightly near the middle of the pronotum. A broad brown stripe usually extends from the front of the head to the posterior margin of the pronotum. General body color ranges from mostly brown to mostly green, but some males tend toward blackish. The most distinctive feature is the pattern on the forewings. The leading edge of the forewing (ventral surface when wings are closed) is green or grayish, whereas the trailing edge is brown to black. These contrasting colors meet in a wavy or crenulate pattern that immediately distinguishes females. Some males, however, have the forewings marked with discrete spots rather than a crenulate design, or are almost entirely dark. The sexes differ markedly in size. Males are 22–31 mm long, females 35–49 mm.

Ecology: Commonly associated with dry grasses of short to moderate height. The males fly readily when disturbed, whereas the females fly awkwardly and often escape by leaping. Both sexes stridulate. Adults are found from July to October.

Similar species: Although usually distinct, this species can resemble the spotted-wing grasshopper, *Orphulella pelidna* (Plate 5). Usually, the handsome grasshopper can be separated by its overall larger size. The lateral ridges on the dorsal surface of the pronotum are compressed near the mid-point, but not as markedly as in the spotted-wing grasshopper. Also, the forewings are not heavily spotted and speckled. The handsome grasshopper also resembles Montezuma's grasshopper, *Syrbula montezuma* (Plate 5), but the black on the lateral lobes of the pronotum and the black hind wings serve to distinguish Montezuma's grasshopper.

MONTEZUMA'S GRASSHOPPER

Syrbula montezuma (Plate 5)

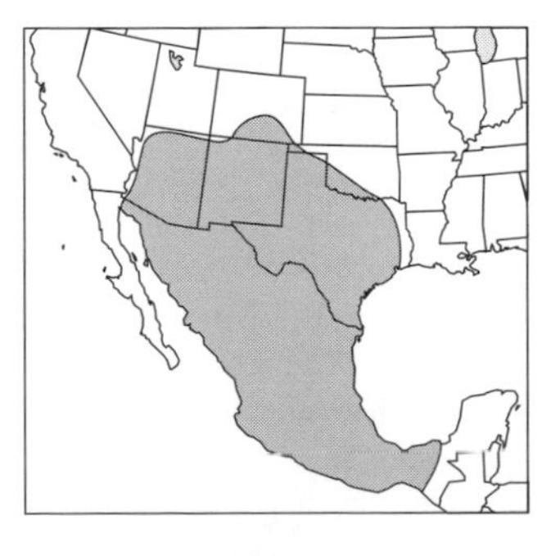

Montezuma's grasshopper

Distribution: Found in the Southwest; occurs principally from Texas to Arizona, but extends as far north as Colorado and south throughout Mexico.

Identification: This species is quite similar to the handsome grasshopper, *Syrbula admirablis*, in appearance: slender body, pronounced slanted face, and long hind legs. Also, in males the slender antennae are expanded at the tip. In males, greenish forms and blackish forms occur. Both sexes tend to have some black markings on the lateral lobes of the pronotum. They tend to lack a crenulate pattern on the forewings. The hind wings are black, an unusual character in this subfamily. Males are 21–32 mm long, females 30–46 mm.

Ecology: Inhabits taller grass vegetation within arid grasslands. Feeds exclusively on grasses. Adults are found from July to October.

Similar species: Montezuma's grasshopper resembles the handsome grasshopper, *Syrbula admirablis* (Plate 5); it is distinguished by the black on the lateral lobes of the pronotum and the black hind wings.

Silent Slantfaced Grasshoppers

Family Acrididae, subfamily Acridinae

This is a very small subfamily in North America, with only one species known from the United States. Several genera and numerous species, however, occur in South America. Acridinae are very similar in appearance to the stridulating slantfaced grasshoppers (subfamily Gomphocerinae), but as the common name suggests, members of this subfamily lack stridulatory pegs on the hind femora of males, and thus do not produce sound.

Grasshoppers in this subfamily have a slanted face (see Fig. 3), as is found in stridulating grasshoppers (Gomphocerinae), and flattened, sword-shaped antennae (see Fig. 2), as are found among some stridulating grasshoppers and spurthroated grasshoppers (subfamily Cyrtacanthacridinae). Acridines lack the spine between the front legs (the prosternal spine) that is found in the spurthroated grasshoppers and lubbers (subfamily Romaleinae). The hind wings are colorless or nearly colorless, lacking the dark band found in the band-winged grasshoppers (subfamily Oedipodinae).

CLIPPED-WING GRASSHOPPER

Metaleptea brevicornis (Plate 7)

Distribution: Found throughout eastern North America west to the Mississippi River, and in the south its distribution extends west into Texas. Its range includes most of Central and South America.

clipped-wing grasshopper

Identification: This slantfaced species is distinguished by the angled tip of the forewings, a distinct contrast from the rounded or pointed wing tips found in other long-winged slantfaced species. Hind wings are not pigmented. Males usually are green above and brown on the sides. Females are more variable, mostly brown or green, but sometimes light brown above and green on the sides. Antennae are markedly sword-shaped. Hind tibiae are brownish. Males are 25–38 mm long, females 36–53 mm.

Ecology: Preferred habitat is tall grasses along ponds and marshes, often in heavily shaded areas. Sometimes occurs in salt marshes. This species is a strong flier, and unlike most grasshoppers, can be attracted to lights. Adults are found from July to October.

Similar species: This species is not readily confused with others. Some band-winged species have wing tips that are similarly angled, but they are not slantfaced species, and they bear pigmented hind wings.

Band-winged Grasshoppers

Family Acrididae, subfamily Oedipodinae

The band-winged grasshoppers tend to be heavy bodied and bear enlarged hind legs. The heads of these grasshoppers can appear enlarged and broadly rounded (see Fig. 3). The orientation of the face is nearly vertical, a distinct contrast to the slantfaced grasshoppers (subfamily Gomphocerinae). Band-winged grasshoppers lack the spine between the front legs (the prosternal spine) that is found in the spurthroated grasshoppers (subfamily Cyrtacanthacridinae) and lubber grasshoppers (subfamily Romaleinae). The band-winged grasshoppers tend to be gray or brown, and often are mottled with darker spots, and the forewings frequently bear distinct or indistinct transverse bands. The pronotum bears ridges, wrinkles, or small tubercles, imparting a rough appearance.

The band-winged grasshoppers usually bear bright colors, but this may not be obvious. The hind wings can be yellow, orange, or reddish basally, with a broad black band crossing near the center of the wing. These colorful hind wings are hidden by the front wings (tegmina) except when in flight. Similarly, the inner face of the hind femora is often yellow, orange, red, or blue, but this is not usually apparent. These species blend exceptionally well with soil. The excellent camouflage provided by the earthy tones, mottling, and stripes makes it difficult to see band-winged grasshoppers until they take flight. They are capable of strong, directed flight and aerial acrobatics.

Males, and sometimes females, produce sound in flight (crepitation). The snapping, crackling, or buzzing sound apparently is made when the membranes between large veins in the hind wings are suddenly popped taut. They do not always crepitate in flight, as sound production is related to mate selection. These grasshoppers can also produce sound while at rest (stridulation) by rubbing the hind femora against the forewings, but the femora lack the stridulatory pegs found in the stridulatory slantfaced grasshoppers (Gomphocerinae).

The oedipodine grasshoppers normally are associated with open, sunny areas, and particularly with bare soil. Thin, overgrazed pastures or barren areas within pastures are the preferred habitat. Not surprisingly, oedipodines are most numerous in the arid western states, where they feed principally on grasses. When disturbed, these grasshoppers fly readily, but alight on soil rather than on plants. Their general color varies slightly, depending on the color of the substrate in their environment. Thus, they can be very difficult to detect when they sit motionless on soil.

One group of oedipodines, the genus *Trimerotopis*, is quite numerous (about 40 species) and difficult to distinguish. Most are exclusively western in distribution. Although some common species are treated here, for more definitive identification refer to *The North American Grasshoppers*, volume 2 (1984) by

Daniel Otte. This is an excellent resource for other band-winged groups that are not treated completely in this field guide.

SPECKLE-WINGED RANGELAND GRASSHOPPER

Arphia conspersa (Plate 8)

Distribution: Found throughout most of the western United States west of the Mississippi River, and in most of western Canada and south into Mexico.

speckle-winged rangeland grasshopper

Identification: Forewings and general body color are grayish brown. Forewings have a light anterior edge that often forms a light stripe along the back. Hind wings are red or yellow, and bordered with black. Abdomen is bright yellow. Inner face of the femur is black basally, and yellow with a black band distally. Hind tibiae are pale yellow or blue. Males are 20–25 mm long, females 25–35 mm.

Ecology: A common element of western grasslands, this species feeds primarily on grasses but also consumes broadleaf plants. Generally overwinters in the nymphal stage, so adults are present fairly early in the summer, but there is considerable geographic variability in the life cycle. Adults often crepitate noisily during flight.

Similar species: The speckle-winged rangeland grasshopper occurs in the same areas as the northwestern red-winged grasshopper, *Arphia pseudonietana* (Plate 8), though the former tends to occur in the adult stage early in the summer, whereas the adult stage of the latter occurs later in the summer. The northwestern red-winged grasshopper also tends to be darker, sometimes black, or more boldly mottled. In California, three other yellow- or orange-winged species (or subspecies) are known.

SOUTHERN YELLOW-WINGED GRASSHOPPER

Arphia granulata (Plate 8)

Distribution: Found in the southeastern states from North Carolina to Mississippi. It is common throughout the Florida peninsula, but elsewhere in the Southeast is often replaced by the autumn yellow-winged grasshopper, *Arphia xanthoptera*.

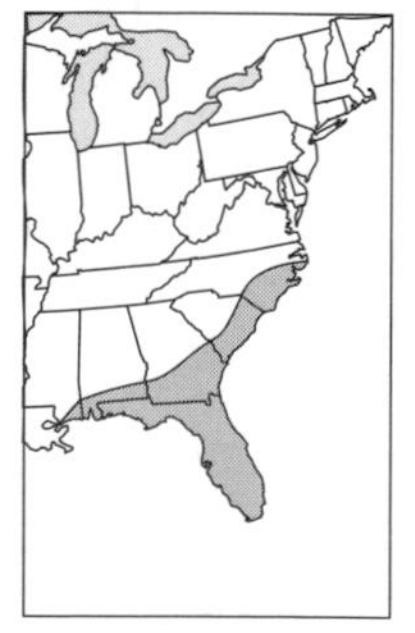
southern yellow-winged grasshopper

Identification: One of several similar-appearing species in the genus. It is light to dark brown in color, often bearing small dark or black speckles on the forewings and elsewhere. Forewings usually bear a narrow, pale-yellow hind margin, which forms a dorsal yellow band when the wings are held at rest. In the field, the most distinctive feature is the bright yellow hind wings. Hind wings also are marked with a curved black band. Hind tibiae are yellowish basally, with a black band separating the basal third of the tibia from the second third. The distal two-thirds of the hind tibiae are mostly pale or yellowish but often contain some additional dark coloration. This portion of the tibiae is not predominantly black, however. The dorsal median ridge on the pronotum is slightly, but distinctly, elevated. Males are 27–33 mm long, females about 30–37 mm.

Ecology: Inhabits brushy fields, open woods, roadsides, and to a lesser degree, grasslands. Presence of the adults is readily apparent because they make short, noisy flights in which they produce a crackling sound (crepitation) and flash their brightly colored wings. They overwinter principally as nymphs, transforming to adults in the early spring, but adults and nymphs can be found nearly any time in Florida.

Similar species: Though similar in appearance to other *Arphia* species, the southern yellow-winged grasshopper is likely confused only with the autumn yellow-winged grasshopper, *Arphia xanthoptera* (Plate 9), and the sulfur-winged grasshopper, *A. sulphurea*, because of its restricted southeastern distribution. The autumn yellow-winged grasshopper is distinguished by its greatly elevated median ridge on the pronotum. It is quite difficult to distinguish the southern yellow-winged grasshopper from the sulfur-winged grasshopper. The dark band of the hind wing is slightly wider in the former than in the latter.

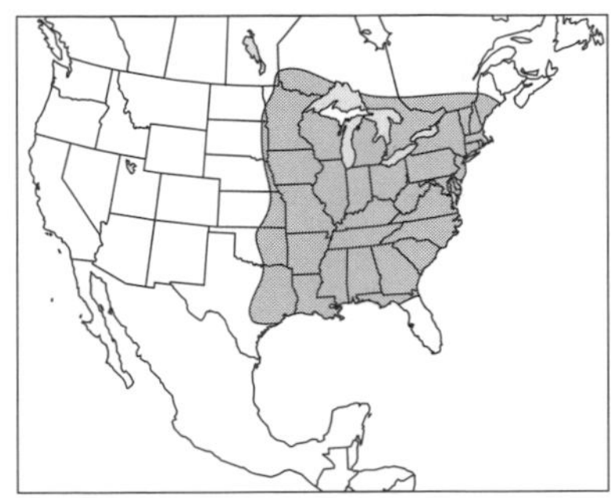

sulfur-winged grasshopper

NORTHWESTERN RED-WINGED GRASSHOPPER

Arphia pseudonietana (Plate 8)

Distribution: Found throughout the western United States west of the Mississippi River, the Great Lakes region, the southern regions of western Canada, and northern Mexico.

Identification: A dark-colored, late-season species. Body color, including abdomen color, varies from mottled to solid black. Hind wings are bright red-orange with a solid black margin. Hind femora are dark with a light ring distally. Hind tibiae, also blackish in color, bear a light ring near the junction with the femur. Abdomen is dark. Males are 30–41 mm long, females 32–47 mm.

northwestern red-winged grasshopper

Ecology: A common late-season element of western grasslands, this species feeds primarily on grasses but also consumes broadleaf plants. Adults often crepitate noisily during flight. Adults are found from July to November.

Similar species: The northwestern red-winged grasshopper occurs in the same areas as the speckle-winged rangeland grasshopper, *Arphia conspersa* (Plate 8), though the latter tends to occur in the adult stage early in the summer, whereas the former occurs later in the summer. The northwestern red-winged grasshopper tends to be darker, sometimes black, or more boldly mottled. In Oregon and northern California, a similar species occurs, Saussure's

Saussure's red-winged grasshopper

red-winged grasshopper, *Arphia saussureana*. This species is distinguished from the northwestern red-winged grasshopper by the yellow or brownish abdomen.

PLAINS YELLOW-WINGED GRASSHOPPER

Arphia simplex (Plate 9)

Distribution: Less widespread than some other *Arphia* species—found only in the central Great Plains from South Dakota to Texas, and south into eastern Mexico. It is found throughout Texas.

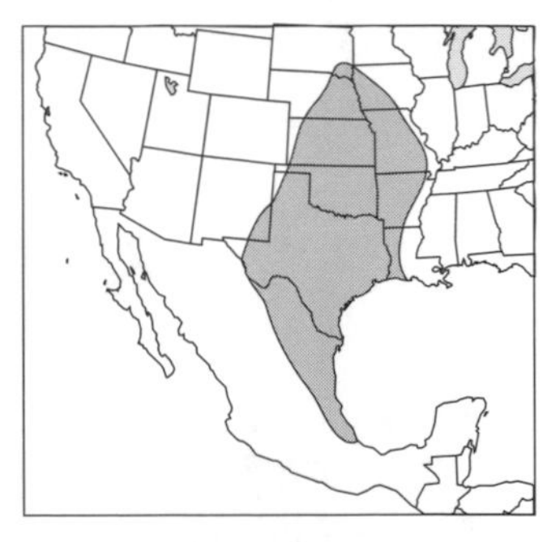
plains yellow-winged grasshopper

Identification: General body color is brown and sometimes speckled. Hind wings are yellow, yellow orange, or orange basally with a narrow black band marginally. Hind tibiae are dark and partly bluish. Males are 28–40 mm long, females 34–44 mm.

Ecology: Inhabits bunchgrass prairie, old fields, open woodlands, and edges of wooded areas. Nymphs overwinter and adults are common in early summer, but both stages can be found in summer months. Crepitation by males is common.

Similar species: The plains yellow-winged grasshopper is separated from the similar and co-occurring northwestern red-winged grasshopper, *Arphia pseudonietana* (Plate 8), and the speckle-winged rangeland grasshopper, *Arphia conspersa* (Plate 8), by the bluish color on the hind tibiae.

AUTUMN YELLOW-WINGED GRASSHOPPER

Arphia xanthoptera (Plate 9)

Distribution: Found widely in the United States, from the East Coast west to western Nebraska and Oklahoma.

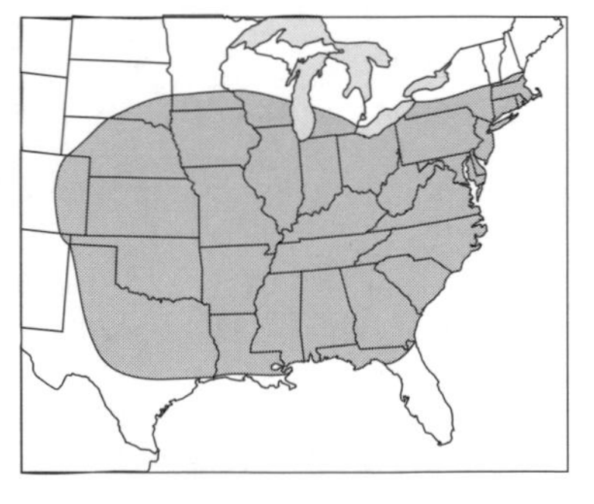
autumn yellow-winged grasshopper

Identification: The largest of the *Arphia* species found in the eastern states; its large size is a distinguishing characteristic. Brown to blackish body with yellow (sometimes orange) hind wings that are marked with a curved black band crossing the wing. Forewings are uniformly colored brownish to blackish. Dorsal median ridge on the pronotum is strongly elevated and arched. Tibiae are mostly dark, including the distal region, with a pale ring in the basal quarter. Males are 31–38 mm long, females 36–46 mm.

Ecology: Habitats of the autumn yellow-winged grasshopper include weedy borders of cultivated fields, brushy fields, and open woods. Nymphs typically occur from April to June, and adults from July to November, but as in other *Arphia* species, the life cycle is quite variable.

Similar species: The median ridge on the pronotum of the autumn yellow-winged grasshopper is distinctly elevated and arched, and this serves as the most reliable diagnostic feature to separate it from other *Arphia* species. The dark distal region of the tibiae is not entirely consistent but is a fairly reliable character to separate the autumn yellow-winged grasshopper from the sulfur-winged grasshopper, *Arphia sulfurea*, and especially from the southern yellow-winged grasshopper, *Arphia granulata* (Plate 8). The forewings of the autumn yellow-winged grasshopper lack the distinct yellowish stripe that is common on the other eastern *Arphia*.

CLEAR-WINGED GRASSHOPPER

Camnula pellucida (Plate 9)

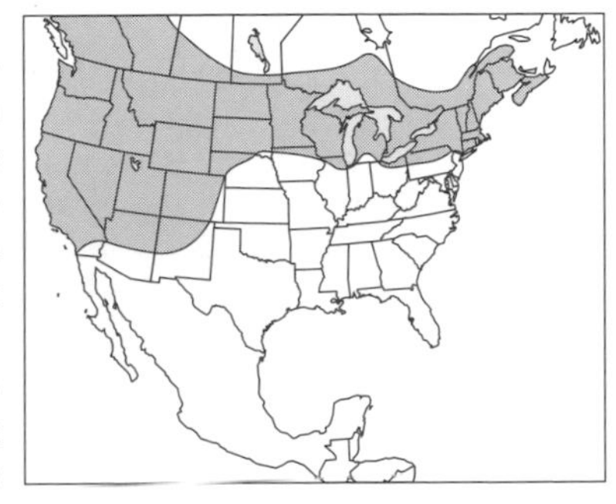

clear-winged grasshopper

Distribution: Primarily western in distribution. Found throughout western North America east to the western margin of the Great Plains, but is also found in the northern states and southern Canada in the region of the Great Lakes, New England, and east to the Maritime Provinces.

Identification: Differs from most other band-winged grasshoppers by having transparent hind wings, lacking bands and other forms of pigmentation. In other respects, however, this yellowish or gray-brown grasshopper is quite typical. Median ridge on the pronotum is slightly elevated and notched. Forewings bear dark round or oval blotches. When viewed from above, amber lines on the forewings converge to form a distinct V shape. Lateral lobes of the pronotum are marked with black. Hind femora normally bear black bands. Hind tibiae are yellowish. Males are 20–25 mm long, females 25–31 mm.

Ecology: Inhabits grassy meadows, often in hilly or mountainous areas. Feeds mostly on grasses, and its abundance and tendency to aggregate when densities are high can lead to significant damage to pastures, cultivated grains, and canola crops. It displays extreme fluctuations in abundance, possibly because it is quite susceptible to a fungus disease. Adults sometimes form aggregations where mating occurs.

Similar species: The clear-winged grasshopper is easily distinguished from Scudder's clear-winged grasshopper, *Trepidulus hyalinus* (Plate 20), another clear-winged species, by the presence of the V dorsally.

PAINTED MEADOW GRASSHOPPER

Chimarocephala pacifica (Plate 9)

painted meadow grasshopper

Distribution: Found in the southern two-thirds of California, and south into Baja California, Mexico.

Identification: A medium-sized grasshopper that occurs in two color forms, either mostly green or mostly brown. Central ridge on the pronotum is elevated. Forewings lack bands. Hind wings are mostly transparent, but have pale yellow basally and smoky gray distally. Hind femora generally are marked with a narrow dark band. Males are 15.5–20.0 mm long, females 20–29 mm.

Ecology: Found in a variety of grassy habitats; adults are most abundant in the spring.

Similar species: The painted meadow grasshopper is distinguished from the *Chortophaga* species, with which it is readily confused, by its more pointed head and western distribution. In and near the Central Valley of California, a similar species occurs, the Central Valley grasshopper, *Chimarocephala elongata*. The hind wing bears the same color pattern as the painted meadow grasshopper, but is translucent rather than transparent.

Central Valley grasshopper

SOUTHERN GREEN-STRIPED GRASSHOPPER

Chortophaga australior (Plate 10)

Distribution: Found in the Southeast from Florida and Georgia to east Texas. Its distribution overlaps that of the northern green-striped grasshopper slightly, but largely replaces the northern species in the coastal south.

southern green-striped grasshopper

Identification: There are two color forms, a green form and a brown form, with intermediates found in both sexes. The principal difference between forms is in the coloring of the head, thorax, and outer face of the hind femora. Median ridge on the pronotum is slightly elevated. An X-shaped mark is present on the dorsal surface of the pronotum in the brown forms. Leading edge of the forewings is marked with 2 to 3 large green or light brown spots, with the balance of the forewings colored dark brown. The most important distinguishing character is the color of the hind wing. Unlike most band-winged species, the southern green-striped grasshopper lacks a bold black transverse band on the hind wing. The dark band is present but greatly muted, reduced to no more than a smoky area in many individuals. Similarly, the yellow color in the basal area of the hind wing is muted to absent. The upper surface of the hind femur usually is marked with about 3 large dark spots; the central or largest spot is triangular in shape when viewed from above. Hind tibiae are brown or bluish green in color. Males are 21–27 mm long, females 29–37 mm.

Ecology: Inhabits open areas. Favored habitats are old fields, heavily grazed pastures, edges of crop fields, and unpaved roadways. Adults and nymphs can be found year-round in Florida, where two generations occur annually.

Similar species: The southern green-striped grasshopper is quite similar to the more widespread northern green-striped grasshopper, *Chortophaga viridifasciata* (Plate 10), but can be distinguished by the color of the hind tibiae. The tibiae are bluish or bluish green in the former but usually yellowish brown or brown in the latter. The posterior angle of the pronotum, viewed from above, can also be used to distinguish the species: the angle is about 90 degrees in the southern green-striped grasshopper, but acute in the northern green-striped grasshopper (see Fig. 6). Species in the genus *Chimarocephala* (Plate 9) resemble *Chortophaga* species, but their geographic distribution distinguishes them; they are found only in California.

NORTHERN GREEN-STRIPED GRASSHOPPER

Chortophaga viridifasciata (Plate 10)

Distribution: Found throughout the eastern states west to the Rocky Mountains, but absent from Florida. Also found in southern Canada and eastern Mexico.

Identification: As in the southern green-striped grasshopper, *Chortophaga australior*, there are two color forms, a green form and a brown form, with intermediates found in both sexes. The principal difference between forms is in the coloring of the head, thorax, and outer face of the hind femora. Abdomen is reddish brown in both forms. Median ridge on the pronotum is slightly elevated. Hind wing is not distinctly pigmented, though the tip is grayish or blackish and the base yellowish. Antennae are red and rather short. Hind tibiae are normally yellowish or brownish, though sometimes tending toward blue. Males are 23–30 mm long, females 28–38 mm.

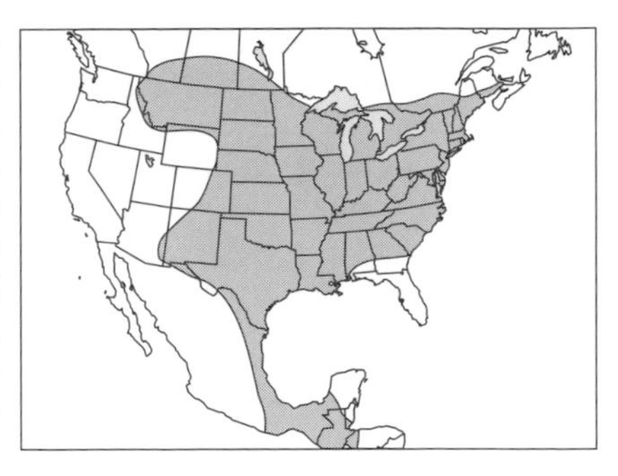

northern green-striped grasshopper

Ecology: Favors wet areas with short grasses, and feeds primarily on grasses. Nymphs overwinter, and adults are present in the spring and early summer. There are reports of adults also overwintering, and in southern areas apparently two generations can occur annually.

Similar species: The northern green-striped grasshopper is quite similar to the southern green-striped grasshopper, *Chortophaga australior* (Plate 10), but can be distinguished by the color of the hind tibiae. The tibiae are yellowish brown or brown in the former but bluish or bluish green in the latter. Generally they can be distinguished by their geographic range, but their distribution overlaps in some southern states. The posterior angle of the pronotum, viewed from above, can be used to distinguish the species: the angle is acute in the northern green-striped grasshopper but about 90 degrees in the southern green-striped grasshopper (see Fig. 6). Species in the genus *Chimarocephala* (Plate 9) resemble *Chortophaga* species, but their geographic distribution distinguishes them; they are found only in California.

CARLINIAN SNAPPER GRASSHOPPER

Circotettix carlinianus (Plate 11)

Distribution: Found widely in western North America from the Canadian Prairie Provinces to northern New Mexico, and from the Pacific Coast states east to the Dakotas.

Identification: Gray to reddish brown, with weak transverse bands crossing the forewings and the outer face of the hind femora. Hind wings bear several unusually thickened veins, and generally are black basally but may be entirely colorless. Hind tibiae are yellowish. Males are 29–40 mm long, females 33–46 mm.

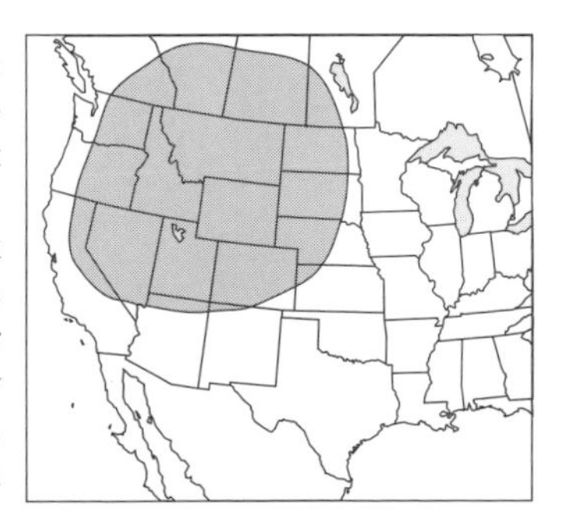

Carlinian snapper grasshopper

Ecology: Habitats include dry hillsides and prairies, dirt roadways, and fallow wheat fields. This grasshopper derives its name from the noisy crackling display produced by males during their hovering flight, though many other species are similarly noisy. Adults are found from June to September.

Similar species: The Carlinian snapper grasshopper is readily distinguished from the wrangler grasshopper, *Circotettix rabula* (Plate 11), by the color pattern of the hind wing: the Carlinian snapper lacks yellow basally. There are five other species in this group, and careful examination of the wing veins is necessary to distinguish among them. Four of the five species have restricted distributions and are found only in California or nearby Nevada: the shasta grasshopper, *Circotettix shastanus*; the rattling grasshopper, *C. crotalum*; the California snapper grasshopper, *C. stenometopus*; and the spotted snapper grasshopper, *C. maculatus.* The fifth species, the undulant-winged grasshopper, *C. undulatus*, is more widespread, and where its distribution overlaps that of the Carlinian snapper and the wrangler grasshopper, it can be distinguished by the presence of blue or gray hind tibiae. Also, the undulant-winged grasshopper sometimes has greenish-blue or blue hind wings.

shasta grasshopper

rattling grasshopper

California snapper grasshopper

spotted snapper grasshopper

undulant-winged grasshopper

WRANGLER GRASSHOPPER

Circotettix rabula (Plate 11)

Distribution: Found throughout the Rocky Mountains Region of western North America, and along the western edge of the Great Plains.

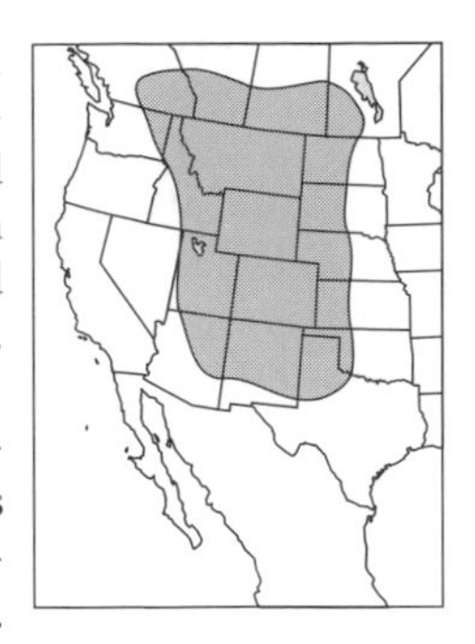
wrangler grasshopper

Identification: A fairly large, gray-colored grasshopper. Forewings are marked with small spots that sometimes coalesce to form bands. Hind wings are pale yellow basally and bear a weak to distinct black band centrally. Lateral margin of the hind wing is concave. Some of the veins of the hind wing are thickened. Hind tibiae are yellow, brown, or bluish. Males are 30–40 mm long, females 34–49 mm.

Ecology: Inhabits sparsely vegetated, eroded, and rocky hillsides and riverbanks. Feeds on low-growing broadleaf weeds and mosses. Males frequently produce noisy crackling displays and make wide circling flights for long periods of time. Adults normally are found from June to September, but both earlier and later in the year in mild southern climates.

Similar species: The wrangler grasshopper is readily distinguished from the Carlinian snapper grasshopper, *Circotettix carlinianus* (Plate 11), by the color pattern of the hind wing: the Carlinian snapper lacks yellow basally. There are five other species in this group, and careful examination of the wing veins is necessary to distinguish among them. Four of the five species have restricted distributions and are found only in California or nearby Nevada: the shasta grasshopper, *Circotettix shastanus*; the rattling grasshopper, *C. crotalum*; the California snapper grasshopper, *C. stenometopus*; and the spotted snapper grasshopper, *C. maculatus*. The fifth species, the undulant-winged grasshopper, *C. undulatus*, is more widespread, and where its distribution overlaps that of the common Carlinian snapper and the wrangler grasshopper, it can be distinguished by the presence of blue or gray hind tibiae. Also, the undulant-winged grasshopper sometimes has greenish-blue or blue hind-wings.

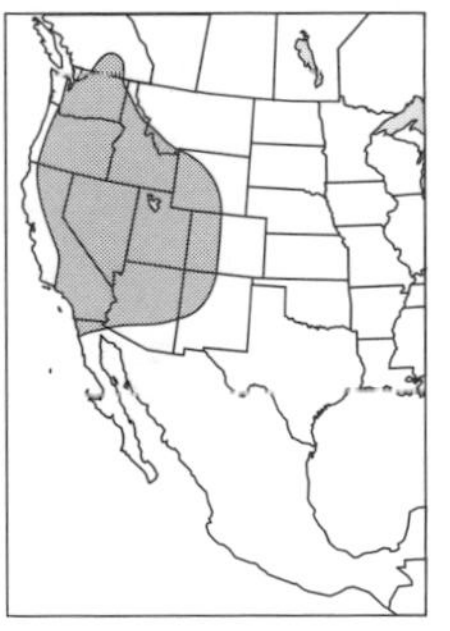
groove-headed grasshopper

GROOVE-HEADED GRASSHOPPER

Conozoa sulcifrons (Plate 10)

Distribution: Found from the Rocky Mountains west to British Columbia and California.

Identification: A grayish-brown species with black bands on the forewings. Wings are long, extending well beyond the tip of the abdomen. Hind wings are yellow basally and with a transverse black band. Hind tibiae are red or orange. Males are 19–27 mm long, females 25–37 mm.

Ecology: Associated with arid environments, occupying habitats that range from sparsely to densely vegetated. A mixed feeder, preferring grasses. Adults are found from July to September throughout the range, but both earlier and later in the year in benign southern climates.

Similar species: Though there are several species in this genus, only the groove-headed grasshopper and the Texas cristate grasshopper, *Conozoa texana*, occur over large geographic areas. The two are distinguished by the color of the hind tibiae: red or orange in the former but bluish or gray in the latter.

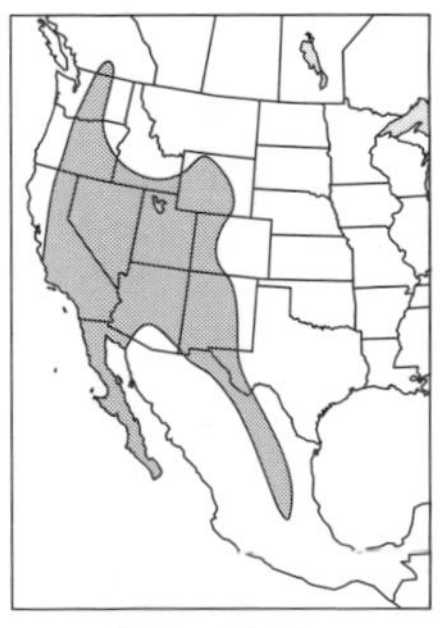
Texas cristate grasshopper

PRONOTAL RANGE GRASSHOPPER

Cratypedes neglectus (Plate 12)

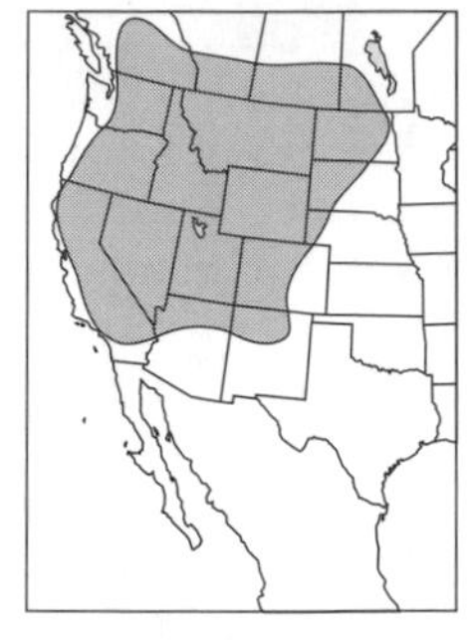
pronotal range grasshopper

Distribution: Found widely in western North America from the Rocky Mountain region to the Pacific Ocean.

Identification: This grayish-brown grasshopper is normally marked with pale dorsal stripes on the forewings that converge on the top of the back to form an elongated V. Forewings bear indistinct broad bands basally and small spots toward the wing tips. Hind wings are yellow with a black band at the margin. Pronotum is rough or bumpy, the median ridge bearing two cuts. Inner face of the hind femora is yellow, orange, or red, and usually bears two dark bands. Hind tibiae are orange red. Males are 24–32 mm long, females 33–45 mm.

Ecology: Favors sunny clearings in forested and brushy areas but is found in diverse habitats. Feeds primarily on grasses. Crepitates loudly. Adults are found from May or June to August.

Similar species: The pronotal range grasshopper is distinguished from the Nevada red-winged grasshopper, *Cratypedes lateritius* (Plate 12), by the widening of the pronotal lobes; they are wider at the bottom than in the middle. In the Nevada red-winged grasshopper the ventral lobes have parallel sides or narrow toward the bottom. The genus *Cratypedes* resembles *Xanthippus* (Plate 21), but *Cratypedes* species lack the large distinct spots along the entire length of the forewings usually so apparent in *Xanthippus.*

NEVADA RED-WINGED GRASSHOPPER

Cratypedes lateritius (Plate 12)

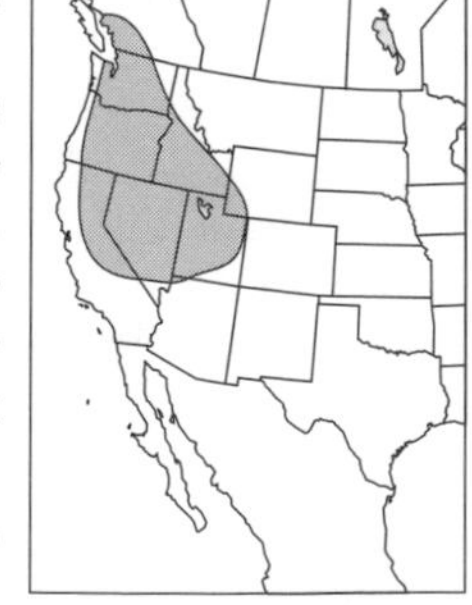
Nevada red-winged grasshopper

Distribution: Found from the western states east to Utah and western Wyoming.

Identification: An unusually variable species. Forewing has irregular dark spots. When the wings are folded, narrow ivory stripes on the forewings converge to form a V, particularly in males. Hind wings are yellow, orangish, pink, or red with a black band. Pronotum is rough or bumpy. Inner face of the femora bears two complete or incomplete black bands, and the background color varies from orange to red. Hind tibiae are yellow or orange. Males are 26–33 mm long, females 33–46 mm.

Ecology: Inhabits open pine woodlands and sagebrush and is common in the Idaho and Nevada desert areas. Adults are found from April or May to September.

Similar species: The Nevada red-winged grasshopper is distinguished from the pronotal range grasshopper, *C. neglectus* (Plate 12), by the shape of the lateral lobes of the pronotum. The lateral lobes narrow toward the bottom or are parallel sided in the former, whereas the lobes are wider at the bottom than in the middle in the latter. The genus *Cratypedes* resembles *Xanthippus* (Plate 21), but *Cratypedes* species lack the large distinct spots along the entire length of the forewings usually so apparent in *Xanthippus.*

HAYDEN'S GRASSHOPPER

Derotmema haydeni (Plate 11)

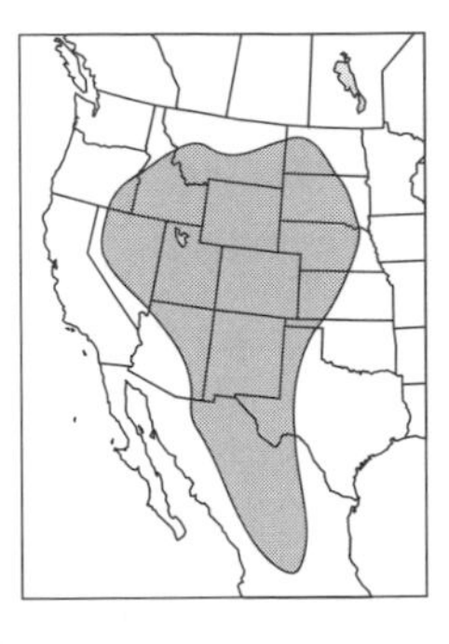

Hayden's grasshopper

Distribution: Found widely in the western United States, from North Dakota to Texas and west to eastern Oregon, Nevada, and Arizona. Its range also extends into central Mexico.

Identification: A brownish-gray species marked by bulbous eyes, slender antennae, and dark markings on the forewings. Median ridge of the prothorax has two deep incisions, resulting in pronounced lobes. Hind wings are yellow or reddish basally with a broad dark band and the apex transparent. Hind tibiae are yellowish brown. Males are 18–24 mm long, females 22–30 mm.

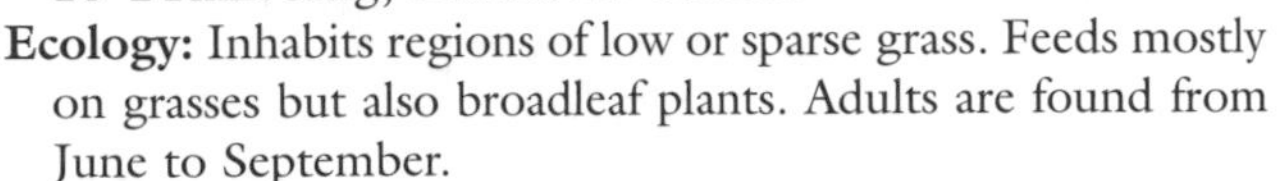

Ecology: Inhabits regions of low or sparse grass. Feeds mostly on grasses but also broadleaf plants. Adults are found from June to September.

Similar species: The bulbous eyes are reminiscent of the genus *Psinidia* (Plate 20) and of other *Derotmema* spp. *Psinidia* spp. have sword-shaped antennae, whereas *Derotmema* spp. have thread-like antennae. There are 5 species in the genus *Derotmema*; Hayden's grasshopper and Saussure's desert grasshopper, *D. saussureanum* (Plate 11), are distinguished from the others by the tall, elevated lobes and deep incisions in the middle ridge of the pronotum. Also, the posterior margin of the pronotum of these two species, when viewed from above, forms a definite angle, whereas in the other three species the posterior margin is rounded. Separating these two species is difficult based on appearance, but they do not overlap in distribution. Although ranging widely in the western states, Hayden's grasshopper is absent from California, where Saussure's desert grasshopper occurs.

SAUSSURE'S DESERT GRASSHOPPER

Derotmema saussureanum (Plate 11)

Saussure's desert grasshopper

Distribution: Found only in southern California and Baja California, Mexico.

Identification: Grayish brown in body color, with slender antennae. Forewings bear a row of dark spots along the leading edge. Hind wing is pale yellow or blue basally, with a black transverse band. Wing apex is transparent. Hind tibiae are yellowish. Males are 19–24 mm long, females 23–28 mm.

Ecology: Inhabits desert flats. Adults are found from June to September.

Similar species: The bulbous eyes are reminiscent of the genus *Psinidia* (Plate 20) and of other *Derotmema* species. *Psinidia* species have sword-shaped antennae, whereas *Derotmema* species have thread-like antennae. There are 5 species in the genus *Derotmema*, and Hayden's grasshopper, *D. haydeni* (Plate 11), and Saussure's desert grasshopper are distinguished from the others by the tall, elevated lobes and deep incisions in the middle ridge of the pronotum. Also, the posterior margin of the pronotum of these two species, when viewed from above, forms a definite angle, whereas in the other three species the posterior margin is rounded. Separating these two species is difficult based on appearance, but they do not overlap in distribution. Although ranging widely in the western states, Hayden's grasshopper is absent from California, where Saussure's desert grasshopper occurs.

CAROLINA GRASSHOPPER

Dissosteira carolina (Plate 13)

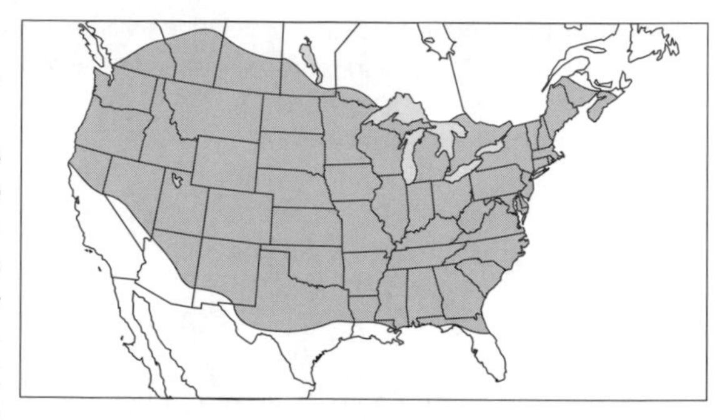

Carolina grasshopper

Distribution: Found throughout the United States.

Identification: Color varies from yellowish gray to reddish brown, and sometimes there are numerous minute dark spots over most of the body. An elevated central ridge is found dorsally on the pronotum; it bears one distinct cut. Hind wings are black except for a marginal yellowish band and smoky gray wing tips. The black hind wings distinguish this species from nearly all other grasshoppers. Hind tibiae are yellow. Males are 32–42 mm long, females 42–58 mm.

Ecology: Associated with barren soil such as unpaved roadways and fallow fields. Feeds both on grasses and on broadleaf plants. This species is a strong flier and is often seen hovering or in the zig-zag, fluttering flight of courtship. It can be mistaken for a butterfly when in flight. Adults are found from May to November, but June to October in northern areas.

Similar species: The Carolina grasshopper is not often confused with other species. The combination of an elevated pronotum and black hind wings is fairly distinctive. However, in the Great Plains region, the High Plains grasshopper, *Dissosteira longipennis* (Plate 13), is similar in appearance. The extent of the black coloration on the hind wing is reduced in the High Plains grasshopper; the distal third of the wing may be smoky or transparent, but not black. Also, the black coloration does not extend to the base of the hind wing, as it does in the Carolina grasshopper. The forewings of the High Plains grasshopper are distinctly spotted; in the Carolina grasshopper the forewings often lack spots or the spots are weak.

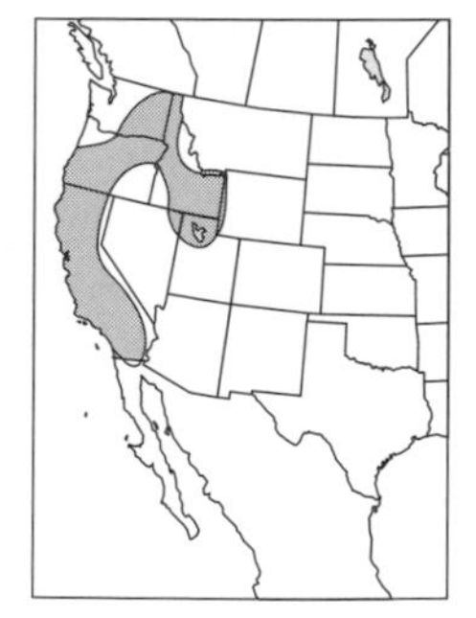

pale-winged grasshopper

West of the Rocky Mountains, particularly in California, two other *Dissosteira* species occur: the pale-winged grasshopper, *D. spurcata*, and the California rose-winged grasshopper, *D. pictipennis* (Plate 13). In both species the medial pronotal ridge is elevated, as is characteristic of this genus, and the body and forewings are heavily mottled with large dark spots. However, the hind wings are not predominantly black. In the pale-winged grasshopper, the hind wing bears only a narrow, faint black band and is otherwise transparent. In the California rose-winged grasshopper, the hind wing bears a broad black band centrally but the wing base is reddish. These western species of *Dissosteira* are easily confused with *Spharagemon*;

however, the hind tibiae of *Dissosteira* species are yellow or bluish, whereas those of *Spharagemon* species are reddish or orange.

HIGH PLAINS GRASSHOPPER

Dissosteira longipennis (Plate 13)

Distribution: Restricted to the central Great Plains, from South Dakota and eastern Wyoming south to Texas.

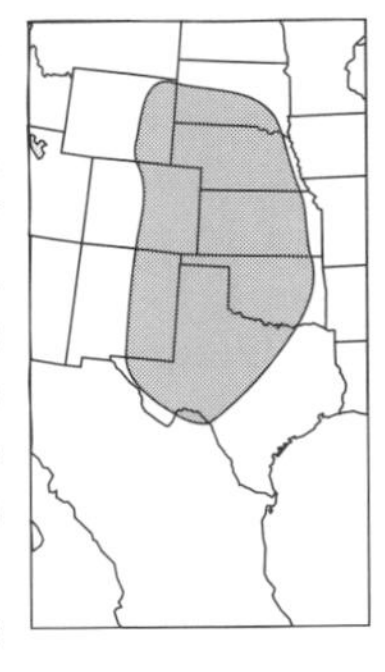

High Plains grasshopper

Identification: This large, yellowish grasshopper is marked by an elevated medial ridge on the pronotum, numerous moderately large spots on the forewings, and extensive black coloration on the hind wings. The pronotal ridge bears one distinct cut. The hind wings, though primarily black, are tan or transparent at the margin and nearly transparent over the distal one-third of the wing. Hind tibiae are yellow. Males are 40–50 mm long, females 48–60 mm.

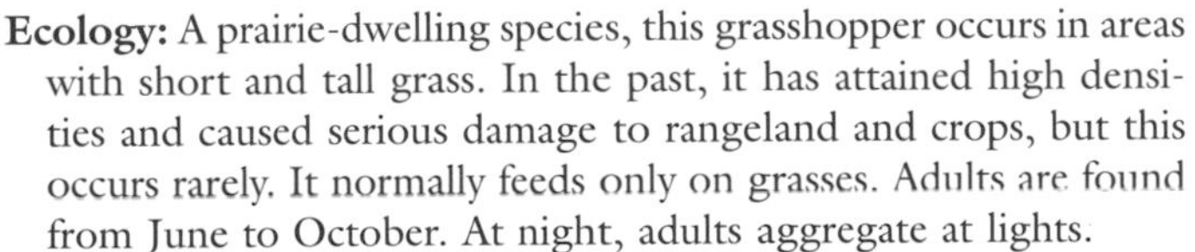

Ecology: A prairie-dwelling species, this grasshopper occurs in areas with short and tall grass. In the past, it has attained high densities and caused serious damage to rangeland and crops, but this occurs rarely. It normally feeds only on grasses. Adults are found from June to October. At night, adults aggregate at lights.

Similar species: The High Plains grasshopper can be confused with the Carolina grasshopper, *Dissosteira carolina* (Plate 13), though only in the Great Plains region. The Carolina grasshopper differs in having the hind wing with black extending nearly to the wing tip and the marginal area of the hind wing yellowish. Also, the forewings of the High Plains grasshopper are mottled with moderately large spots, whereas in the Carolina grasshopper spots are minute or absent.

West of the Rocky Mountains, particularly in California, two other *Dissosteira* species occur: the pale-winged grasshopper, *D. spurcata*, and the California rose-winged grasshopper, *D. pictipennis* (Plate 13). In both species the medial pronotal ridge is elevated, as is characteristic of this genus, and the body and forewings are heavily mottled with large dark spots. However, the hind wings are not predominately black. In the pale-winged grasshopper, the hind wing bears only a narrow, faint black band and is otherwise transparent. In the California rose-winged grasshopper, the hind wing bears a broad black band centrally but the wing base is reddish. These western species of *Dissosteira* are easily confused with *Spharagemon*; however, the hind tibae of *Dissosteira* species are yellow or bluish, whereas those of *Spharagemon* species are reddish or orange.

CALIFORNIA ROSE-WINGED GRASSHOPPER

Dissosteira pictipennis (Plate 13)

Distribution: Principally found in California and Oregon, but occasionally is seen from southern Canada to northern Mexico.

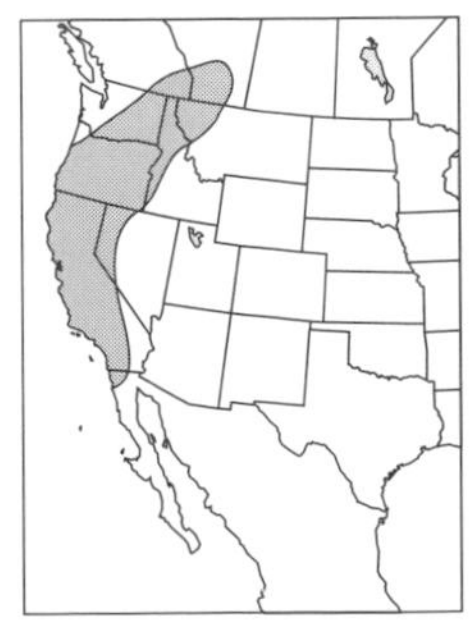

California rose-winged grasshopper

Identification: Somewhat variable in appearance, this small, grayish or brownish species usually is marked with three dark bands on the forewings and four bands on the outer face of the hind femora. The bands sometimes are absent. More distinctive is the deeply cut pronotal ridge that is common among *Dissosteira*. The hind wing is dark pink to rosy red basally with a dark band centrally, and transparent at the tip. Hind tibiae are gray or blue but with a yellowish band basally. In some specimens, the pronotum is marked with a light-colored X. Males are 21–29 mm long, females 29–37 mm.

Ecology: Inhabits high-elevation grass and grass-oak environments. Adults are found from June to October.

Similar species: The California rose-winged grasshopper is easily distinguished from the Carolina grasshopper, *Dissosteira carolina* (Plate 13), the High Plains grasshopper, *D. longipennis* (Plate 13), and the pale-winged grasshopper, *D. spurcata*, by the reddish hind wing. However, it may be confused with *Spharagemon* species (Plates 16, 17). The *Dissosteira* species are distinguished from *Spharagemon* by the yellow or bluish hind tibiae; the tibiae are never reddish or orange as is usually the case in *Spharagemon*.

DUSKY GRASSHOPPER

Encoptolophus costalis (Plate 15)

Distribution: Found throughout the Great Plains and Rocky Mountains regions from Alberta and Manitobas in Canada south to Arizona and Texas, and central Mexico.

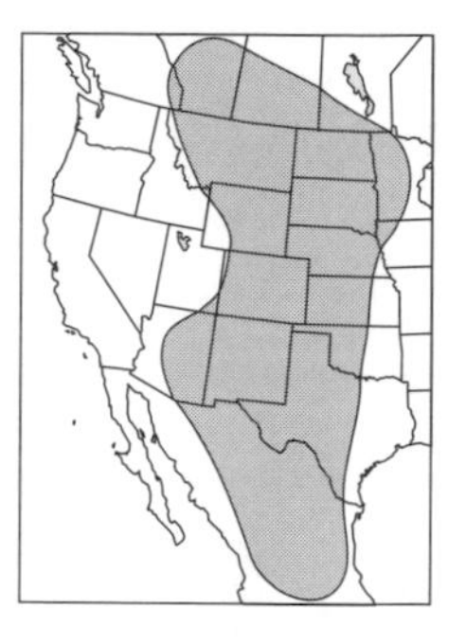

dusky grasshopper

Identification: Usually grayish brown, bearing solid dark bands on the forewings. Some females are greenish. Median ridge of the thorax is pronounced. When viewed from above, the pronotum is marked with a light-colored X. Hind wings are nearly transparent but smoky at the tip. Hind femora are marked with dark bands. Hind tibiae are bright blue. Males are 16–24 mm long, females 22–31 mm.

Ecology: Inhabits grasslands and arid environments. Feeds preferentially on grasses but also accepts broadleaf plants. Adults are commonly found from May to September, but both nymphs and adults overwinter successfully in southern areas.

Similar species: Formerly considered synonymous with the clouded grasshopper, *Encoptolophus sordidus*, this species is now separated because of different geographic range and mating behavior. The clouded grasshopper is more eastern in distribution, ranging from Maine and North Carolina in the east, to the central Great Plains in the west. The distal two-thirds of the hind tibiae are brown, in contrast to the blue tibiae of the dusky grasshopper. To differentiate the dusky grasshopper from the southwestern dusky grasshopper, *Encoptolophus subgracilis* (Plate 15), one should observe the pronotum from above. The posterior angle of the pronotum in the dusky grasshopper is 90 degrees or less, whereas in the southwestern dusky grasshopper the angle is greater than 90 degrees (see Fig. 6).

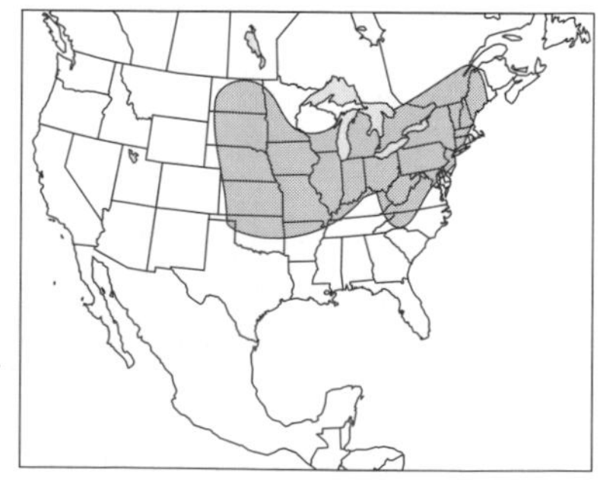

clouded grasshopper

SOUTHWESTERN DUSKY GRASSHOPPER

Encoptolophus subgracilis (Plate 15)

Distribution: Principally found in the southwestern states, though it occurs as far north as South Dakota, and south through Mexico to Central America.

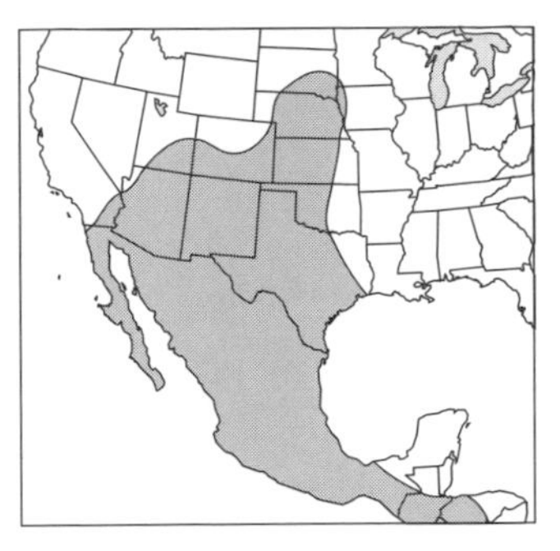

southwestern dusky grasshopper

Identification: Usually grayish brown, bearing solid dark bands on the forewings. Some females are greenish. Median ridge of the thorax is raised but not pronounced. When viewed from above, the pronotum is marked with a light-colored X. Hind wings are nearly transparent but are sometimes greenish yellow at the base and usually smoky at the tip. Hind femora are marked with dark bands. Hind tibiae are bright blue. Males are 18–26 mm long, females 24–33 mm.

Ecology: Inhabits grasslands, including desert grasslands. Feeds principally on grasses. Adults are most common from May to September, but both nymphs and adults overwinter successfully in southern areas.

Similar species: This species is quite similar to the dusky grasshopper, *Encoptolophus costalis* (Plate 15). To differentiate, examine the pronotum from above. The posterior angle of the pronotum in the southwestern dusky grasshopper is greater than 90 degrees, whereas in the dusky grasshopper it is 90 degrees or less (see Fig. 6). Also, the median ridge on the pronotum is not as elevated as in the dusky grasshopper.

MAGNIFICENT GRASSHOPPER

Hadrotettix magnificus (Plate 14)

Distribution: Found from Colorado and Nebraska south to central Mexico.

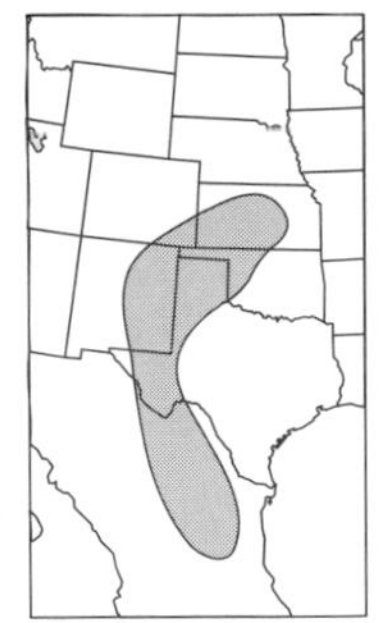

magnificent grasshopper

Identification: This large, brownish or grayish grasshopper is distinctly marked with three broad blackish bands crossing the forewings. Hind wings are pale yellow basally and bear a distinct black band. Antennae are long and black. Hind femora bear a dark band on the outer face; the inner face is predominantly blue black but brown or orange at the base of the femora. Hind tibiae are orange or red. Males are 40–49 mm long, females 44–53 mm.

Ecology: Inhabits arid grasslands and deserts but is relatively uncommon everywhere. Adults are found from July to October.

Similar species: The magnificent grasshopper is similar in appearance to the three-banded range grasshopper, *Hadrotettix trifasciatus* (Plate 14). In the former, the basal region of the inner face of the hind femora is orange or brown, whereas in the latter it is blue.

THREE-BANDED RANGE GRASSHOPPER

Hadrotettix trifasciatus (Plate 14)

Distribution: Found widely in the Great Plains region from southern Canada to northern Mexico.

Identification: This large, tan or grayish grasshopper is distinctly marked with three broad blackish bands crossing the forewings. Hind wings are pale yellow basally and bear a distinct black band. Antennae are long and black. Hind femora bear a dark band on the outer face; the inner face is bluish black to the base of the femora. Hind tibiae are orange or red. Males are 26–37 mm long, females 36–50 mm.

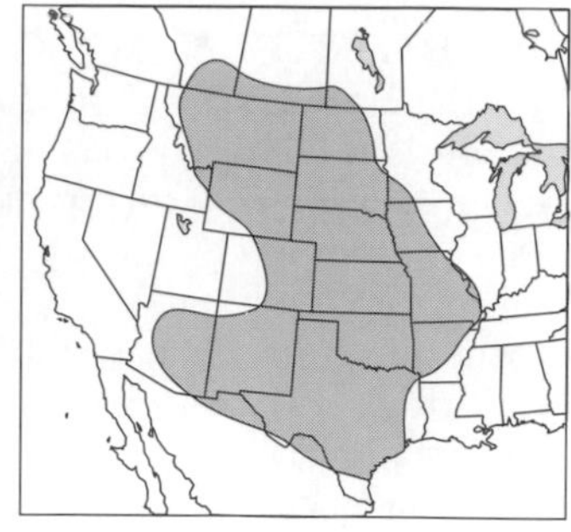

three-banded range grasshopper

Ecology: Inhabits arid portions of grasslands and is not uncommon. Though grasses are consumed, broadleaf plants are preferred. Adults are found from June to September.

Similar species: The three-banded range grasshopper is similar in appearance to the magnificent grasshopper, *Hadrotettix magnificus* (Plate 14). In the former the basal region of the inner face of the hind femora is blue, whereas in the latter it is brown or orange.

ARROYO GRASSHOPPER

Heliastus benjamini (Plate 12)

Distribution: Found only in the southwestern states of Arizona, New Mexico, and Texas, and south to southern Mexico.

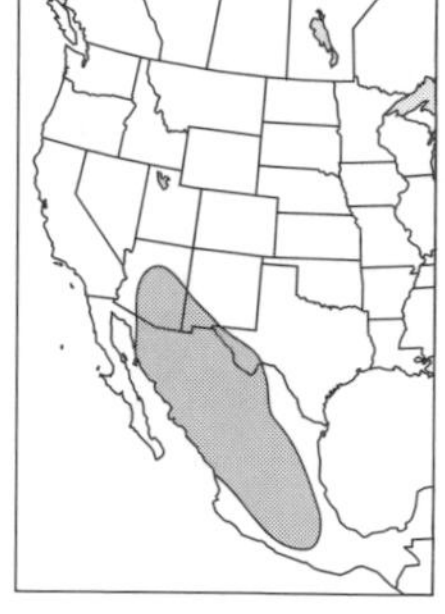

arroyo grasshopper

Identification: This small, brownish grasshopper has discrete dark bands on the forewings. Hind wings are rosy red basally with a clouded region or weak dark band centrally. Lateral lobes of the pronotum are broadly rounded. Hind femora bear two incomplete dark bands on the outer face; the inner face is marked with two black and three yellow bands. Hind tibiae are yellowish basally with a dark band, and reddish distally. Males are 18–23 mm long, females 25–33 mm.

Ecology: Frequents dry streambeds in desert regions, where it is reported to feed on weeds and on algae growing around pools of water. Occasionally found in grasslands. Adults are found from July to November.

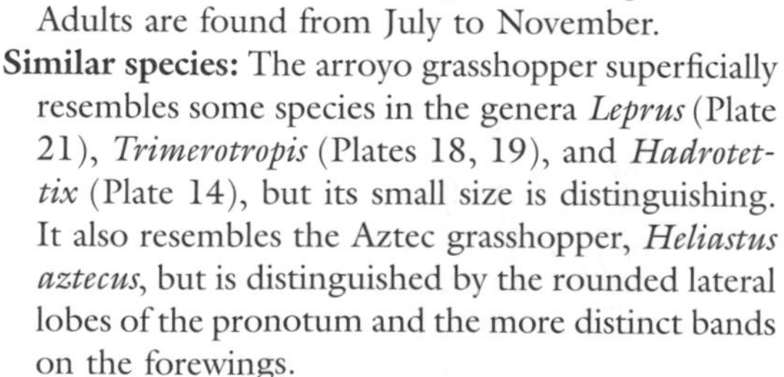

Similar species: The arroyo grasshopper superficially resembles some species in the genera *Leprus* (Plate 21), *Trimerotropis* (Plates 18, 19), and *Hadrotettix* (Plate 14), but its small size is distinguishing. It also resembles the Aztec grasshopper, *Heliastus aztecus*, but is distinguished by the rounded lateral lobes of the pronotum and the more distinct bands on the forewings.

Aztec grasshopper

WRINKLED GRASSHOPPER

Hippiscus ocelote (Plate 15)

Distribution: Found widely east of the Rocky Mountains except for the northern states and Canada. Occasionally found in the desert southwest and Mexico.

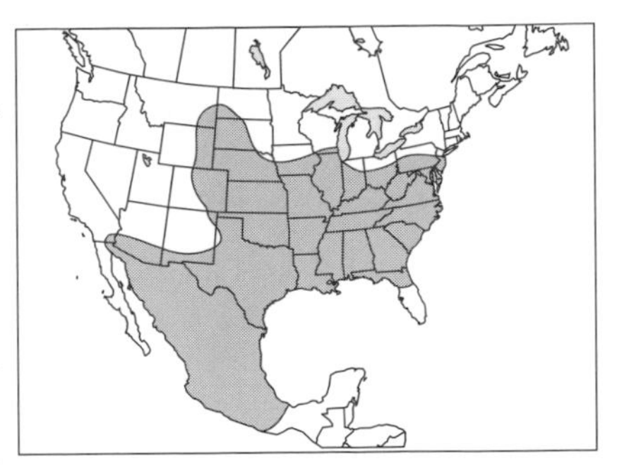
wrinkled grasshopper

Identification: This is a large, heavy-bodied species, and is gray and brown. Pronotum is usually rough or wrinkled, the basis of the name, and often has a light X-shaped mark dorsally, especially in males. Forewings bear large dark spots and light bands; the latter converge at the tips to form a light-colored V dorsally when the wings are closed. Hind wings are usually pale pinkish or orangish basally, but sometimes tending toward yellow. They also have a broad dark band centrally but with the tip poorly pigmented. Hind tibiae are yellow. Males are 28–40 mm long, females 39–53 mm.

Ecology: Inhabits pastures with thin or low-growing grass. Feeds on grass, and is an occasional pasture pest. Females are poor fliers, but males are active. Adults are found from July to October, and during the winter in some southern areas.

Similar species: In the Great Plains, the wrinkled grasshopper can be confused with the red-shanked grasshopper, *Xanthippus corallipes* (Plate 21). The pink or orange hind wing serves to distinguish it from the red-shanked grasshopper, which has a yellow hind wing.

APACHE GRASSHOPPER

Hippopedon capito (Plate 12)

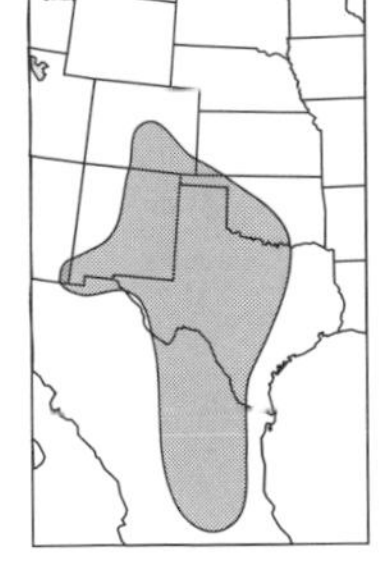
Apache grasshopper

Distribution: Found in the southwestern states, from Colorado and Oklahoma south into central Mexico.

Identification: Largely gray with black markings. Forewings have large spots or bands, and a black band may be present on the face beneath the antennae. Hind wings are yellow basally with a black band centrally. Wings are long. When the folded wings are viewed from above, a light colored V is evident. Median ridge of the pronotum is elevated near the head but reduced on the posterior half. Hind tibiae are yellowish. Males are 20–27 mm long, females 24–33 mm.

Ecology: Inhabits areas of hilly prairie. It is uncommon. Adults are found from July to October.

Similar species: The Apache grasshopper resembles some of the *Trachyrhachys* species (Plate 17) but is distinguished by the reduced median ridge on the posterior region of the pronotum. It can be distinguished from Rehn's slender grasshopper, *Hippopedon gracilipes*, by the presence of the black band on the hind wing. This latter species is quite restricted in distribution, occurring only in southern Arizona, southernmost California, and Baja California, Mexico.

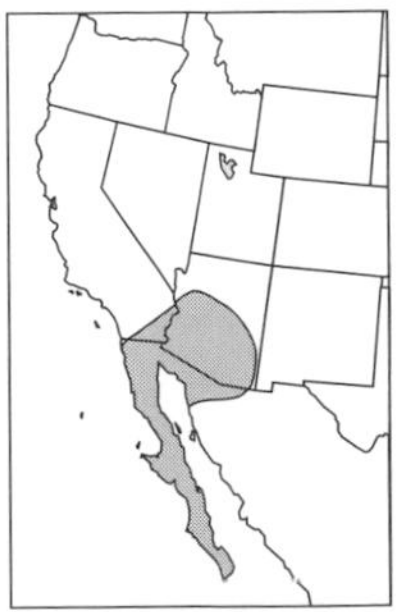
Rehn's slender grasshopper

SAUSSURE'S BLUE-WINGED GRASSHOPPER

Leprus intermedius (Plate 21)

Distribution: Found widely in the southwest from Colorado and Texas west to California. It also occurs in northwestern Mexico.

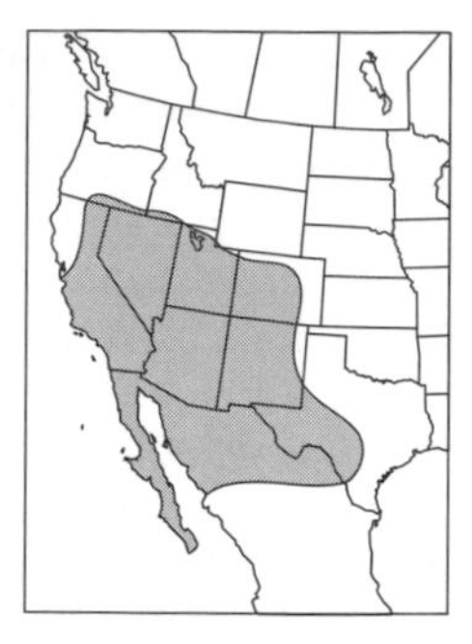

Saussure's blue-winged grasshopper

Identification: The blue hind wings of this species immediately suggest its identity, though this is apparent only when the grasshopper is in flight. It is large, brownish gray, and the forewings are marked with three broad black bands or spots separated by narrow lighter areas. Hind wings are blue or turquoise basally, and bear a black band that sometimes is quite wide, occupying most of the hind wing. Hind tibiae are blue. Males are 24–40 mm long, females 35–49 mm.

Ecology: This species is sometimes locally abundant, though it is not generally common. It inhabits arid, barren areas, particularly hillsides, where it feeds on grass. Adults generally are found from April to October, but occurrence is variable.

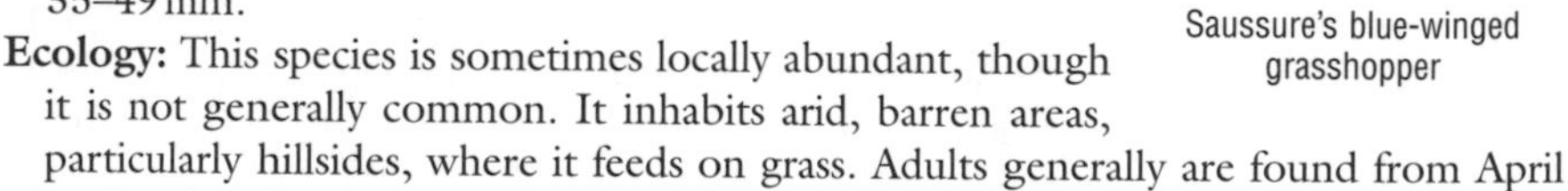

Similar species: This species is superficially similar to *Hadrotettix* species (Plate 14), but the blue hind wings distinguish it. Saussure's blue-winged grasshopper is also similar to Wheeler's blue-winged grasshopper, *Leprus wheeleri*, though in this latter species the base of the hind wings is greenish yellow and the hind tibiae pale blue or grayish.

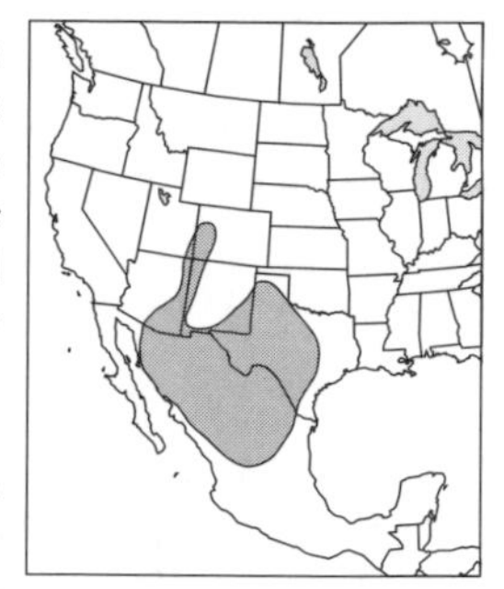

Wheeler's blue-winged grasshopper

Other grasshoppers with blue wings include the blue-winged grasshopper, *Trimerotropis cyaneipennis* (Plate 18), and the undulant-winged grasshopper, *Circotettix undulatus*. The former has irregular dark bands on the forewings, usually consisting of numerous small spots, and a broad light band separating the dark bands. *Circotettix* species have unusually thickened veins in the hind wing.

PLATT RANGE GRASSHOPPER

Mestobregma plattei (Plate 15)

Distribution: Found on the Great Plains from Montana and the Dakotas south to Texas and west to Arizona. It also occurs in northernmost Mexico.

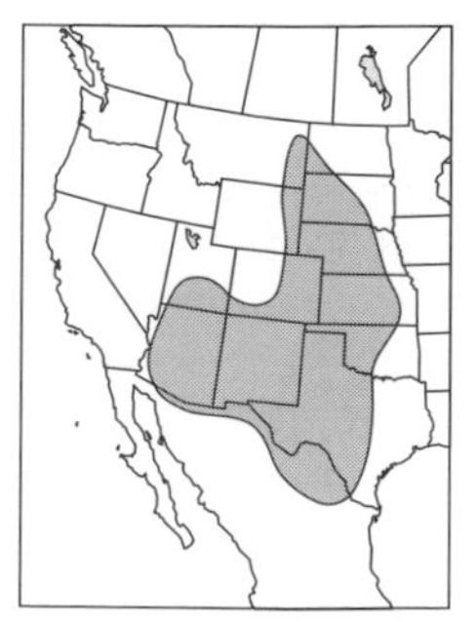

Platt range grasshopper

Identification: A grayish or brownish species, distinctively marked with a curved, dark band on the lateral lobes of the pronotum, and two or three broad bands crossing the forewings. Most specimens bear a dark stripe on the face that connects with the lower part of each eye. Hind wings are brightly colored basally, and may be yellow, orange, or rosy red; they also bear a curved black band. Hind tibiae are blue or yellow. Males are 22–32 mm long, females 27–40 mm.

Ecology: Favors arid areas and regions with sparse grass. In the Southwest, it sometimes is abundant. Its dietary habits are poorly known, though it has been observed to feed on shrubs. Adults are found from June to October.

Similar species: This medium-sized species is fairly easy to distinguish from other band-winged species; it is similar to other *Mestobregma* species, but those species that occur in the United States are less common. The narrow-fronted grasshopper, *M. impexum*, is quite similar, and is distinguished from the Platt range grasshopper by the weak, irregular black bands on the forewings. The dirt-colored grasshopper, *M. terricolor*, is more easily distinguished because the bands on the forewings are abbreviated, failing to cross the forewings. Also, this species lacks a black line on the face. Superficially, *Mestobregma* species resemble species in the genera *Metator* (Plate 14) and *Derotmema* (Plate 11).

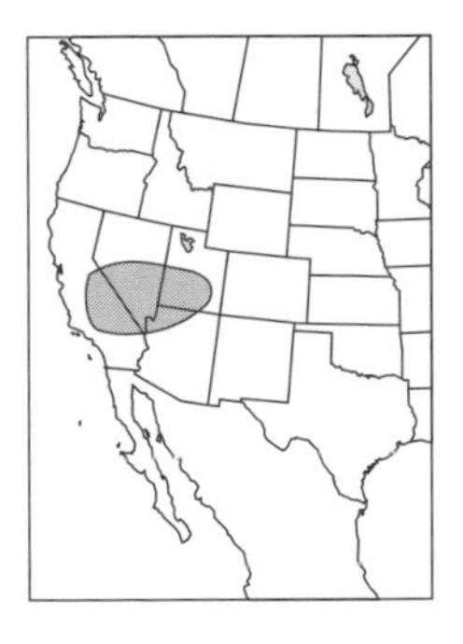

narrow-fronted grasshopper

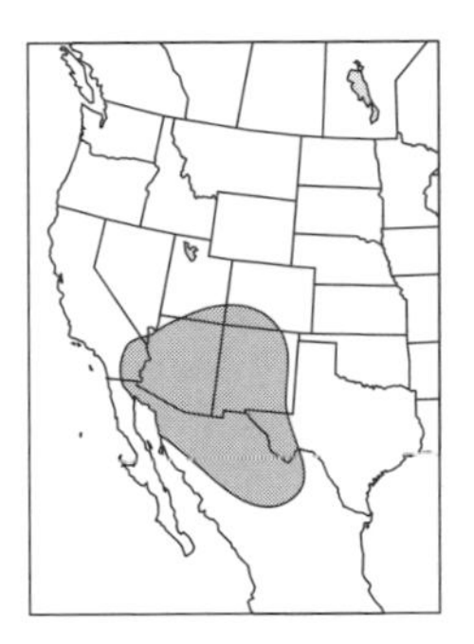

dirt-colored grasshopper

BLUE-LEGGED GRASSHOPPER

Metator pardalinus (Plate 14)

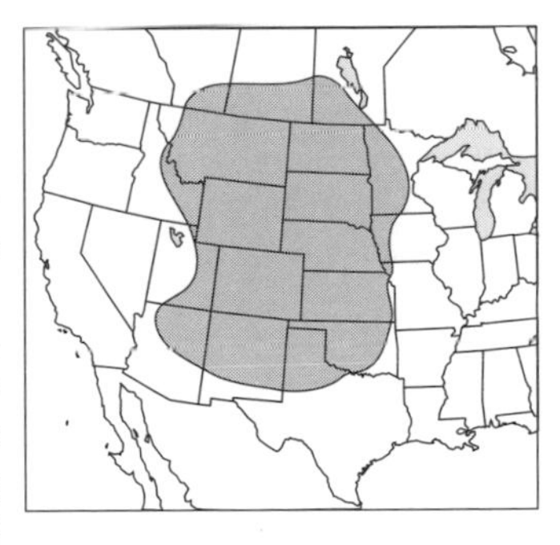

blue-legged grasshopper

Distribution: Found throughout the Rocky Mountains and Great Plains regions of the United States and southern Canada.

Identification: Yellowish brown or grayish brown, but well marked with blackish spots on the forewings. Pale stripes on the forewings converge to form a V when viewed from above. Hind wing is quite variable in color, ranging from yellow to orange and rose, and bears a broad dark band. Inner face of the hind femora and the hind tibiae is blue. Males are 26–38 mm long, females 32–45 mm.

Ecology: Inhabits a variety of grassy areas ranging from arid shortgrass prairie to wet meadows. Feeds on grasses, and occasionally is abundant enough to be considered damaging to rangeland. Adults are found from June or July to September.

Similar species: The blue-legged grasshopper resembles some *Trachyrhachys* species (Plate 17) in general appearance, but the light-colored V on the forewings serves to distinguish it. It is similar to the Apache grasshopper, *Hippopedon capito* (Plate 12), but the latter has yellowish hind tibiae, whereas the blue-legged grasshopper has blue hind tibiae. The

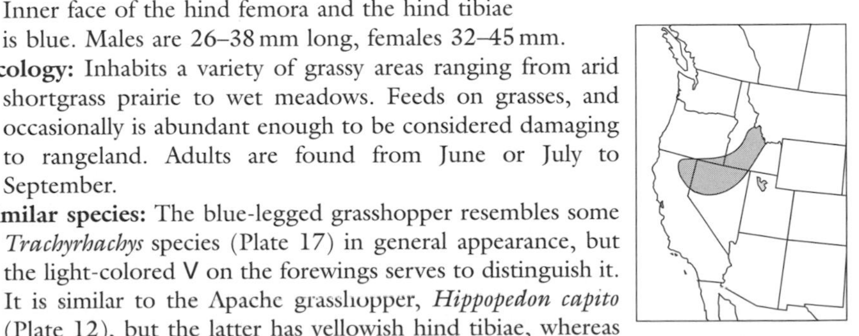

Nevada band-winged grasshopper

Nevada band-winged grasshopper, *Metator nevadensis*, is a similar-appearing species that occurs in the northern Great Basin region of Nevada and adjacent areas. It is distinguished by the smaller number of black spots and by the weakly marked V on the forewings.

ORANGE-WINGED GRASSHOPPER

Pardalophora phoenicoptera (Plate 20)

Distribution: Found throughout the eastern United States, except for the northern region.

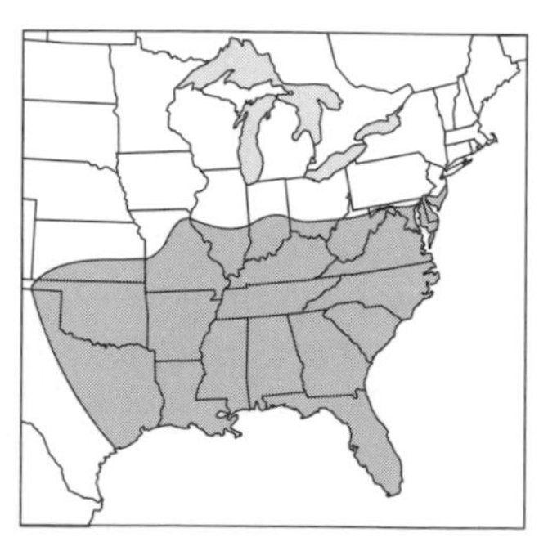

orange-winged grasshopper

Identification: A large, gray and brown grasshopper with large dark spots on the forewings. Forewings bear a light brown or gold diagonal line that forms a V when the grasshopper is viewed from above. Some individuals bear an infusion of green on the head, thorax, and hind femora. The basis of the name is the bright orange or rose-colored hind wing, which also bears a broad, curved black line crossing centrally. Tip of the hind wing is smoky. Inner face of the hind femora is bright blue and black, and orange distally. Hind tibiae are orange. Males are 36–42 mm long, females 45–55 mm.

Ecology: Prefers an open habitat such as old fields and sandy areas, but may also be found in tall grass, brush, and wooded areas if plant density is low. Becomes obvious early in the season because the large nymphs overwinter and the adults are present in the spring, when few other grasshoppers are mature. In the north, adults are found from June to July, but in Florida they are found from April to July. The male is an active flier, the heavy-bodied female tending to remain on the soil. Sound production may occur on the ground (stridulation) or in flight (crepitation), but this is not a particularly noisy species.

Similar species: Other *Pardalophora* species are easily mistaken for the orange-winged grasshopper. Saussure's grasshopper, *P. saussurei*, is found in the south-central Great Plains, but the basal area of the hind wings is yellowish. The coral-winged grasshopper, *P. apiculata*, and Haldeman's grasshopper, *P. haldemani*, are widespread, and distinguished from the orange-winged grasshopper by the lack of blue color at the base of the hind femora. The coral-winged grasshopper is distinguished from Haldeman's grasshopper by the spotting pattern of the forewing: the former has a small number of large or contiguous spots, whereas the latter has numerous spots.

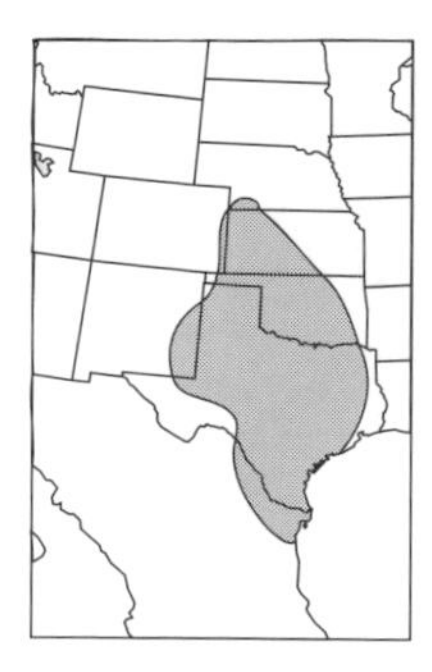

Saussure's grasshopper

coral-winged grasshopper

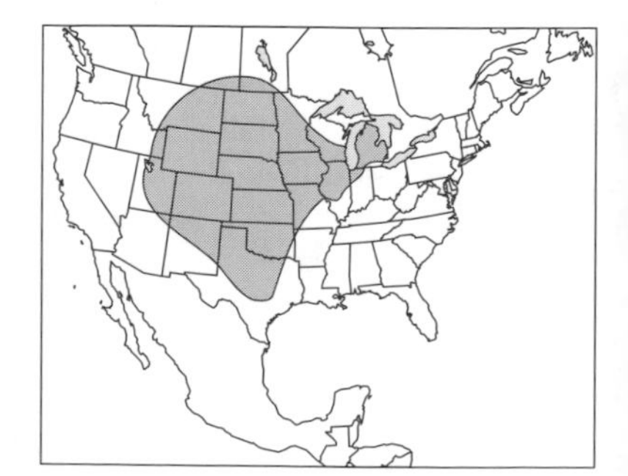

Haldeman's grasshopper

CAUDELL'S LONGHORN GRASSHOPPER

Psinidia amplicornis (Plate 20)

Distribution: Known only from Texas.

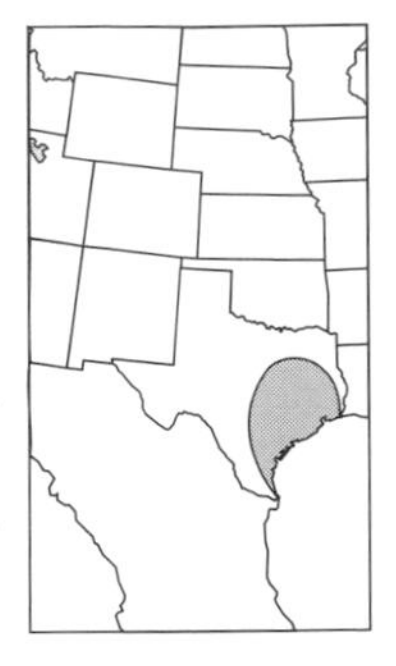

Caudell's longhorn grasshopper

Identification: A grayish-brown species, small and narrow-bodied. Long, sword-shaped antennae are diagnostic. Forewings are weakly spotted. Hind wings are orange with a dark black band. Tip of the hind wing is dark in males but only smoky in females. Males are 22–27 mm long, females, 23–34 mm.

Ecology: Inhabits barren patches of sand in either grassy or wooded habitats.

Similar species: Caudell's longhorn grasshopper is quite similar in appearance to the longhorn band-winged grasshopper, *Psinidia fenestralis* (Plate 20). It is distinguished by the flatter, sword-shaped antennae. Also, the lower posterior margin of the lateral lobe of the pronotum is rounded, whereas in longhorn band-winged grasshopper it is angled sharply. The geographic distribution of these species does not overlap.

LONGHORN BAND-WINGED GRASSHOPPER

Psinidia fenestralis (Plate 20)

Distribution: Found throughout the United States east of the Mississippi River.

longhorn band-winged grasshopper

Identification: A small, thin-bodied species, somewhat recognizable by its relatively long antennae. Antennal segments are slightly flattened and the basal segments slightly larger. General coloration is usually gray and brown, but ranges from yellowish to blackish; overall color tends to match the habitat. A narrow yellowish stripe runs from the back of the eye onto the prothorax. Forewings tend to be marked on the leading edge with alternating light and dark spots. Hind wings bear an unusually wide, curved black band centrally. Basal region of the hind wings is usually orange but sometimes rose or yellow in color. Tip of the hind wing is normally smoky in females but blackened in males. Hind tibiae are yellowish but bear a black band. Males are 20–32 mm long, females, 26–36 mm.

Ecology: Inhabits open grassy areas, particularly barren patches of sand within this general habitat. Sandy areas along the margins of lakes and rivers are commonly inhabited. When disturbed, these grasshoppers fly only a short distance and alight on bare soil, where they blend in remarkably well with the background, becoming almost invisible. Males sometimes produce sound while flying. Adults are found from June to October in most of the range, but small numbers of nymphs and adults can be found throughout the year in Florida.

Similar species: Though similar in appearance to Caudell's longhorn grasshopper, *Psinidia amplicornis* (Plate 20), the longhorn band-winged grasshopper has less flattened antennae, and the posterior angle of the lateral lobe of the pronotum is sharply angled.

BOLL'S GRASSHOPPER

Spharagemon bolli (Plate 16)

Distribution: Found widely in the United States, from the Atlantic Ocean to the Rocky Mountains. The exception is in the peninsular region of Florida, where it is replaced by the crepitating grasshopper, *Spharagemon crepitans.*

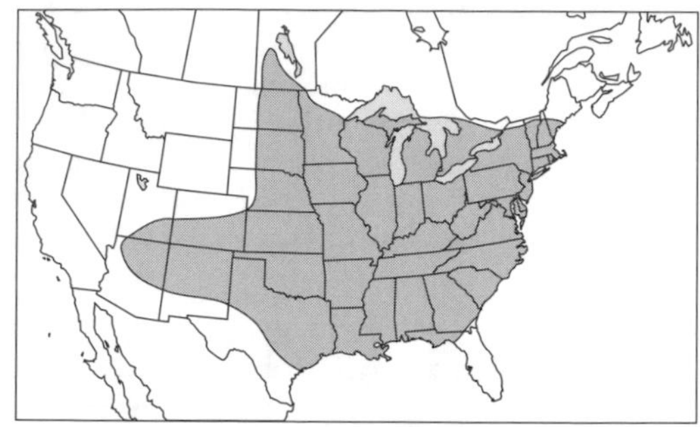

Boll's grasshopper

Identification: A grayish or reddish-brown species, often covered with minute dark spots that blend together on the forewings to form diffuse, broad, transverse bands. Hind wings bear a curved black transverse band centrally, are pale yellow basally, and transparent or smoky at the tip. Dorsal ridge of the pronotum is slightly elevated. Outer face of the hind femora is weakly or indistinctly banded, but the inner face bears alternating black and pale yellow bands. Hind tibiae are yellowish basally and reddish orange distally, with a narrow black band separating the yellow and orange. Males are 30–37 mm long, females 35–45 mm.

Ecology: Preferred habitat is open, sunny woods, although sometimes it is found along the margins. Males crepitate and stridulate frequently, and often hover about one meter above the ground while displaying. Adults are found from June or July to September.

Similar species: The narrowness of the black band on the hind tibiae is useful in separating this species from the similar crepitating grasshopper, *Spharagemon crepitans* (Plate 16), which has a broader black tibial band.

MOTTLED SAND GRASSHOPPER

Spharagemon collare (Plate 16)

Distribution: Found across the northern half of the United States, and in southern Canada. In the Rocky Mountains region it descends farther south, to New Mexico and Arizona.

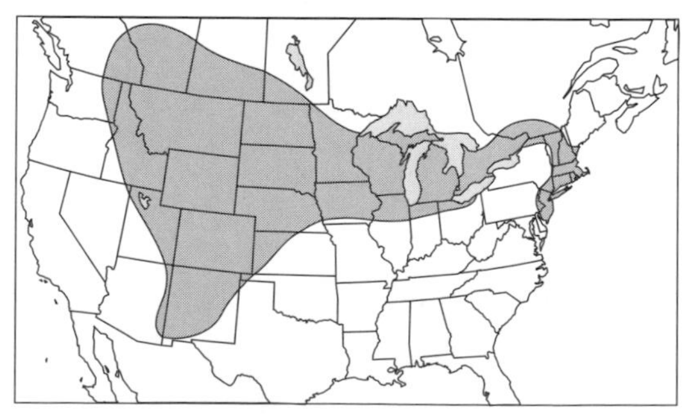

mottled sand grasshopper

Identification: Yellowish brown to dark gray in body color and marked with numerous dark speckles. Very pronounced, elevated median pronotal ridge behind the head; indeed, a better name for this species might be the "collared sand grasshopper." Forewing color is variable and sometimes bears 3 to 4 irregular dark bands. Hind wing is pale yellow basally and bears a black band centrally, and a transparent wing tip. Inner face of the hind femora is yellow with black bands. Hind tibiae are orange or red. Males are 23–31 mm long, females 27–37 mm.

Ecology: Inhabits sandy areas including rangeland with sparse vegetation. Prefers to feed on grasses. Males can produce noisy displays. Adults are found from July to September.

Similar species: In most respects, the mottled sand grasshopper is a slightly smaller, more northern version of the ridgeback sand grasshopper, *Spharagemon cristatum* (Plate 16). These species can usually be distinguished by their differences in size and geographic range. However, the dorsal, posterior angle of the pronotum is also diagnostic: it forms nearly a right angle in the mottled sand grasshopper but is acute in the ridgeback sand grasshopper (see Fig. 6).

Other similar species that have such elevated pronotal ridges are those in the genera *Dissosteira* (Plate 13) and *Arphia* (Plates 8, 9). *Dissosteira* species have black hind wings, however, and *Arphia* species tend to be uniformly dark in body color, so they are readily distinguished.

CREPITATING GRASSHOPPER

Spharagemon crepitans (Plate 16)

Distribution: Found almost exclusively in Florida, although a few specimens have been collected from southern Georgia.

crepitating grasshopper

Identification: A grayish-brown or reddish-brown grasshopper. Forewings sometimes have diffuse, broad dark bands, but usually they are lacking. Hind wings are pale yellow basally but are crossed by a wide, curved black band. Tip of the hind wing is smoky or colorless. Central ridge of the pronotum is slightly elevated. Hind tibiae are yellowish basally and reddish orange distally, with a broad black band centrally. Males are 30–38 mm long, females 37–45 mm.

Ecology: Occurs in diverse habitats. Surprisingly, it may be found in oak woods, in shaded areas not typically inhabited by grasshoppers. Adults are found from July to September.

Similar Species: The transverse black band of the hind wing is wider and located more centrally than the corresponding band in a very similar species, Boll's grasshopper, *Spharagemon bolli* (Plate 16). The width of the black band on the hind tibiae is similar to the width of the orange distal portion, considerably wider than in Boll's grasshopper.

RIDGEBACK SAND GRASSHOPPER

Spharagemon cristatum (Plate 16)

Distribution: Found in the southeastern states and in the southern Great Plains. In northern states, it is replaced by the very similar mottled sand grasshopper, *Spharagemon collare.*

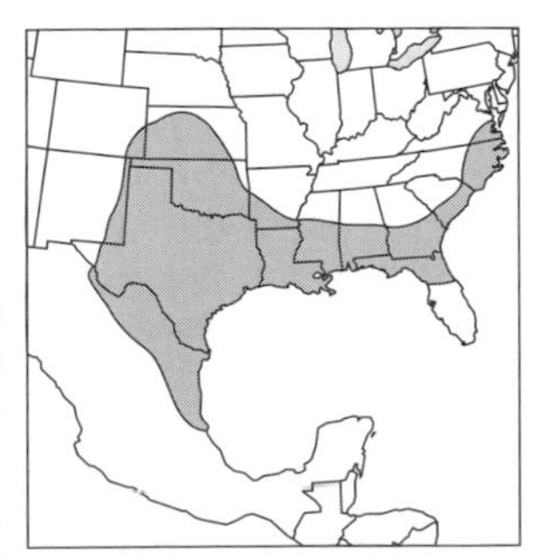

ridgeback sand grasshopper

Identification: A grayish or brownish grasshopper, heavily mottled with black spots. Speckling on the forewings can be aggregated into irregular transverse bands and black wing tips. Ridge on top of the pronotum is sharply elevated, and is the basis for the name. Hind wings are yellow basally, with a curved black band centrally, and a colorless or smoky wing tip. Hind femora are speckled brown on the outer face, but the inner face bears alternating bands of black and yellow. Hind tibiae are pale yellow basally but principally dark orange or red in color. Males are 29–39 mm long, females 34–45 mm.

Ecology: Inhabits fallow crop fields, old fields, margins of woods, and sandy roadsides. Adults are found from May to October.

Similar species: The ridgeback sand grasshopper was formerly considered to be a variant of the mottled sand grasshopper, *Sphagaremon collare* (Plate 16). These two species, easily distinguished from most other grasshoppers by their greatly elevated pronotal ridge, are nearly indistinguishable from each other based on appearance. The larger body size and restricted distribution assist in identifying the ridgeback sand grasshopper, though the mating behavior of these two species also is different. The dorsal, posterior angle of the pronotum can be used to distinguish these species: it forms nearly a right angle in the mottled sand grasshopper but is acute in the ridgeback sand grasshopper (see Fig. 6).

Other similar species that have such elevated pronotal ridges are those in the genera *Dissosteira* (Plate 13) and *Arphia* (Plates 8, 9). *Dissosteira* species have black hind wings, however, and *Arphia* species tend to be uniformly dark in body color, so they are readily distinguished.

SAY'S GRASSHOPPER

Spharagemon equale (Plate 17)

Distribution: Found widely in the central United States, principally in the Great Plains and Rocky Mountains regions, and into southern Canada.

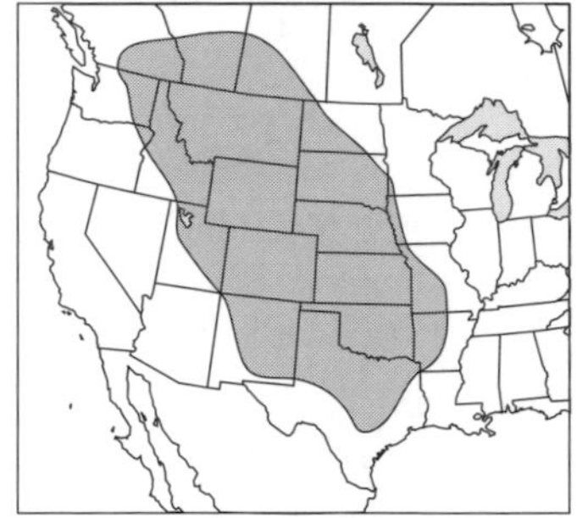

Say's grasshopper

Identification: Grayish brown in color with darker bands crossing the forewings. Lacks the elevated pronotal ridge found in some other *Spharagemon* species. Hind wings are typical of the group: pale yellow basally, a black transverse band centrally, and a transparent wing tip. Hind tibiae are orange. Males are 29–43 mm long, females 36–47 mm.

Ecology: Inhabits grasslands with relatively little vegetation, where it feeds preferentially on grasses. Adults are found from July to November.

Similar species: Within the range inhabited by Say's grasshopper there commonly occurs a similar species, the campestral grasshopper, *Spharagemon campestris.* The campestral grasshopper is distinguished by the presence of two notches or cuts in the medial pronotal ridge; only a single cut is present in Say's grasshopper.

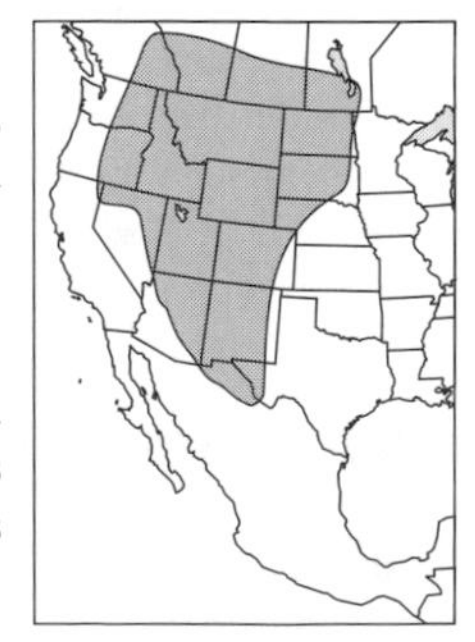

campestral grasshopper

MARBLED GRASSHOPPER

Spharagemon marmorata (Plate 17)

Distribution: Found primarily along the Atlantic and Gulf coasts. It also can be found in the United States and Canada in the vicinity of the Great Lakes.

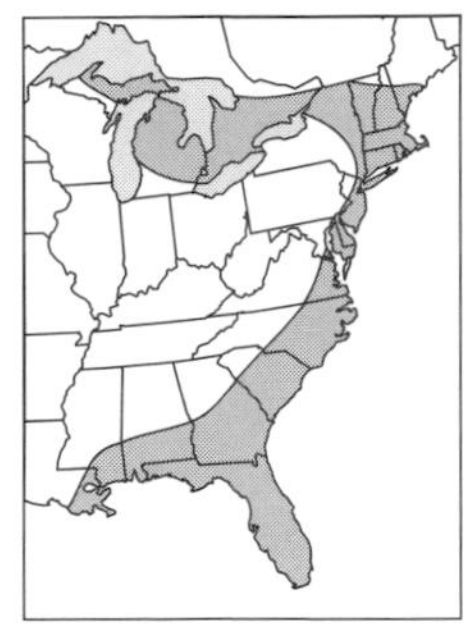

marbled grasshopper

Identification: Gray and brown in general color, but has well-marked, blackish transverse bands on the leading edge of the forewings that merge into a solid black trailing edge. Hind wings bear an unusually wide, curved black band centrally. Basal region of the hind wings is orange yellow or dark yellow in color. Tip of the hind wing is smoky or partially blackened. Hind tibiae are orange or red, with a pale yellowish ring basally. Males are 26–31 mm long, females 27–35 mm.

Ecology: Frequents open sandy areas. Sand dunes along beaches, disturbed areas of pastures, and sunny, sandy areas of open woods are common habitats. Males crepitate loudly during their lengthy zig-zag flights. While on the ground, they also stridulate and make complicated leg-lifting movements as part of their courtship ritual. Both the adult and nymphal stages can be found throughout the year in Florida.

Similar Species: The banding pattern of the forewings helps distinguish the marbled grasshopper from the longhorn band-winged grasshopper, *Psinidia fenestralis* (Plate 20), a co-occurring species in sandy habitats. Tibia color also differs, with the longhorn band-winged grasshopper having yellow and black hind tibiae.

FINNED GRASSHOPPER

Trachyrhachys aspera (Plate 17)

finned grasshopper

Distribution: Found in eastern Colorado, New Mexico, and adjacent areas, and is also known from central Mexico.

Identification: A medium-sized grayish-brown grasshopper with large black spots or bands on the forewings. Elevated central ridge on the pronotum bears two notches or cuts. Face is usually marked with dark stripes beneath the eyes. Antennae are slender. Forewings are long. Hind wings are pale yellow or greenish yellow basally; a black band may occupy the distal portion of the wing, or the wing tip may be transparent. Hind femora are markedly finned (the basis for its name), the upper fin collapsing near the midpoint. Hind tibiae are pale blue or blue gray, becoming black distally. Males are 16–23 mm long, females 21–27 mm.

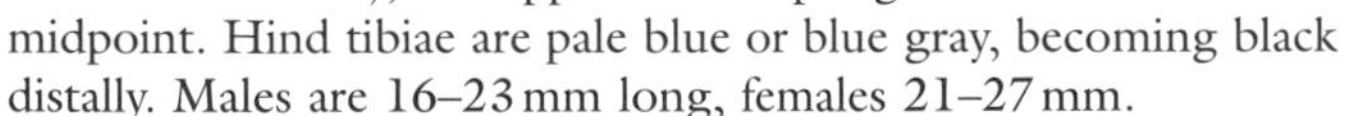

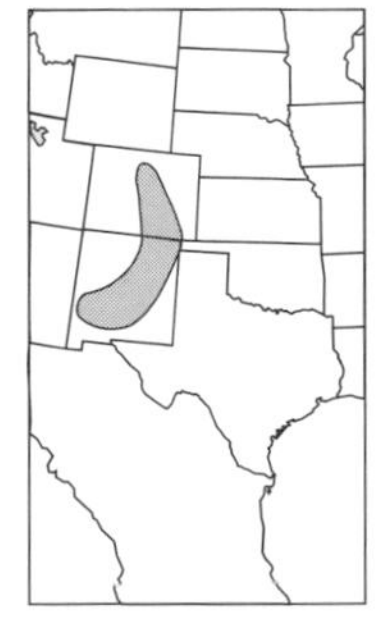

crowned grasshopper

Ecology: Inhabits shortgrass prairie and feeds on grasses. Adults are found from August to October.

Similar species: The finned grasshopper can be distinguished from the Kiowa rangeland grasshopper, *Trachyrhachys kiowa* (Plate 17), by the posterior angle of the lateral lobes of the pronotum. In the latter, the posterior angle is acute, whereas in the former it is nearly a right angle (see Fig. 7). Also, the restricted geographic range of the finned grasshopper makes separation easy in most areas.

The crowned grasshopper, *Trachyrhachys coronata*, may be found in the same areas occupied by finned grasshopper. It is distinguished from the other *Trachyrhachys* species by the presence of 6 to 8 large dark spots distributed over the forewings; the other species have fewer spots aggregated into 3 to 4 bands. Due to the spotting pattern, the crowned grasshopper resembles Haldeman's grasshopper, *Pardalophora haldemani*, and the red-shanked grasshopper, *Xanthippus corallipes* (Plate 21), but is much smaller, measuring only 22–25 mm in males and 26–32 mm in females.

KIOWA RANGELAND GRASSHOPPER

Trachyrhachys kiowa (Plate 17)

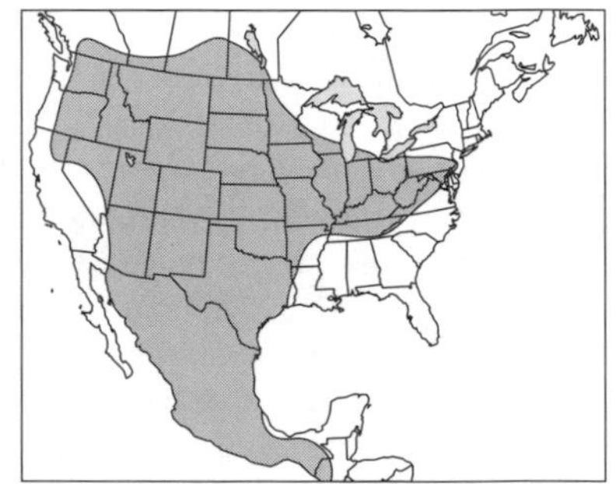

Kiowa rangeland grasshopper

Distribution: Found widely in North America, though it is predominately a western species. It is absent from most of the southeastern states and from New England. It also occurs in southwestern Canada and throughout Mexico.

Identification: A medium-sized grayish-brown or greenish-yellow grasshopper, marked by large black spots or bands on the forewings. More diagnostic is the elevated central ridge on the pronotum that bears two notches or cuts. Head is relatively large and elevated above the pronotum. Face is usually marked with dark stripes beneath the eyes. Antennae are slender. Forewings are long. Hind wings are quite variable in color. Often they are pale yellow basally with an incomplete black band and the tip of the wing transparent; however, in northern areas color is lacking, and other regional variations are known. Hind tibiae are pale blue or blue gray. Males are 20–25 mm long, females 23–30 mm.

Ecology: Inhabits sandy or barren areas, where it feeds on grasses. Sometimes it is abundant enough to cause damage to rangeland grasses. Adults are found from June to November.

Similar species: The Kiowa rangeland grasshopper resembles the finned grasshopper, *Trachyrhachys aspera* (Plate 17). It can be distinguished by the posterior angle of the lateral lobes of the pronotum. In the former the posterior angle is acute, whereas in the latter it is nearly a right angle (see Fig. 7).

SCUDDER'S CLEAR-WINGED GRASSHOPPER

Trepidulus hyalinus (Plate 20)

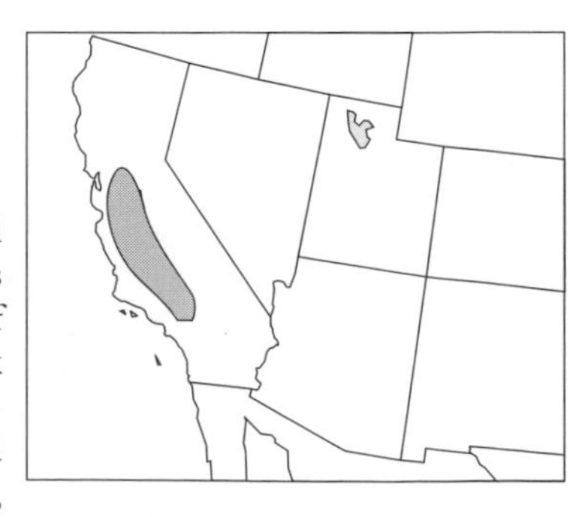

Scudder's clear-winged grasshopper

Distribution: Known only from California.

Identification: A grayish-yellow, thin-bodied species. Forewings often bear dark markings along the anterior edge, and the lateral lobe of the pronotum is crossed by a narrow black band, but there are few distinct markings elsewhere. Hind wings are transparent. Hind tibiae are yellow. Males are 21–24 mm long, females 27–31 mm.

Ecology: Inhabits dry grassy areas. Adults are found from July to September.

Similar species: The thin-bodied genus *Trepidulus* possesses large eyes, somewhat resembling *Psinidia* (Plate 20) and *Derotmema* (Plate 11). It is distinguished from these genera by the lack of a dark band crossing the hind wing. The presence of transparent hind wings also suggests the clear-winged grasshopper, *Camnula pellucida* (Plate 9), but the clear-winged grasshopper is easily distinguished from Scudder's clear-winged grasshopper by the presence of the V dorsally. Also found in the United States is the shy rose-winged grasshopper, *Trepidulus rosaceus*, but this latter species bears reddish or pinkish coloration on the hindwing.

shy rose-winged grasshopper

CALIFORNIA BAND-WINGED GRASSHOPPER

Trimerotropis californica (Plate 18)

Distribution: Found throughout the southwestern states from Oregon and Idaho to western Texas.

Identification: A typical *Trimerotropis* species. Grayish or brownish, with two distinct dark bands on the forewings. Hind wing is yellow basally and bears a narrow black band centrally. Hind tibiae are reddish or orange. A distinguishing character is the presence of a small extension or tooth on the lower posterior margin of the lateral lobe of the pronotum. Males are 25–35 mm long, females 30–43 mm.

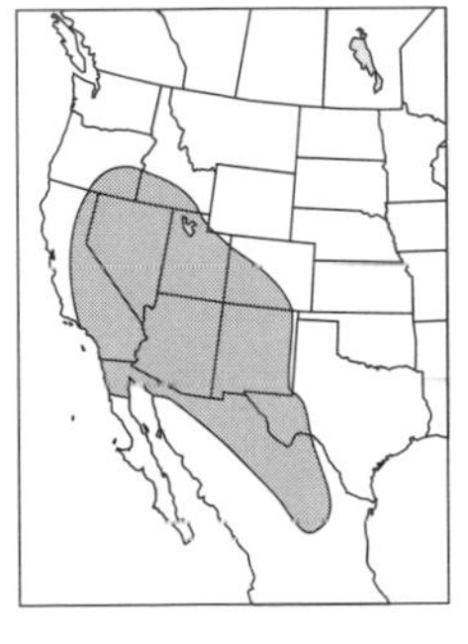

California band-winged grasshopper

Ecology: Inhabits arid, rocky ground where it feeds on grasses. Males crepitate freely, and while on the soil display femur shaking. Adults are found from June to October.

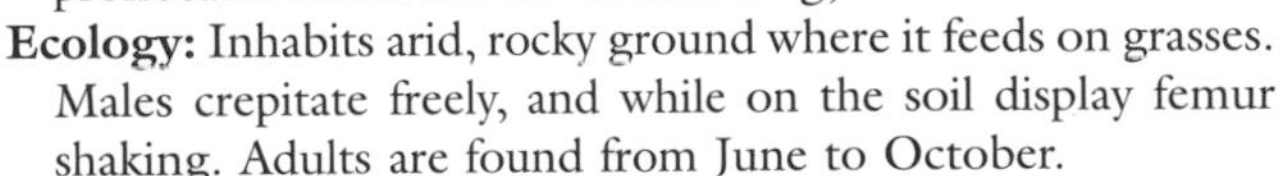

Similar species: Although it resembles several *Trimerotropis* species, the California band-winged grasshopper is most easily confused with the broad-banded grasshopper, *T. latifasciata* (Plate 19). The presence of the tooth on the pronotal lobe (see Fig. 8) distinguishes the California band-winged grasshopper.

BLUE-WINGED GRASSHOPPER

Trimerotropis cyaneipennis (Plate 18)

Distribution: Found widely in the western states, from California, Oregon, and Washington east to Colorado and west Texas.

Identification: This species derives its name from the bluish or bluish-green hind wings that are found in most individuals. In addition to the bluish basal portion of the hind wing, a black central band is also present, and the wing tip is transparent. In New Mexico, yellow-winged specimens are known, but even these have a greenish cast basally. Gray in overall body color. Forewings are marked with dark spots that form irregular transverse bands. Outer face of the hind femora bears two dark bands; the inner face is mostly black but with two light bands. Hind tibiae are blue or bluish green. Males are 25–35 mm long, females 28–40 mm.

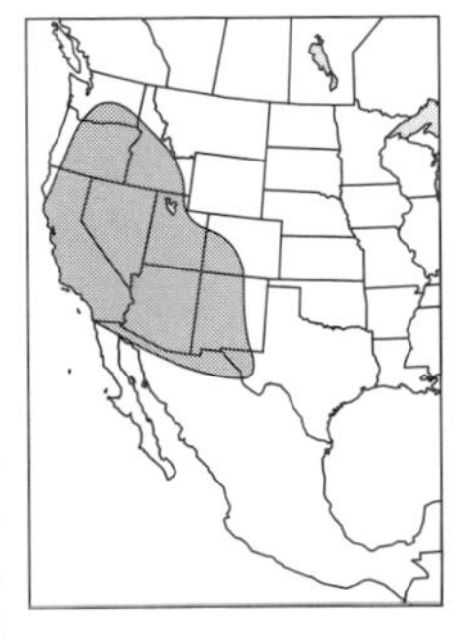

blue-winged grasshopper

Ecology: Dwells in rocky areas of mountainous regions. Adults are found from July to October.

Similar species: The taxonomic status of this species is uncertain. Its status relative to brother's band-winged grasshopper, *Trimerotropis fratercula*, needs further study, as the two species seem to hybridize in central New Mexico. Generally, they can be distinguished by the color of the hind tibiae: blue or bluish green in the blue-winged grasshopper; yellowish brown or faintly greenish in brother's band-winged grasshopper.

brother's band-winged grasshopper

Other common blue-winged species are *Leprus* species (Plate 21) and the undulant-winged grasshopper, *Circotettix undulatus*. The dark bands of the forewings are broad, dark, and smooth edged in *Leprus* spp. The hind wings of *Circotettix* species (Plate 11) bear unusually thickened wing veins.

BROAD-BANDED GRASSHOPPER

Trimerotropis latifasciata (Plate 19)

Distribution: Found throughout most of western North America from southern Canada to central Mexico.

Identification: Grayish or brownish, with two distinct dark bands on the forewings. Hind wing is yellow basally and bears a black band centrally, which may be either narrow or wide. Inner face of the hind femora has at least two pale bands of either tan or reddish color. Hind tibiae are reddish or orange. Males are 30–41 mm long, females 35–49 mm.

Ecology: Various habitats include dry prairie, shrubby, and barren areas. Adults are found from June to October.

Similar species: Though the broad-banded grasshopper resembles several species of *Trimerotropis*, it is most easily confused with the California band-winged grasshopper, *T. californica* (Plate 18). The lack of a pronotal tooth (see Fig. 8) distinguishes the former from the latter. In specimens with a broad black band on the hind wing, the broad-banded grasshopper can be confused with the black-winged grasshopper, *Trimerotropis melanoptera* (Plate 19), but can be distinguished by the presence of at least two pale bands on the inner face of the hind femora.

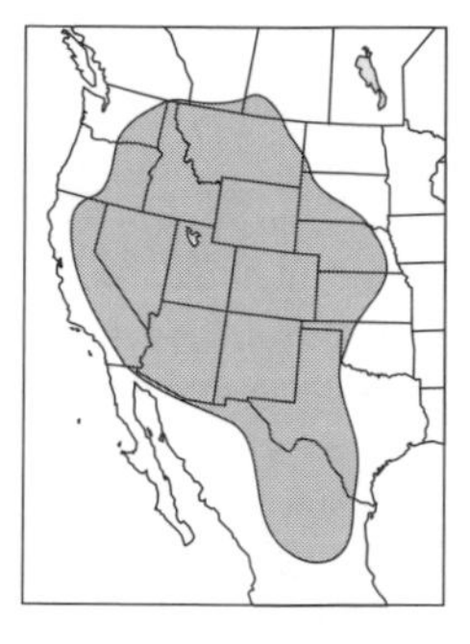
broad-banded grasshopper

BLACK-WINGED GRASSHOPPER

Trimerotropis melanoptera (Plate 19)

Distribution: Found in a small region of the Southwest, from Arizona and Colorado south to central Mexico.

Identification: Grayish or brownish, with two distinct dark bands on the forewings. Hind wing is yellow basally and bears a very wide black band centrally. Often nearly the entire wing is blackened. Inner face of the hind femur is mostly black, bearing only a single pale band of reddish or orange color. Hind tibiae are reddish or orange. Males are 32–40 mm long, females 37–50 mm.

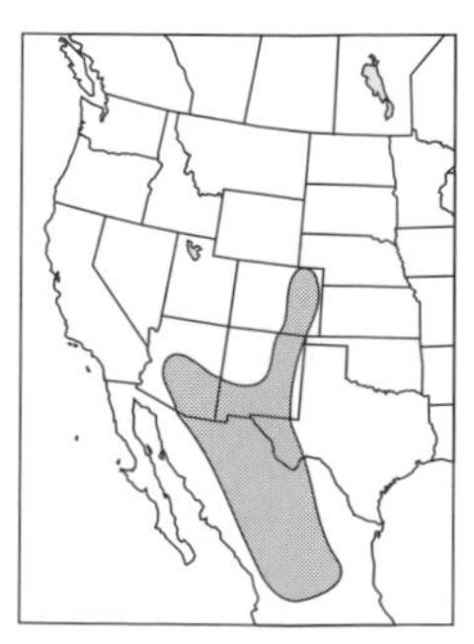
black-winged grasshopper

Ecology: Inhabits shortgrass prairie, often in the most barren areas. Adults are found from June to November.

Similar species: A broad black band on the hind wing is found in other genera and species, particularly *Dissosteira* species (Plate 13) and some specimens of the broad-banded grasshopper, *Trimerotropis latifasciata* (Plate 19). The failure of the black wingband to cross the wing entirely serves to distinguish *Dissosteira*. The presence of only a single pale band on the inner face of the hind femora distinguishes the black-winged grasshopper from the broad-banded grasshopper.

SEASIDE GRASSHOPPER

Trimerotropis maritima (Plate 18)

Distribution: Found throughout the eastern United States west to the Rocky Mountains.

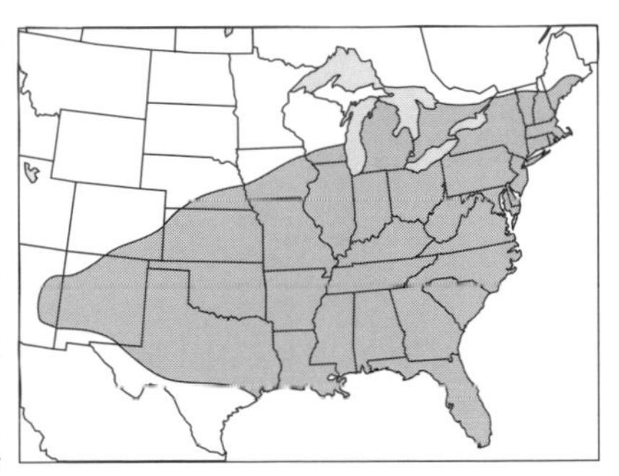

seaside grasshopper

Identification: Light gray to dark grayish brown, with numerous small brown speckles over most of the body and weakly to strongly marked wide transverse bands on the forewings. Hind wings are pale yellow basally and marked with a curved black transverse band centrally. Tip of the hind wing is transparent. Central ridge on the pronotum is barely elevated. Outer face of the hind femora is gray and brown, with only weak evidence of bands. Inner face, however, is pale yellow with three black bands. Hind tibiae are yellow to red. Males are 29–33 mm long, females 30–40 mm.

Ecology: A sand-loving grasshopper, found in arid, barren areas. A common resident of ocean, lake, and river margins, the seaside grasshopper also frequents fallow crop fields and sandy roadways. When disturbed, it is likely to crouch motionless, blending well with its sandy background. It is a strong flier, however, and can travel long distances. Males display sound production in flight (crepitation) and on the ground (stridulation) during their courtship ritual. Adults are found from July to October.

Similar species: Though the genus *Trimerotropis* is quite large, it is almost entirely western in distribution. Thus, any light-colored *Trimerotropis* species found east of the Great Plains is likely to be the seaside grasshopper. The very dark-colored crackling forest grasshopper, *Trimerotropis verruculata*, also occurs in the northeastern states, but these two species differ markedly in appearance.

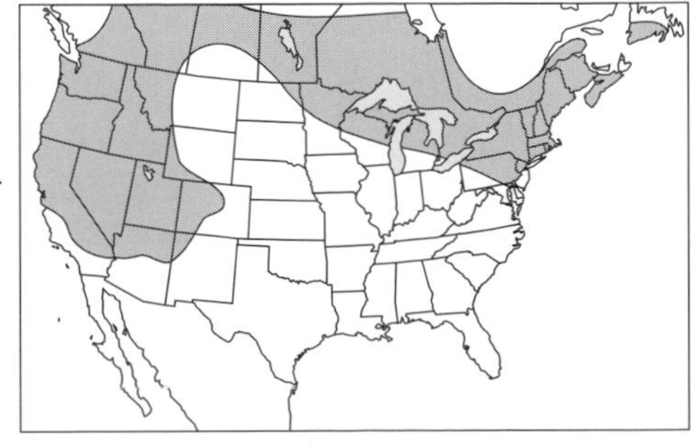

crackling forest grasshopper

PALLID-WINGED GRASSHOPPER

Trimerotropis pallidipennis (Plate 19)

Distribution: Found throughout the western states and Mexico, and though absent from Central America it is found in South America.

Identification: Grayish or brownish, with two distinct dark bands on the forewings. Hind wing is whitish yellow or yellow basally, and bears a narrow or wide black band centrally. The black band is not as wide as in some species, however, not exceeding about one-third the width of the hind wing. Hind tibiae are yellowish or brown. Males are 25–35 mm long, females 30–45 mm.

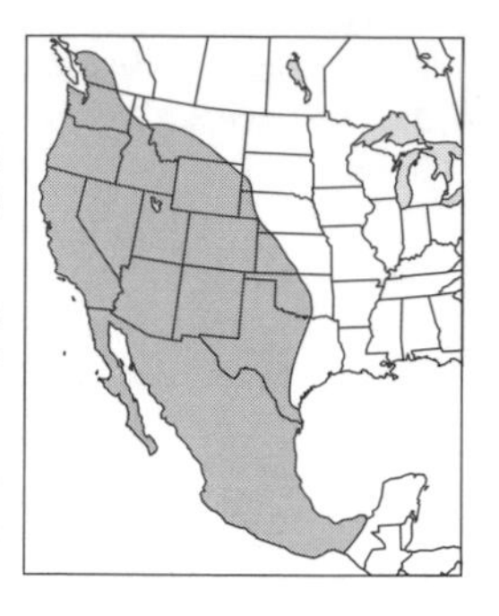

pallid-winged grasshopper

Ecology: Inhabits grasslands and deserts, but not high elevations, and prefers to feed on grasses. Its mating displays are noisy. In northern areas, adults are found from June to October. In southern areas, 2 to 3 generations occur, so adults are found year-round. It is often attracted to lights at night.

Similar species: This common species resembles many other *Trimerotropis* species (Plates 18, 19) and is separated by the combination of characters listed above.

GREAT CRESTED GRASSHOPPER

Tropidolophus formosus (Plate 21)

Distribution: Found on the southern Great Plains, from southeastern Wyoming, eastern Colorado, and western Nebraska south to southern Arizona, western Texas, and northern Mexico.

Identification: This is one of the most spectacular and recognizable grasshoppers. A fairly large, green grasshopper, it has a high, arching, toothed pronotal crest. Head is broadly rounded. Body often is marked with brown spots ringed with yellow, including about 6 spots on each forewing. Forewings extend beyond the tip of the abdomen in males, but do not reach the tip of the abdomen in females. Hind wings are orange and normally bear an incomplete brown band. Males are 35–46 mm long, females 38–50 mm.

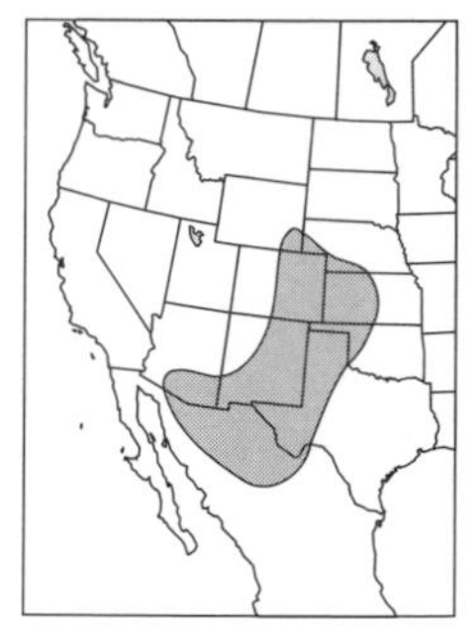

great crested grasshopper

Ecology: Inhabits grasslands and desert grasslands, where it feeds principally on plants in the Malvaceae family. Males produce noisy mating displays both in flight (crepitation) and on the ground (stridulation). Females do not fly. Adults are found from July to October.

Similar species: This species is likely confused only with the green fool grasshopper, *Acrolophitus hirtipes* (Plate 6), though they are only superficially similar because of their green color and arched thorax. The green fool grasshopper is easily distinguished because it has a pointed head and lacks teeth on the pronotal arch. Also, the hind wings of the green fool grasshopper are pale greenish yellow with a black band.

RED-SHANKED GRASSHOPPER

Xanthippus corallipes (Plate 21)

Distribution: Found throughout the western states and southwestern Canada, south to southern Mexico.

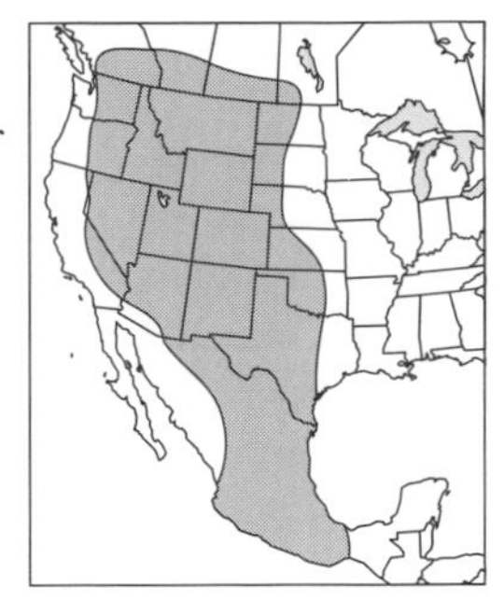
red-shanked grasshopper

Identification: A large, grayish-brown grasshopper. Though variable in appearance, it is characteristically marked by large, clearly defined dark spots over the entire length of the forewings. Pale stripes on the forewings converge posteriorly, forming a V when viewed from above. Hind wings are yellow, pink, rose, or orange basally, and bear a transverse black band centrally. Central ridge of the pronotum is low and cut by two notches; the area between the notches is particularly low and indistinct. Pronotum is unusually rough or bumpy. Inner face of the hind femora is normally bright orange or red, though sometimes dark blue. Hind tibiae are orange. Males are 25–48 mm long, females 26–65 mm.

Ecology: This adaptable species is found over a wide range of altitudes and in numerous plant communities. It feeds on grasses. Because it tends to overwinter as a nearly full-grown immature, it is readily detected in the spring when other grasshoppers are in the egg stage or present as small nymphs. Adults are found from May to June. In Canada, and elsewhere at high altitudes, it is reported to require two years to complete its development.

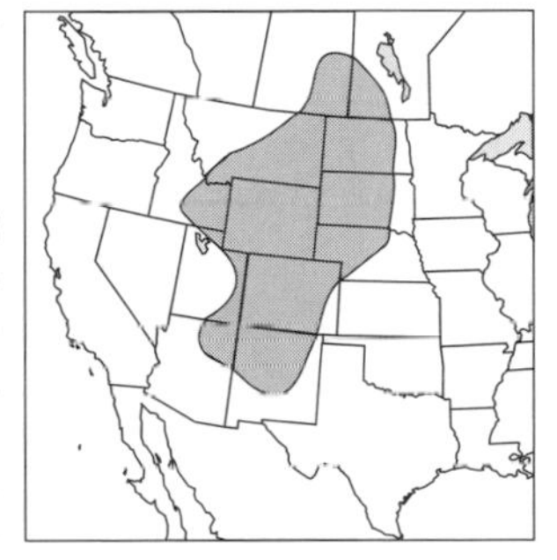
Montana band-winged grasshopper

Similar species: The Montana band-winged grasshopper, *Xanthippus montanus*, is a similar-appearing species that is distinguished by the less discrete spots on the forewings, as are grasshoppers of the genus *Cratypedes* (Plate 12). In the genus *Pardalophora*, including the orange-winged grasshopper, *Pardalophora phoenicoptera* (Plate 20), and in the wrinkled grasshopper, *Hippiscus ocelote* (Plate 15), the middle ridge of the pronotum is cut by only a single notch.

Spurthroated Grasshoppers

Family Acrididae, subfamily Cyrtacanthacridinae

The treatment of this subfamily varies among authors. Some use the subfamily name Cyrtacanthacridinae to include a great number of genera and species. Others restrict the name to a smaller group, the bird grasshoppers, and divide the remaining species into two or more subfamilies, often the Melanoplinae (*Melanoplus* and close relatives) and the Romaleinae (lubber grasshoppers). The lubbers are also sometimes treated as a separate family, the Romaleidae. We take an intermediate approach, placing *Melanoplus*, *Schistocerca*, and their relatives in the subfamily Cyrtacanthacridinae (spurthroated grasshoppers) but treating the lubber grasshoppers as a separate subfamily (Romaleinae). Although the lubbers bear a prosternal spine like the spurthroated grasshoppers, they are fairly distinct. The lubber grasshoppers are large and heavy bodied; in addition, they have a spine on both the inner and outer surface at the tip of the hind tibiae. The other grasshoppers may have a moveable spur, which resembles a spine, but only the lubbers have immovable spines at this position.

The spurthroated grasshoppers, as the common name suggests, bear a spine (the prosternal spine, or "spur") ventrally and immediately in front of, or between, the front legs. The antennae usually are thread-like, not flattened or sword-shaped. The head usually is not especially large, and these grasshoppers do not appear to be especially heavy-bodied. In most genera, the head has a vertical orientation, but in some groups the face is slanted (see Fig. 3). These grasshoppers may be wingless, or may bear abbreviated or long wings. The forewings (tegmina) are pigmented but lack the transverse bands common in the band-winged grasshoppers (subfamily Oedipodinae). The hind wings are not pigmented. The flying ability of spurthroated grasshoppers varies greatly, even within a single genus. The genus *Schistocerca* contains especially long-winged, strong fliers, and they are sometimes called bird grasshoppers in recognition of their large size and strong flying abilities. Spurthroated grasshoppers do not make sounds during flight (crepitation), as do the band-winged grasshoppers (subfamily Oedipodinae). Nor do they stridulate, as do the stridulating slantfaced grasshoppers (subfamily Gomphocerinae).

The subfamily Cyrtacanthacridinae is large and diverse. The largest group is the genus *Melanoplus*, with at least 200 species in North America. It is sometimes difficult to distinguish among *Melanoplus* species, and to a lesser degree among *Schistocerca* species; the sexual or terminal abdominal structures of the males are commonly used to distinguish among similar-appearing species. There are several characters of value in distinguishing among males: the forked, two-lobed furcula; the supra-anal plate; and the subgenital plate (see Fig. 9). In many cases, it is not possible to distinguish among females. To aid in identification of *Melanoplus* species, each species account includes a description of the diagnostic features of the male and a figure illustrating these features. Comparisons of many similiar *Schistocerca* species are included in the account of the leather-colored bird grasshopper, *S. alutacea*.

The habitat of these grasshoppers is highly variable, but they are primarily phytophagous. Although some frequent trees or the undergrowth of dense woods, most are found in open grassy or weedy areas. The dietary habits vary from narrow to wide. Species may specialize on grasses, forbs, shrubs, or trees, may feed freely among all these plant types, or may scavenge for insect cadavers or other sources of nutrition including dead plant material.

RUSSIAN-THISTLE GRASSHOPPER

Aeoloplides turnbulli (Plate 22)

Distribution: Found in the Rocky Mountains region and in the western portion of the Great Plains.

Identification: A medium-sized, robust species. Usually greenish and tan, but in some specimens the green is replaced by brown. A green stripe extends from the top of the head across the upper surface of the pronotum. On the pronotum, this stripe expands and is usually yellow centrally. A dark stripe crosses the lateral lobe of the pronotum. Forewings are pale

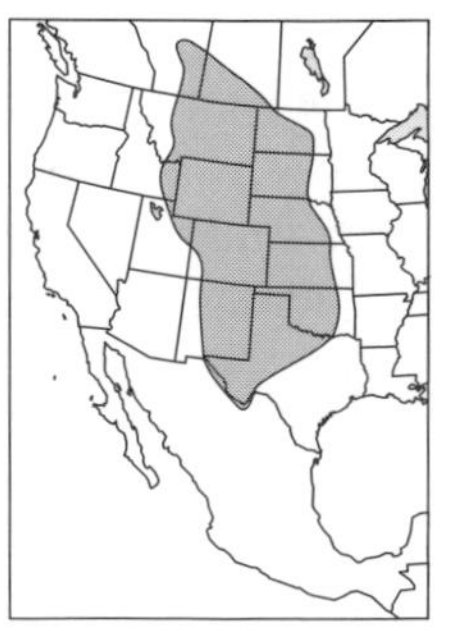

Russian-thistle grasshopper

green or brown and may be shorter or longer than the abdomen. Hind femora bear three dark bands. Hind tibiae are bluish green. Cerci are narrow and sharply pointed. Tip of the abdomen bears a small protuberance. Males are 16–20 mm long, females 19–25 mm.

Ecology: Feeds principally on Chenopodiaceae such as Russian thistle, kochia, lamb's-quarters, and saltbush. These are not particularly valuable plants for ranchers, so this grasshopper can be considered beneficial in that it suppresses these plants. Occasionally, the Russian-thistle grasshopper attains very high densities, and under such conditions is reported to damage sugarbeet crops, but this is very unusual. Eggs commence hatching relatively early, in midspring. Adults are found from June to October.

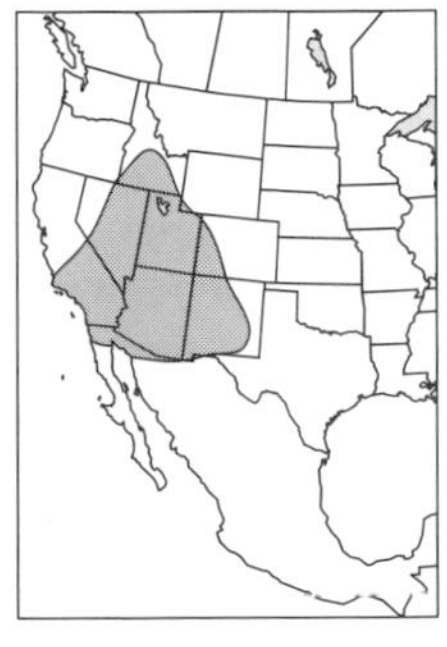

narrow-winged bush grasshopper

Similar species: The Russian-thistle grasshopper is easily confused with species in the genus *Melanoplus.* It can be distinguished, however, by the shape of the spine protruding from between the front legs. The spine is more slender and sharply conical in the Russian-thistle grasshopper than in *Melanoplus.* Another *Aeoloplides* species, the narrow-winged bush grasshopper, *A. tenuipennis,* replaces the Russian-thistle grasshopper in the southwestern states. It can be distinguished by the purplish hind tibiae and a projecting wedge-shaped structure near the base of the hind femora.

LINEAR-WINGED GRASSHOPPER

Aptenopedes sphenarioides (Plate 22)

Distribution: Found throughout Florida and in adjacent southeastern states. Often abundant.

linear-winged grasshopper

Identification: This species is usually green, but sometimes tends toward purplish brown. It has a strongly slanted face, and superficially resembles slantfaced grasshoppers (subfamilies Gomphocerinae and Acridinae). It bears the ventral spine or spur, however, that marks the subfamily Cyrtacanthacridinae. In the adult stage, wings are reduced to small, elongate linear pads, which is the basis of its name. Thus, this species is flightless. A stripe that is light on one side and dark on the other is usually present on each side, running from the top of the eye to the tip of the wing pads. A similar stripe may occur on top, down the center of the back, though this is more frequent among males than females. Hind tibiae are bluish green. Females are considerably larger and more robust than males. Males are16–25 mm long, females 22–33 mm.

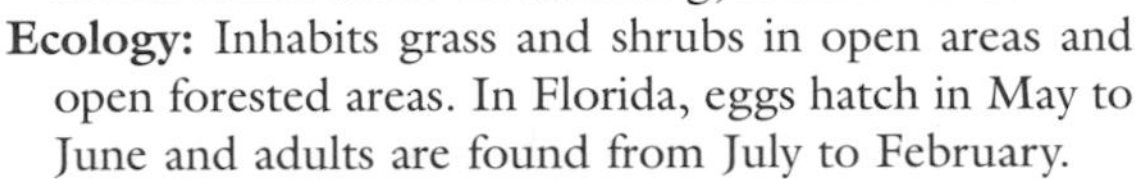

Ecology: Inhabits grass and shrubs in open areas and open forested areas. In Florida, eggs hatch in May to June and adults are found from July to February.

Florida wingless grasshopper

Similar species: Osceola's grasshopper, *Hesperotettix osceola,* bears a strong resemblance to the linear-winged grasshopper because the body forms and color patterns are similar. However, the short but obvious wings of adult Osceola's grasshopper should serve to distinguish it from the linear-winged grasshopper, which bears only a small flap or pad rather than a wing. The Florida wingless grasshopper, *Aptenopedes aptera*, is also similar, but completely lacks wings, and even wing pads.

SLOW MOUNTAIN GRASSHOPPER

Bradynotes obesa (Plate 22)

Distribution: Found widely in western states and into southernmost Canada, but nowhere is it plentiful.

slow mountain grasshopper

Identification: A robust, wingless species, brownish above and lighter below. It bears contrasting light and dark markings, including a light-colored stripe on the top of the abdomen in some individuals. The face sometimes bears red markings. Hind femora are greenish with black of variable intensity on the outer face, and reddish on the lower surface. Prosternal spine near the base of the front legs is weakly developed. Hind tibiae are blackish at the base and red distally, or entirely reddish. This small species is 18–25 mm long.

In males, the furcula is absent or nearly so. The cerci expand from the base and then taper markedly in the basal half of the cerci. The distal half of the cerci is long and finger-like, tapering only slightly.

Ecology: This sluggish species inhabits rocky slopes. In Arizona, it is reported to feed on *Potentilla.*

FUZZY OLIVE-GREEN GRASSHOPPER

Campylacantha olivacea (Plate 22)

Distribution: Found widely in the south-central United States, from South Dakota to Texas, and from New Mexico to South Carolina.

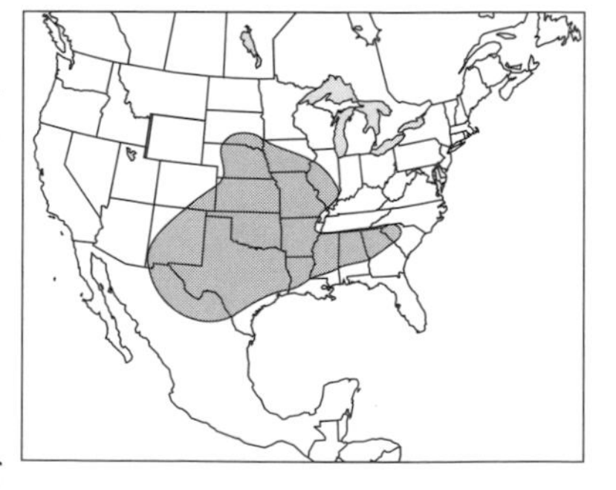

fuzzy olive-green grasshopper

Identification: Normally uniform olive green, though sometimes gray or brownish. The numerous hairs on the body (particularly the pronotum and legs) are the basis for its name, though some magnification is required to observe the hairs. Forewings do not attain the tip of the body, often covering about half of the abdomen. Hind tibiae are yellowish green or orange. Males are 21–25 mm long, females 28–35 mm.

In males, the tip of the abdomen is slightly enlarged. The furcula is barely visible. Cerci are moderate in length and taper to a blunt point.

Ecology: Commonly associated with thick grasses and weeds such as those found along fences and the edges of fields. Feeds on broadleaf weeds such as sunflower and lamb's-quarters.

PICTURED GRASSHOPPER

Dactylotum bicolor (Plate 22)

Distribution: Found along the western edge of the Great Plains from Montana to northern Mexico, and west to Arizona.

Identification: The contrasting reddish orange, yellow or white, and blue or black coloration makes this species immediately identifiable. Because of this pattern, it is sometimes called the barber-pole grasshopper. The color pattern is variable, but in general the forewings are black with yellow veins, the body is banded with black and yellow, and the dorsal areas of the thorax and abdomen bear spots of orange. Hind femora are black and orange. Hind tibiae are yellow and green. Wings are small. Males are about 24 mm long, females about 32 mm.

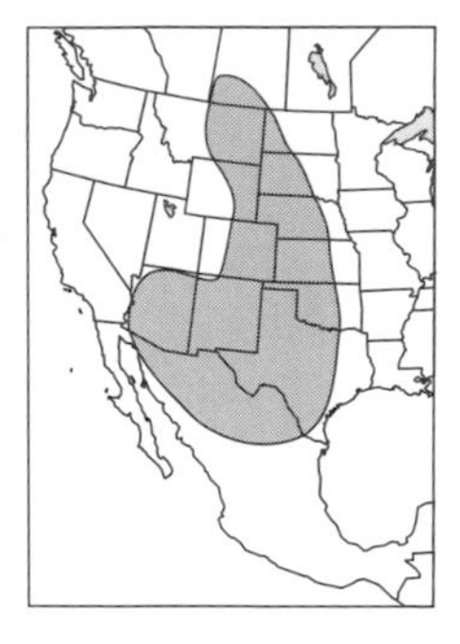

pictured grasshopper

Ecology: Inhabits areas of sparse vegetation, where it feeds on low-growing, broadleaf plants. Eggs commence hatching relatively late, often in early summer. Adults are found from August to September.

LITTLE EASTERN GRASSHOPPER

Eotettix pusillus (Plate 23)

Distribution: Found in several southeastern states from North Carolina to Alabama.

little eastern grasshopper

Identification: This pretty grasshopper is strikingly small. It is metallic yellowish green, brownish green, or reddish brown. Behind each eye, the lateral lobe of the pronotum bears a black spot that does not extend completely to the posterior margin. Hind femora and tibiae are reddish gold or greenish gold in color. Most abdominal segments are partially black, which results in a black vertical banding pattern. The forewings of this flightless species are almost round, an important distinguishing character, and are shorter than the pronotum. The nymphal stage differs considerably, possessing a black body with red and gold markings on its head. Males are 10–15 mm long, females 15.5–23 mm.

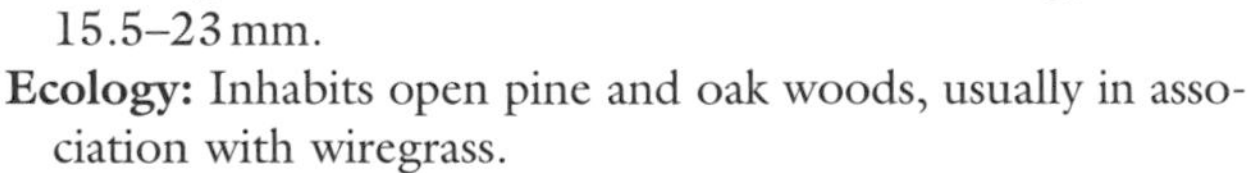

Ecology: Inhabits open pine and oak woods, usually in association with wiregrass.

Similar species: The black spot on the lateral lobe of the pronotum does not extend to the posterior margin, as occurs in the swamp eastern grasshopper, *Eotettix palustris*, and the handsome Florida grasshopper, *Eotettix signatus* (Plate 23). The hind tibiae of these other *Eotettix* species are reddish, whereas in the little eastern grasshopper they tend toward gold.

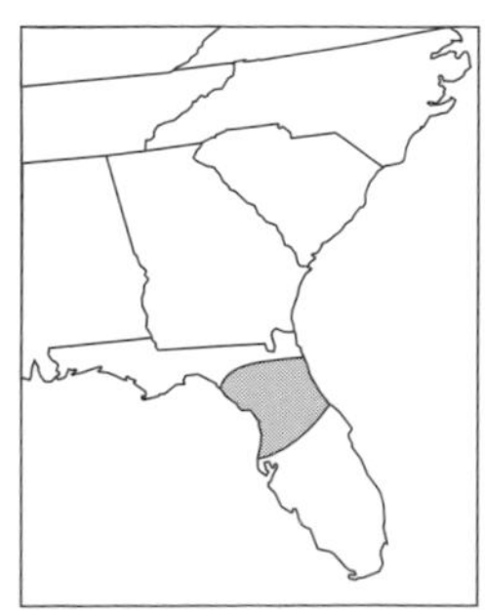

swamp eastern grasshopper

HANDSOME FLORIDA GRASSHOPPER

Eotettix signatus (Plate 23)

Distribution: Found throughout peninsular Florida, but apparently not outside Florida.

Identification: Metallic yellowish green or bluish green. The forewings of this flightless species are elongate oval and as long as, or longer than, the pronotum. Each side is marked with a black stripe running from the eye to the posterior edge of the pronotum. Hind tibiae are red. Males are 18–23 mm long, females 19.5–32 mm.

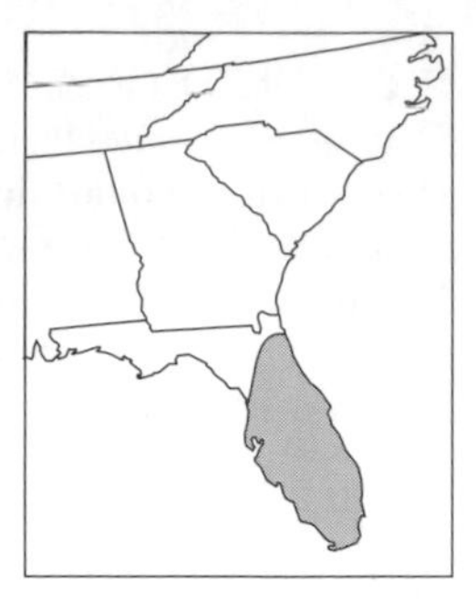

handsome Florida grasshopper

Ecology: Occurs in varied habitats. It can be found in moist or fairly dry areas, in open pine forests or on prairie. Sometimes it is abundant adjacent to ponds.

Similar species: The length and shape of the forewings serve to distinguish the handsome Florida grasshopper from the swamp eastern grasshopper, *Eotettix palustris*, and the little eastern grasshopper, *Eotettix pusillus* (Plate 23). The latter two species have round (little eastern) to oval (swamp eastern) forewings that are shorter than the pronotum.

LITTLE WINGLESS GRASSHOPPER

Gymnoscirtetes pusillus (Plate 23)

Distribution: Found in Georgia and Florida.

Identification: A small grasshopper, greenish yellow to tan with a black-and-white lateral stripe running from each eye to about the midpoint of the abdomen or beyond. The adult is wingless. Males are 12.5–17 mm long, females 17–25 mm.

In males, the furcula is barely visible. Cerci taper gradually to a point, but the upper edge is almost straight. Tip of the subgenital plate is slightly extended into a tubercle about as high as it is wide.

little wingless grasshopper

Ecology: Occurs in wet areas of pine forests, where it inhabits the undergrowth, or adjacent to ponds. This is an agile species that easily eludes capture.

Similar species: The little wingless grasshopper greatly resembles Morse's wingless grasshopper, *Gymnoscirtetes morsei*. In the former the tubercle at the tip of the subgenital plate in males is about as broad as high, whereas in the latter the tubercle is twice as high as wide. The lateral margins of the subgenital plate are not elevated, as they are in Morse's wingless grasshopper. Also, the cercus of the little wingless grasshopper is relatively straight, with only the tip curved slightly downward, whereas in Morse's wingless grasshopper it is strongly curved.

Morse's wingless grasshopper

WESTERN GRASS-GREEN GRASSHOPPER

Hesperotettix speciosus (Plate 23)

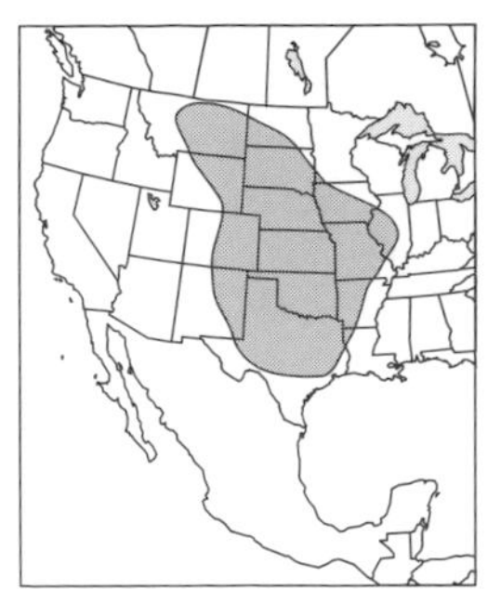

western grass-green grasshopper

Distribution: Found throughout the Great Plains region of the United States.

Identification: An attractive, uniformly colored grasshopper. Normally, it is grass green, though some individuals are tinged with red or purple. It is heavy-bodied. Antennae are red. Pronotum is roughly sculptured and has a pinkish or reddish stripe on the top. Forewings usually are shorter than the abdomen. Hind tibiae are green. Males are 20–26 mm long, females 28–37 mm.

Ecology: Inhabits weedy areas, where if feeds on broadleaf plants, particularly sunflower.

Similar species: In Florida and nearby states, a very similar species, the Florida purple-striped grasshopper, *Hesperotettix floridensis*, can be found. The forewings of the Florida purple-striped grasshopper are abbreviated and broadly rounded, whereas those of the western grass-green grasshopper are bluntly pointed.

Florida purple-striped grasshopper

MEADOW PURPLE-STRIPED GRASSHOPPER

Hesperotettix viridis (Plate 23)

Distribution: Found throughout the United States.

meadow purple-striped grasshopper

Identification: This is a colorful species, although the eastern forms are less striking than those found in western states. It is principally green, but the forewings often are colored by a broad stripe of reddish purple edged with white. Purplish coloration may be found along the top of the abdomen and sometimes on the side of the pronotum. A short black-and-white stripe occurs behind each eye. The light stripe on the upper surface of the pronotum is bordered by dark stripes. Forewings may be long, reaching the tip of the abdomen or slightly beyond, or may be abbreviated, extending about two-thirds the length of the abdomen. Hind tibiae are blue. Males are 16–22 mm long; females 20–30 mm.

Ecology: Inhabits weedy and brushy locations, particularly dry habitats. In the western states, it is also known as "snakeweed grasshopper," a reflection of its taste for snakeweed, rabbitbrush, ragweed, and other noxious plants. From the rancher's perspective, this grasshopper could be beneficial to rangeland productivity. Though it feeds readily on weeds, it only occasionally becomes sufficiently abundant to curb their growth.

Similar species: The forewings are always longer than the pronotum in this species, a character that distinguishes it from Osceola's grasshopper, *Hesperotettix osceola*. The meadow purple-striped grasshopper is quite variable in appearance, so its specific status is subject to debate and it has acquired several names.

Osceola's grasshopper

CUDWEED GRASSHOPPER

Hypochlora alba (Plate 23)

Distribution: Found in the Great Plains region of Canada and the United States from southern portions of the Prairie Provinces south to Texas.

Identification: Pale grayish green, though close examination reveals numerous small brown spots over the surface of the body. A dark stripe extends from the back of the eye along the side of the pronotum. Wings normally are short and pointed. Hind tibiae are bluish green. Males are about 15 mm long, females 20 mm.

In males, the cerci are broad basally but slender at the tip. The tip of the abdomen bears a short, blunt protuberance.

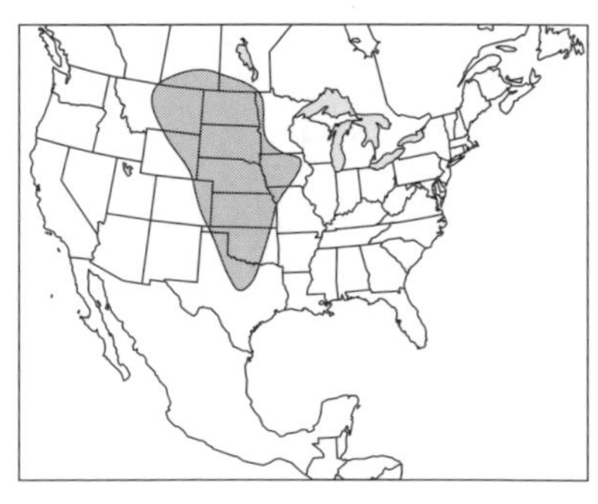

cudweed grasshopper

Ecology: This species is normally associated with cudweed sagewort, *Artemisia ludoviciana*, though other plants in this genus are eaten in lesser amounts. This degree of specificity is unusual in grasshoppers. Other grasshoppers do not survive if fed only this plant. Adults are found from August to October.

CATTAIL TOOTHPICK GRASSHOPPER

Leptysma marginicollis (Plate 30)

Distribution: Found from Maryland and Florida in the east, and west to New Mexico and Colorado.

Identification: A slender, elongate grasshopper with a very pointed head and flattened, sword-shaped antennae. It superficially resembles stridulating slant faced grasshoppers (subfamily Gomphocerinae), but is easily distinguished by the presence of the spine between the base of the front legs. It is usually brownish, with a whitish, yellow, or brown stripe from the eye to the base of the front legs. The head is as long as, or longer than, the pronotum. Dorsally, the body may also be reddish or pinkish. Forewings are sharply pointed, extending 3 to 5 mm beyond the tip of the abdomen. Males are 28–38 mm long, females 31–45 mm.

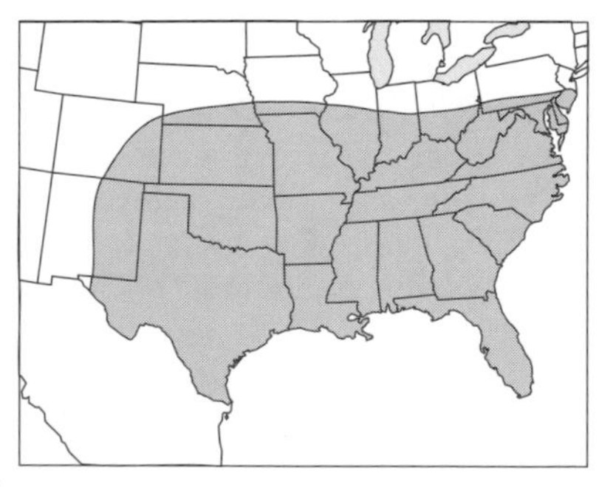

cattail toothpick grasshopper

Ecology: Inhabits wet areas, and is usually found on emergent vegetation such as cattails and sedges. Adults fly readily when disturbed but never alight on soil, usually landing on the stem of emergent vegetation and moving quickly to the side opposite the source of disturbance. In Florida, nymphs and adults can be found year-round.

Similar species: This species is easily confused with the glassy-winged toothpick grasshopper, *Stenacris vitreipennis* (Plate 30), but in the former the head is as long as, or longer than, the pronotum, whereas in the latter the head is shorter than the pronotum. The antennal segments are considerably wider in the cattail toothpick grasshopper. This species can be distinguished from the long-headed toothpick grasshopper, *Achurum carinatum* (Plate 1), the stubby toothpick grasshopper, *Achurum minimipenne*, Sumichrast's toothpick grasshopper, *Achurum sumichrasti*, and the Wyoming toothpick grasshopper, *Paropomala wyomingensis* (Plate 3), by the presence of a spine between the base of the front legs.

NARROW-WINGED SPURTHROATED GRASSHOPPER

Melanoplus angustipennis (Plate 24, Fig. 25)

Distribution: Found centrally in the United States over a wide area extending from the Appalachian Mountains in the east to the Rocky Mountains in the west. In southern Canada, it occurs over a similar range.

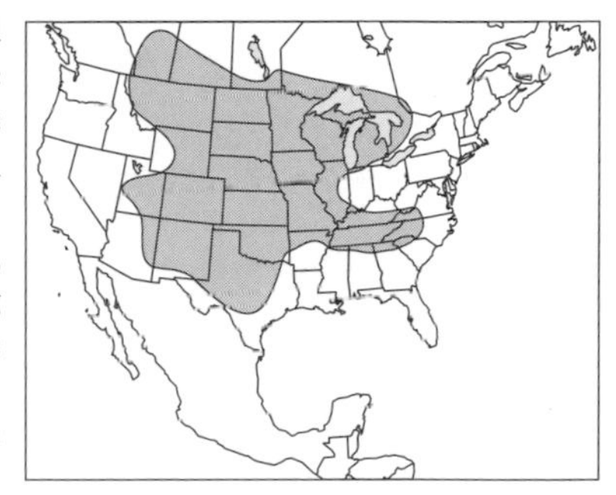

narrow-winged spurthroated grasshopper

Identification: Reddish brownish above, lighter below, with a dark stripe extending from behind the eye about two-thirds of the distance across the lateral lobe of the pronotum. Forewings are about the length of the abdomen or slightly longer and are not distinctly marked. Hind femora are yellowish brown. Hind tibiae are greenish blue or dull red. Males are 20–24 mm long, females 21–27 mm.

In males, the cerci are short and constricted near the middle. The tip is expanded and broadly rounded, and slightly depressed. The extensions of the furcula are divergent, and no more than one-third the length of the supra-anal plate. The supra-anal plate is narrowed and pointed at the tip. The subgenital plate is slightly grooved at the tip.

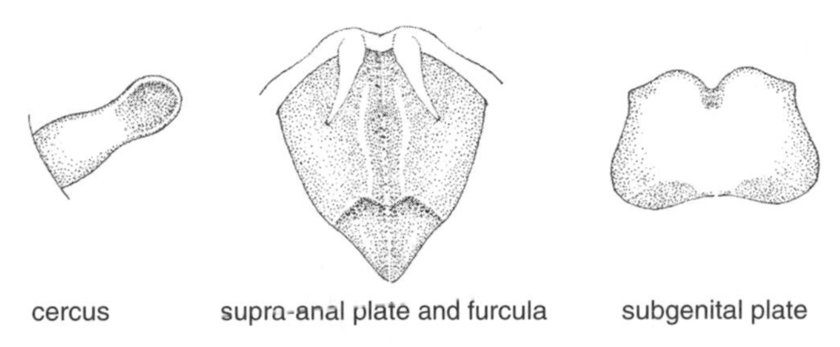

Figure 25. Diagnostic features of male narrow-winged spurthroated grasshopper

Ecology: Inhabits a number of grasslands, particularly sandy, arid regions. Feeds both on grasses and on broadleaf plants, preferring the latter. Though sometimes numerous, it rarely is considered to be a pest. Eggs hatch relatively early, in mid-spring. Adults are found from July to October.

ARIZONA SPURTHROATED GRASSHOPPER

Melanoplus arizonae (Plate 24, Fig. 26)

Distribution: Southwestern in distribution, from western Kansas and Oklahoma west to Utah and Arizona, and south into northern Mexico.

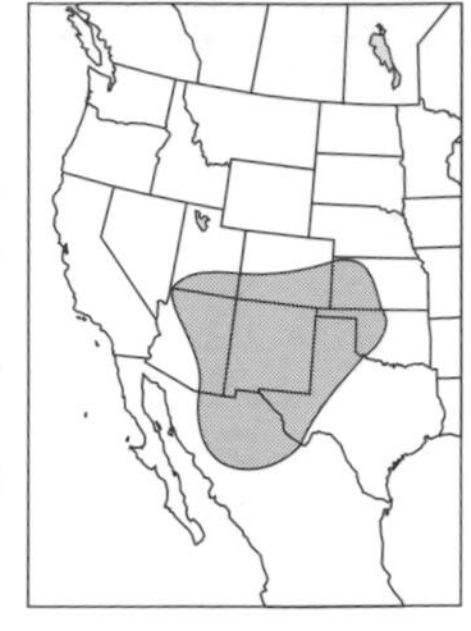
Arizona spurthroateded grasshopper

Identification: A brownish grasshopper with dark markings and paler color below. Wings extend to the tip of the abdomen or beyond. Hind tibiae are blue. Males are about 25 mm long, females 30 mm.

In males, the subgenital plate of the male is slightly prolonged at the tip, but not notched. The furcula is medium in size. The cerci taper to a blunt point.

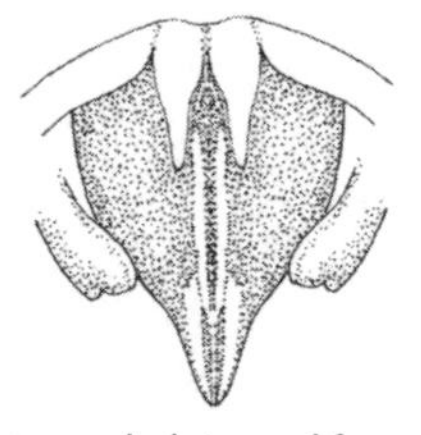
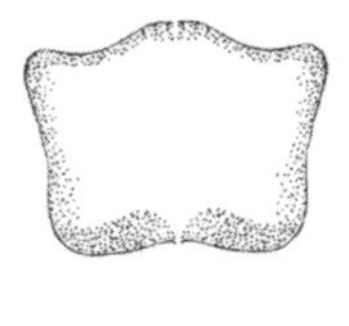

Figure 26. Diagnostic features of male Arizona spurthroated grasshopper

Ecology: Inhabits grasslands, including desert grasslands.

TWO-SPINED SPURTHROATED GRASSHOPPER

Melanoplus bispinosus (Plate 24, Fig. 27)

Distribution: Found in the southeastern states and west to Texas and Kansas.

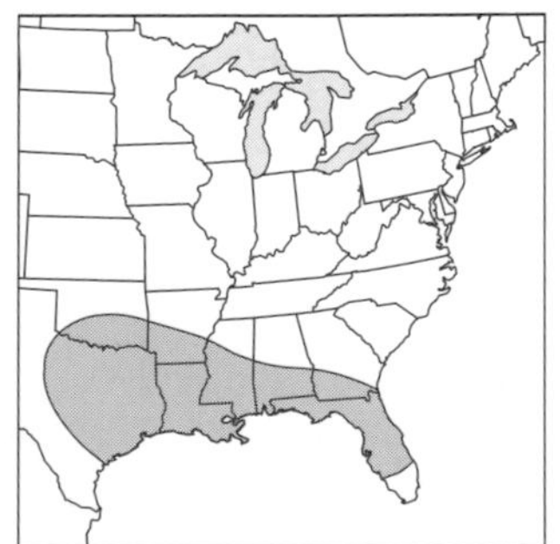
two-spined spurthroated grasshopper

Identification: A medium-sized grasshopper, grayish brown to reddish brown. A dark stripe extends from the eye onto the lateral lobe of the pronotum. Forewings are marked with a row of dark spots centrally, and they extend to the tip of the abdomen or beyond. Hind femora bear large dark spots or diffuse bands. Hind tibiae are bluish green or blue. Males are 25–30 mm long, females 26–32 mm.

In males, the extensions of the furcula are slightly divergent and extend to about one-half the length of the supra-anal plate. The presence of large "spine-like" furcula apparently is the basis for the name of this grasshopper. The cerci are elongate, narrowed at the middle and rounded distally; the outer face of the tip is grooved or recessed.

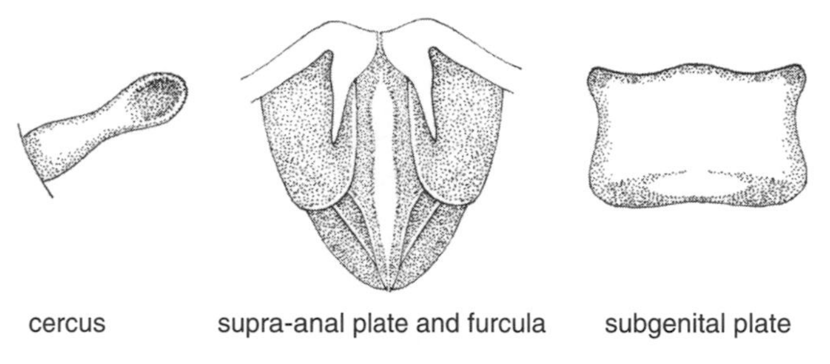

Figure 27. Diagnostic features of male two-spined spurthroated grasshopper

Ecology: Frequents pastures, crop fields, and roadsides. Adults are found all summer and well into the autumn.

TWO-STRIPED GRASSHOPPER

Melanoplus bivittatus (Plate 24, Fig. 28)

Distribution: Found over most of the United States, though it is absent from the southernmost areas. In Canada, it occurs only in southern regions.

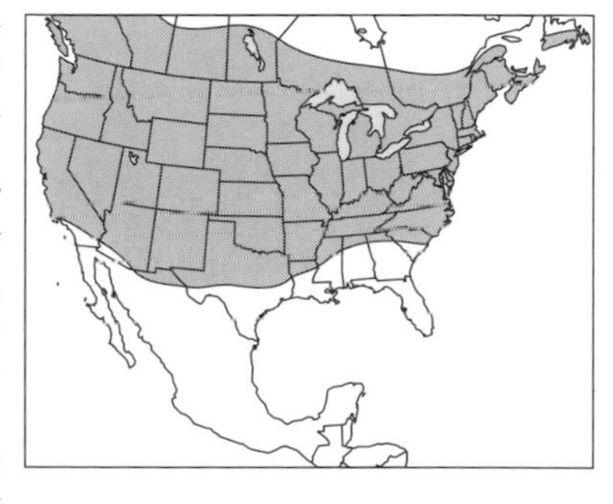

two-striped grasshopper

Identification: This large, robust grasshopper derives its name from the two pale yellow stripes extending from the back of the eyes, across the lateral lobes of the pronotum, to the tips of the forewings. The stripes converge to form a V at the tip of the wings. Over most of the length of the body, the stripes are bordered by black, especially below the stripes. Background color is olive green or yellowish green above, and yellow below. Wings normally are about the length of the abdomen. Hind femora are yellowish but bear a dark stripe. Hind tibiae are variable in color. Males are 23–25 mm long, females 29–45 mm.

In males, the cerci are fairly short, broad, and boot-shaped. The extensions of the furcula are short, triangular, and widely spaced.

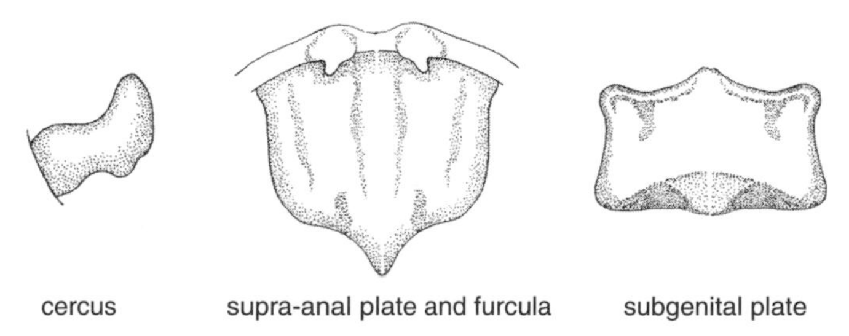

Figure 28. Diagnostic features of male two-striped grasshopper

Ecology: This adaptable species is found in many habitats but prefers tall, lush vegetation. It commonly is associated with disturbed sites, including agricultural areas. It is one of the most significant grasshopper pests of cultivated crops. It feeds both on broadleaf plants and on grasses, but prefers broadleaf plants. Life cycle is completed in a single year in most locations, but at high elevations in Canada a second year is required. In most areas, eggs commence hatching relatively early, in mid-spring. Adults are found from July to September.

NORTHERN SPURTHROATED GRASSHOPPER

Melanoplus borealis (Plate 24, Fig. 29)

Distribution: As suggested by its name, this species is northern in distribution. Found throughout nearly all of Canada and in Alaska, and south to the central regions of the United States. The exception is in the Rocky Mountains region where, like many northern species, it extends farther south because of the cool weather associated with higher altitudes.

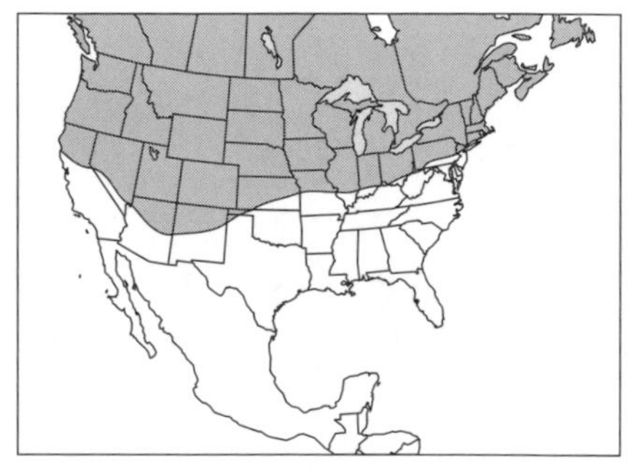

northern spurthroated grasshopper

Identification: Reddish brown or greenish yellow above, and dull yellow below. A dark stripe extends from the back of the eye onto the lateral lobe of the pronotum. In long-winged forms, the forewings extend to the tip of the abdomen and are free of markings. Short-winged forms also occur, and sometimes predominate. Hind femora are brown or yellow. Hind tibiae are usually red but sometimes yellow. Males are 15–21 mm long, females 17–26 mm.

In males, the cerci are broad basally, tapering strongly in the basal half, and less so in the distal half. The tip is broadly rounded and slightly angled upward. The extensions of the furcula diverge, taper to points, and extend over about one-half the supra-anal plate.

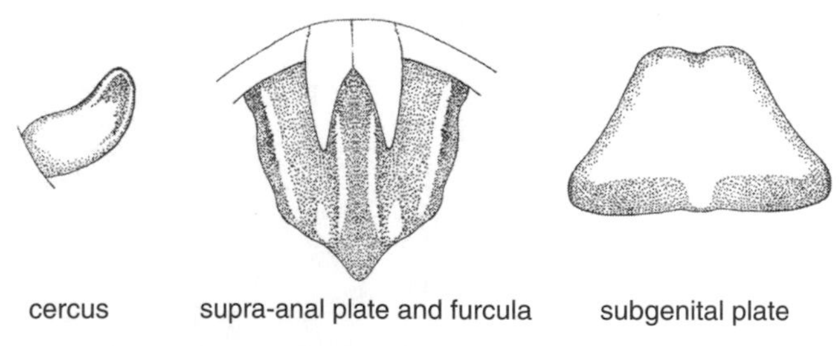

Figure 29. Diagnostic features of male northern spurthroated grasshopper

Ecology: Associated with lush vegetation: often found in moist areas and in the vicinity of lakes and streams. It occurs at nearly all altitudes, but is often the dominant species at high altitudes. Eggs remain in the soil for two years before hatching, resulting in a two-year life cycle. Eggs commence hatching relatively late, in early summer. Adults are found from July to September.

SAGEBRUSH GRASSHOPPER

Melanoplus bowditchi (Plate 24, Fig. 30)

Distribution: Found in the Rocky Mountains region of North America, extending east to the central Great Plains.

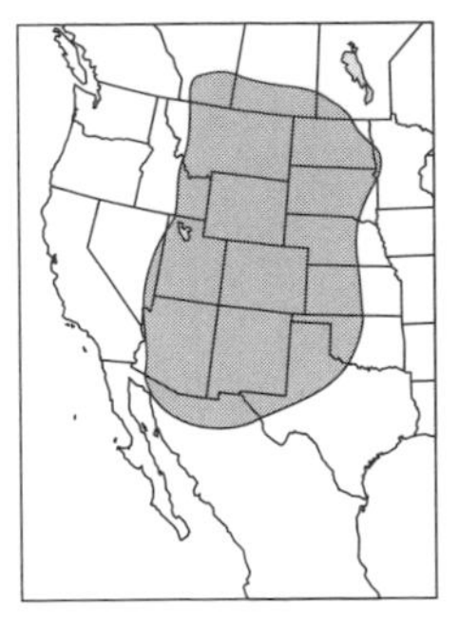

sagebrush grasshopper

Identification: A medium-sized grasshopper, bluish gray to reddish brown. A dark stripe extends from the back of the eye across the lateral lobes of the pronotum. Forewings are not well marked, though they bear a few dark spots along the midline. Hind femora are yellow with two dark, irregular bands. Hind tibiae are bluish gray. Males are 19–23 mm long, females 22–26 mm.

In males, the cerci are wide basally, curving slightly and tapering to a blunt point. The furcula is about one-third or one-half the length of the supra-anal plate, divergent, and blunt at the tips.

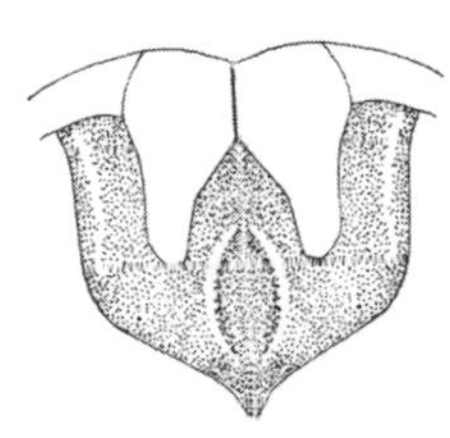

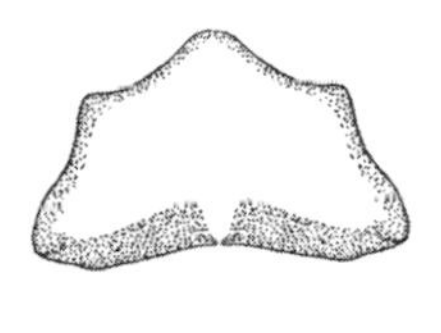

Figure 30. Diagnostic features of male sagebrush grasshopper and yellowish spurthroated grasshopper

Ecology: Most often encountered on the plains, but also occurs in foothills and mountain valleys. Feeds only on sagebrush (genus *Artemesia*). Eggs commence hatching in late spring. Adults are found from July to September.

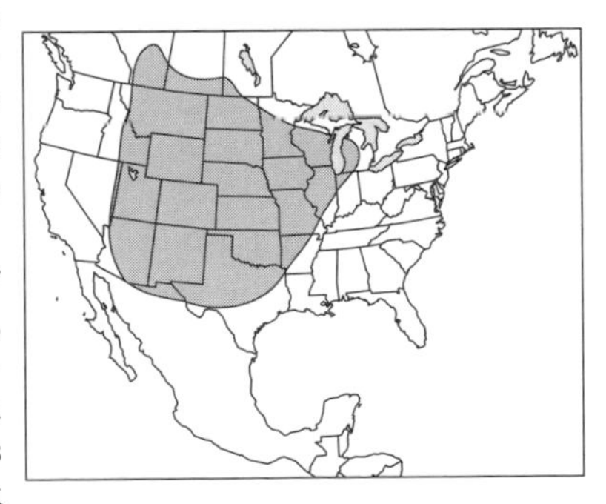

yellowish spurthroated grasshopper

Similar species. This species is quite similar to the yellowish spurthroated grasshopper, *Melanoplus flavidus.* They are difficult to distinguish reliably based on appearance, but in the sagebrush grasshopper the forewings are more regularly spotted, and the dark stripe extending from the eye across the lateral lobes of the pronotum is darker and more discrete.

LITTLE PASTURE SPURTHROATED GRASSHOPPER

Melanoplus confusus (Plate 26, Fig. 31)

Distribution: Found throughout most of the northern region of the United States and southern Canada. In the Great Plains region, the range extends south to northern Texas.

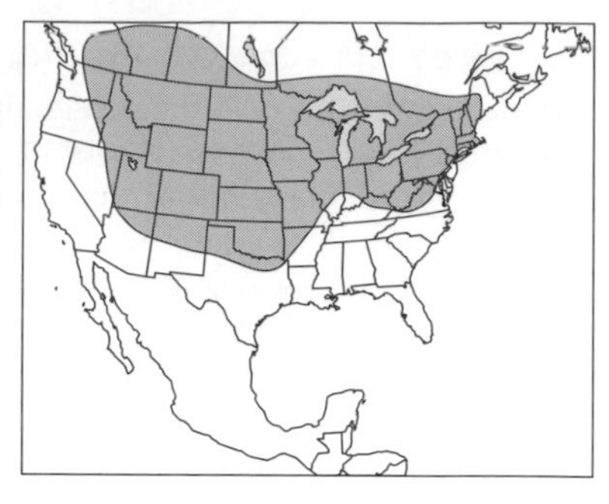

little pasture spurthroated grasshopper

Identification: A yellowish-brown grasshopper; tends to be darker above and lighter below. Dark stripes extend back from the eyes onto the lateral lobes of the pronotum, and from the top of the head dorsally onto the pronotum. Forewings extend beyond the tip of the abdomen and are uniformly colored. Hind femora are brownish yellow. Hind tibiae are usually pale blue but sometimes reddish or yellowish. Males are 16–20 mm long, females 19–27 mm.

In males, the cerci are distinctive: boot-shaped, with the base rather rectangular, the "heel" small, and the "toe" bluntly rounded and curved inward. The extensions of the furcula are short, only about one-fourth the length of the supra-anal plate, and widely separated.

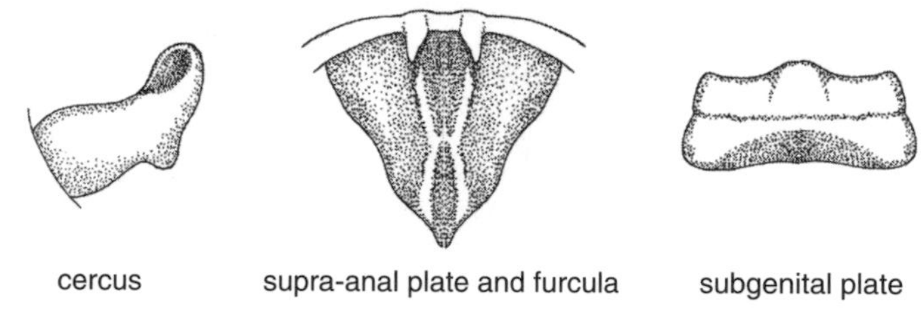

Figure 31. Diagnostic features of male little pasture spurthroated grasshopper

Ecology: Inhabits a variety of habitats, such as pastures, roadsides, and open woods. Feeds on grasses, sedges, and broadleaf plants, though favoring the latter. Among the species that overwinter as eggs, it is generally the first to hatch (in early spring) and to reach maturity (in early summer). Adults are found primarily in June to July.

DAVIS'S OAK GRASSHOPPER

Melanoplus davisi (Plate 26, Fig. 32)

Distribution: Found in northern Florida; apparently does not occur in adjacent states.

Davis's oak grasshopper

Identification: This species is considered large for a short-winged *Melanoplus*, but otherwise is indistinct. In general, it is brown, olive, or yellowish brown above and yellowish below, with a black stripe extending from the eye onto the pronotum. Forewings are oval, shorter than the length of the pronotum, and overlap above. Hind femora are reddish yellow, with two bands often present in males but absent in females. Hind tibiae are red. Males are 18–25 mm long, females 25–32 mm.

In males, the furcula consists only of minute, rounded appendages. The cerci are broad and short, turning upward distally to a flattened, blunt tip.

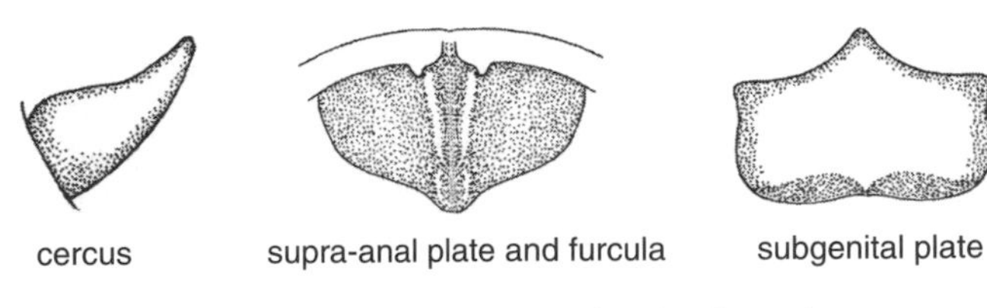

Figure 32. Diagnostic features of male Davis's oak grasshopper

Ecology: Dry pine woods is the favored habitat, where it feeds on understory, particularly low-growing oak. Adults are found in late summer and autumn.

Similar species: Davis's oak grasshopper shares the overlapping forewing character with Scudder's short-winged grasshopper, *Melanoplus scudderi* (Plate 29), and several other short-winged *Melanoplus* species. They can be distinguished based on the shape of the male and furcula cerci: the furcula of Davis's oak grasshopper appears to be minute rounded appendages, whereas in Scudder's short-winged grasshopper they are pointed. The tip of the cerci is much narrower in Davis's than in Scudder's.

DAWSON'S SPURTHROATED GRASSHOPPER

Melanoplus dawsoni (Plate 25, Fig. 33)

Distribution: A northern species, occurring widely in southern Canada and the northern states, though absent from the West Coast. It also occurs south in the Rocky Mountains region to New Mexico. Sometimes it is the most numerous grasshopper in the species assemblage.

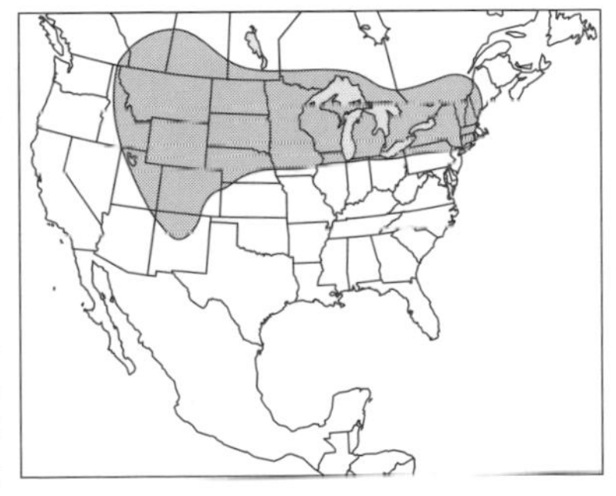

Dawson's spurthroated grasshopper

Identification: A small grasshopper, grayish brown above and yellowish below. The dorsal surface of the abdomen bears transverse light stripes. A black stripe extends from the back of the eye onto the lateral lobes of the pronotum. Femora are variable in length. When long, they extend to the tip of the abdomen. When short, they are about the length of the pronotum, are oval, and taper to a blunt point. Hind femora are yellowish with weak dark bands. Hind tibiae are red. Males are 14–19 mm long, females 19–25 mm.

In males, the cerci taper strongly in the basal half, then expand slightly to the bluntly rounded tip. The base is about twice the width of the apical region. The

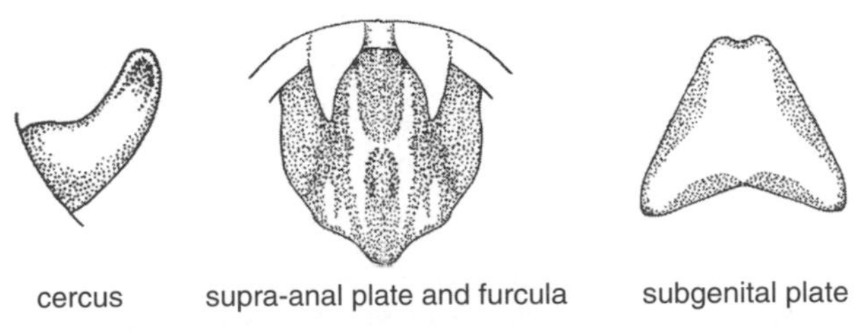

Figure 33. Diagnostic features of male Dawson's spurthroated grasshopper

upper edge of the cerci is strongly concave, the lower edge slightly convex. The furcula bears elongate extensions that extend about half the length of the supra-anal plate and taper to narrow points.

Ecology: Occurs in a wide variety of habitats, including grassy, shrubby, and wooded areas, though there is a preference for dry soils. Feeds mostly on broadleaf plants. Eggs hatch in late spring. Adults are found from July to October.

DEVASTATING GRASSHOPPER

Melanoplus devastator (Plate 25, Fig. 34)

Distribution: A Pacific Coast species occurring from Washington south through California.

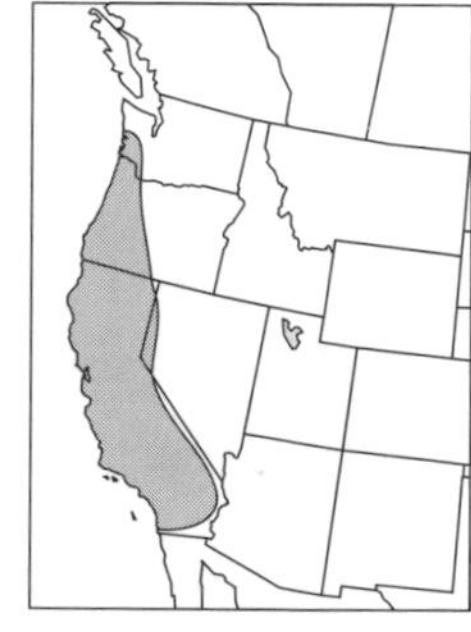

devastating grasshopper

Identification: Yellowish or brownish, with black on the lateral lobes of the pronotum and outer face of the hind femora. Forewings extend to about the tip of the abdomen and are marked with a series of small spots centrally. Hind femora bear three complete or incomplete bands. Hind tibiae are red or blue. Males are about 22 mm long, females about 25 mm.

In males, the cerci taper gradually to a broadly rounded point. The upper edge of the cerci is slightly concave, the lower edge slightly convex. The furcula bears elongate extensions that extend about half the length of the supra-anal plate and taper to narrow points.

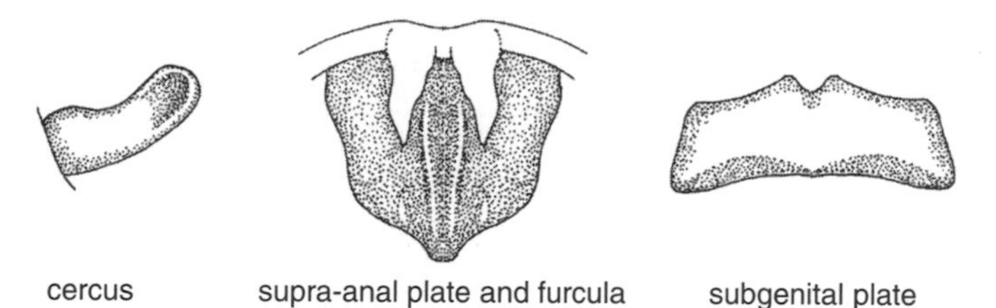

Figure 34. Diagnostic features of male devastating grasshopper

Ecology: Inhabits the semi-arid foothills, where it feeds on grasses and broadleaf plants. Adults sometimes descend from the foothills into agricultural areas where they cause considerable damage to crops. This tendency to cause severe injury is the basis for its name. Eggs commence hatching in late spring, and grasshoppers begin to attain the adult stage starting in July. Adults are found from July to November, but do not reproduce in the hot, dry summer months, waiting until late September or later to deposit eggs. This delayed egg production is an adaptation to the climate that prevails in much of this grasshopper's habitat.

DIFFERENTIAL GRASSHOPPER

Melanoplus differentialis (Plate 25, Fig. 35)

Distribution: Found over much of the United States, though it is largely absent from the northeastern, southeastern, and extreme northwestern states. Within this range, it is most commonly found in the central states from Indiana to Colorado and south to Texas. It also occurs in northern Mexico.

Identification: One of the largest members of the genus; though variable in color, it is relatively easy to distinguish. Forewings and pronotum are uniform, lacking distinctive marks. Body is often olive green with contrasting black markings, though sometimes yellow or black. Hind femora bear black herringbone markings on the outer face, a feature that is evident even in the black members of this species. Hind tibiae are yellow. Males are 28–37 mm long, females 34–50 mm.

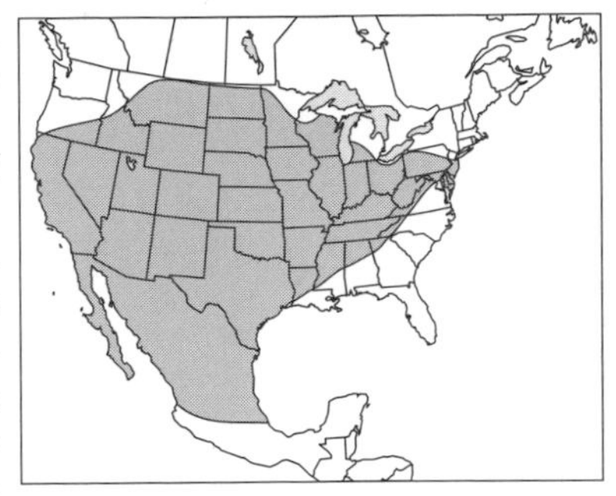

differential grasshopper

In males, the cerci are large and boot-shaped. The furcula is barely visible.

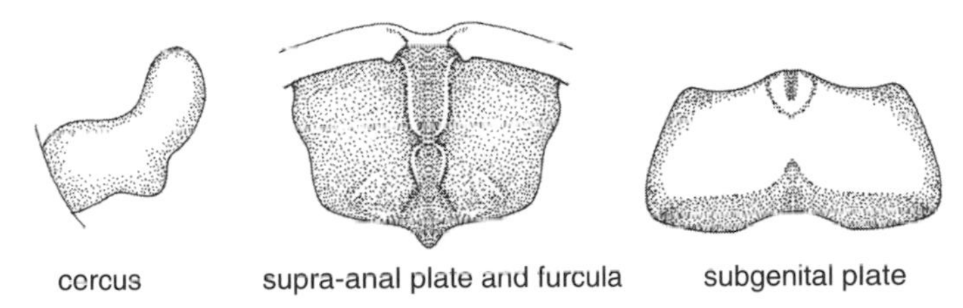

Figure 35. Diagnostic features of male differential grasshopper

Ecology: Favors areas with lush vegetation. It is common in moist cultivated crop areas and on disturbed sites. The differential grasshopper is a significant crop pest in the midwestern states. It feeds both on grasses and on broadleaf plants, but it prefers the latter. Eggs hatch in late spring. Both nymphs and adults display a tendency to aggregate. Adults are found from July to October.

HUCKLEBERRY SPURTHROATED GRASSHOPPER

Melanoplus fasciatus (Plate 25, Fig. 36)

Distribution: Northern in distribution, occurring in most of Canada and in Alaska, south to northern New Mexico and northern Alabama, but absent from the northwestern states.

Identification: A small but heavy-bodied species, grayish brown or reddish brown dorsally and yellow ventrally. A black stripe extends from the back of the eye onto the lateral lobe of the pronotum. Forewings are reddish brown, sometimes bearing a few speckles. They normally extend over about two-thirds of the

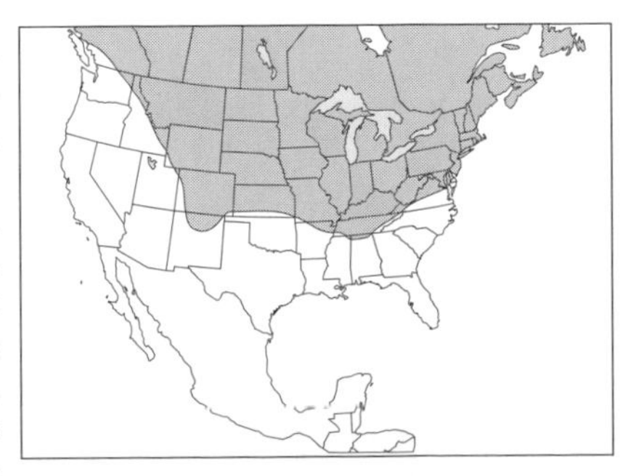

huckleberry spurthroated grasshopper

abdomen in females, and three-fourths in males, but long-winged individuals also occur. Hind femora are yellowish brown and bear two or three dark angled bands on the outer face. Hind tibiae normally are red. Males are 17–20 mm long, females 20–25 mm.

In males, the cerci are only moderately wide and slightly narrowed at the midpoint. The tips are rounded and curve inward. The furcula is very small. The tip of the subgenital plate is extended.

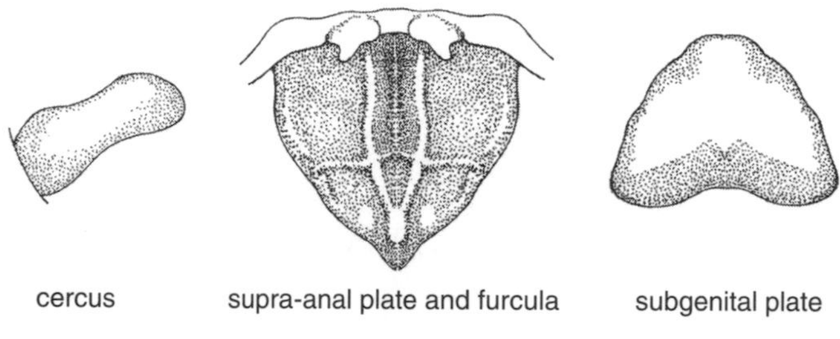

Figure 36. Diagnostic features of male huckleberry spurthroated grasshopper

Ecology: Preferred habitat is open woods and shrubby areas. As suggested by its name it is often found in association with huckleberry, but also with blueberry, bearberry, and heath. Feeds on broadleaf plants. It is rather sedentary, moving only short distances when disturbed. Though completing its life cycle in one year over most of the range, two years is necessary in the far north.

RED-LEGGED GRASSHOPPER

Melanoplus femurrubrum (Plate 26, Fig. 37)

Distribution: One of the most common and adaptable species. Found throughout the United States except for the coastal plain region of the Southeast, where it is replaced by a closely related species, the southern red-legged grasshopper, *Melanoplus propinquus.*

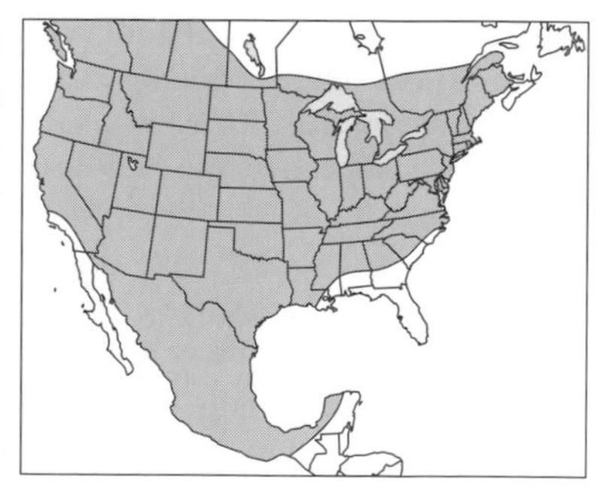

red-legged grasshopper

Identification: A small- to medium-sized grasshopper, reddish brown above and yellowish below. In some populations, the adults are reddish purple, or more infrequently, bluish. A black stripe extends from the back of the eye onto the lateral lobes of the pronotum. Forewings are narrow and uniform in color, extending beyond the tip of the abdomen. Hind femora are reddish brown and usually bear two dark bands. Hind tibiae are usually red but sometimes yellowish green. Males are 17–24 mm long, females 18–30 mm.

In males, the cerci are broad at the base, strongly narrowed in the basal one-half, and then only slighty narrowed toward the tip. The tip is truncated diagonally and longer along the upper edge. The extensions of the furcula are narrow and pointed, extending about half the length of the supra-anal plate. The tip of the abdomen is slightly bulbous, with the upper lip of the subgenital plate depressed.

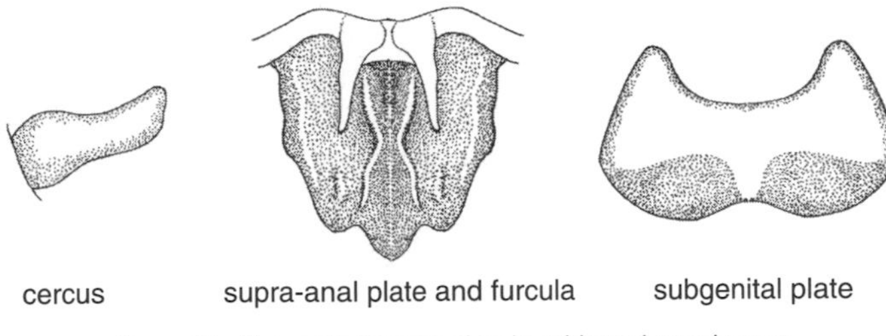

Figure 37. Diagnostic features of male red-legged grasshopper

Ecology: Prefers areas of thick vegetation; in arid areas it is found mostly near water. Often is abundant in disturbed areas such as fallow fields, roadside vegetation, and along fencerows and irrigation ditches. It is difficult to find on arid rangeland. Feeds readily on broadleaf plants, but also accepts grasses, and sometimes becomes a pest of crops. At high densities, it displays a slight tendency to aggregate and disperse. It serves as an intermediate host for the intestinal tapeworms that are acquired by birds such as quail and turkeys when they consume the grasshoppers. Eggs hatch in late spring. Adults are found from July to October.

Similar species: The southern red-legged grasshopper, *Melanoplus propinquus*, is closely related to the red-legged grasshopper. It is found in the southeastern states from North Carolina to Louisiana. It can be distinguished by the shape of the cerci: in the southern red-legged grasshopper, the tips are rounded instead of truncated diagonally, as in the red-legged grasshopper. In all other respects, they are quite similar in appearance.

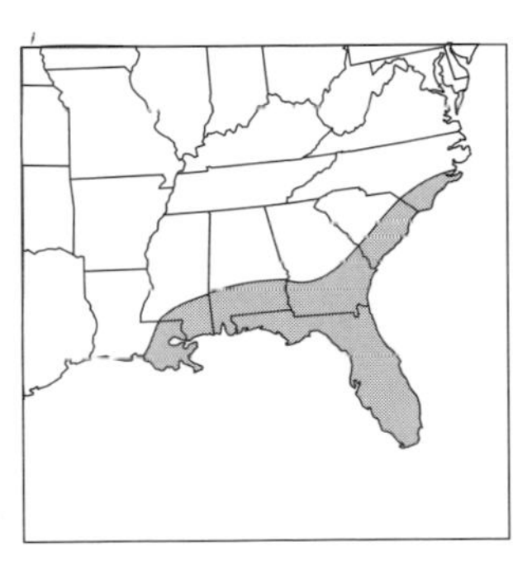

southern red-legged grasshopper

FOEDUS GRASSHOPPER

Melanoplus foedus (Plate 26, Fig. 38)

Distribution: Found widely in the Northwest. Its northern distribution is limited to southern portions of Canada's western provinces, its eastern boundary is Illinois, and it occurs south to Arizona and New Mexico.

Identification: Reddish, yellowish, or greenish-brown above; lighter, often yellowish, below. A dark stripe extends from behind the eye onto the lateral lobes of the pronotum, but it is not distinct. Forewings are grayish brown, usually with a few small spots, and extend beyond the tip of the abdomen. Hind femora yellowish with a weak dark stripe along the upper edge. Hind tibiae are usually red. Males are 22–26 mm long, females 25–32 mm.

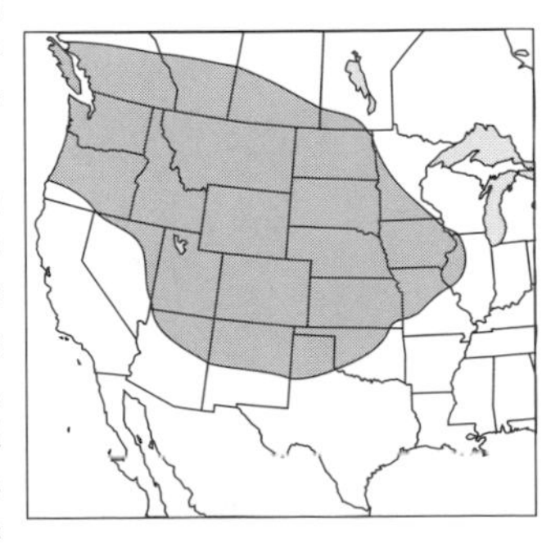

foedus grasshopper

In males, the cerci are symmetrical, constricted at the midpoint, and spoon-shaped at the tip. They are bent inward at the tip. The extensions of the furcula are broad basally but short, extending not even half the length of the supra-anal plate.

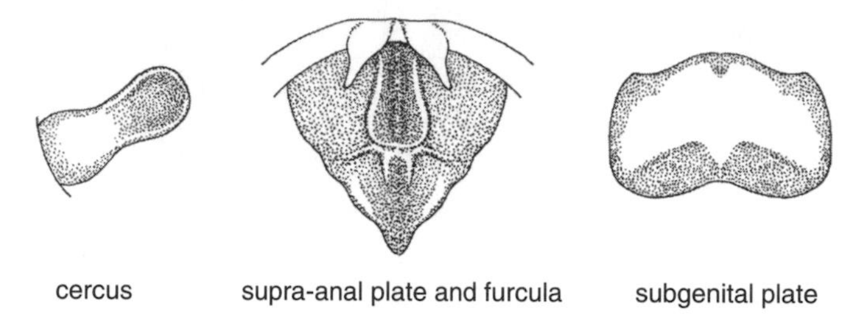

Figure 38. Diagnostic features of male foedus grasshopper and Packard's grasshopper

Ecology: Common in western prairie habitat and along rivers and streams. Feeds on numerous broadleaf plants and occasionally on grasses. Adults are found from July to September.

Similar species: Easily confused with Packard's grasshopper, *Melanoplus packardii* (Plate 28), the foedus grasshopper often can be distinguished by its lighter body coloration, particularly the absence of a dark but diffuse stripe extending from the top of the head across the dorsal region of the pronotum.

GLADSTON'S SPURTHROATED GRASSHOPPER

Melanoplus gladstoni (Plate 26, Fig. 39)

Distribution: Found along the western edge of the Great Plains from Alberta south to Mexico, and eastward into Minnesota and Iowa.

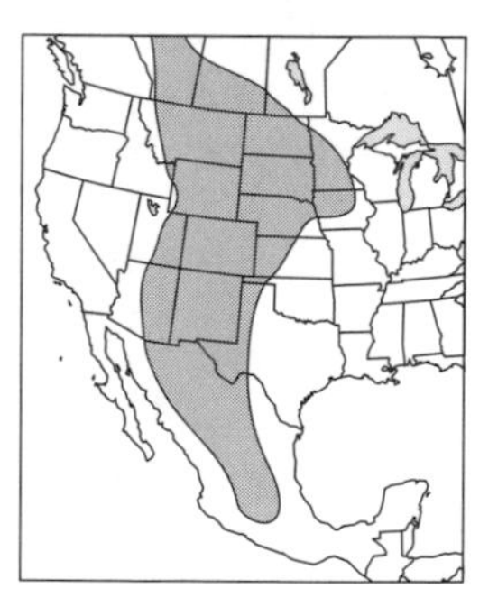

Gladston's spurthroated grasshopper

Identification: A robust species, yellowish brown or brownish gray and lighter below. Lateral lobe of the pronotum bears a black stripe. Forewings are spotted and extend beyond the tip of the abdomen. Abdomen is marked with black and yellow bands. Hind femora are yellowish brown and bear two oblique dark bands. Hind tibiae are normally red tinged with purple, but sometimes greenish. Males are 17–21 mm long, females 19–24 mm.

In males, the cerci are slightly constricted near the midpoint, grooved at the tip, and curve inward. The furcula is short and the extensions widely separated.

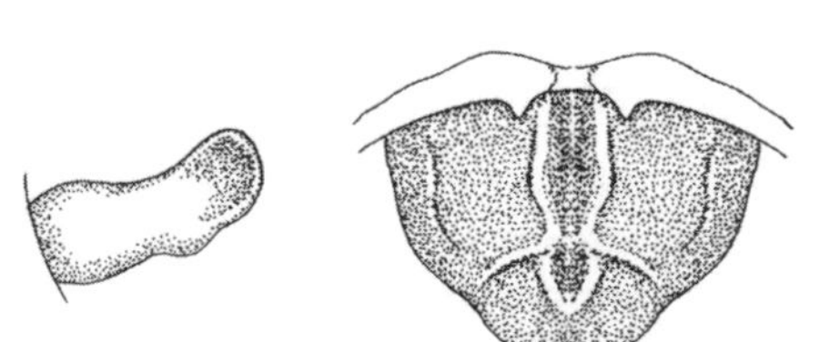

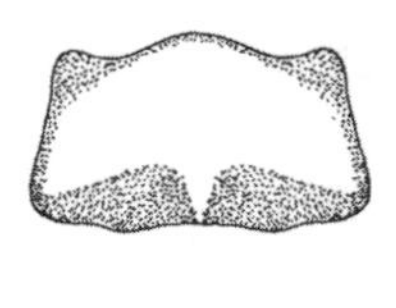

cercus   supra-anal plate and furcula   subgenital plate

Figure 39. Diagnostic features of male Gladston's spurthroated grasshopper

Ecology: Inhabits open dry uplands and dry pastures, usually with sandy or gravelly soil. Feeds both on grasses and on broadleaf plants. It is a late-maturing species, with eggs hatching in June and adults present from August or September to November.

ARROW-WEED SPURTHROATED GRASSHOPPER

Melanoplus herbaceus (Plate 27, Fig. 40)

Distribution: Found from southern California to western Texas and northwestern Mexico.

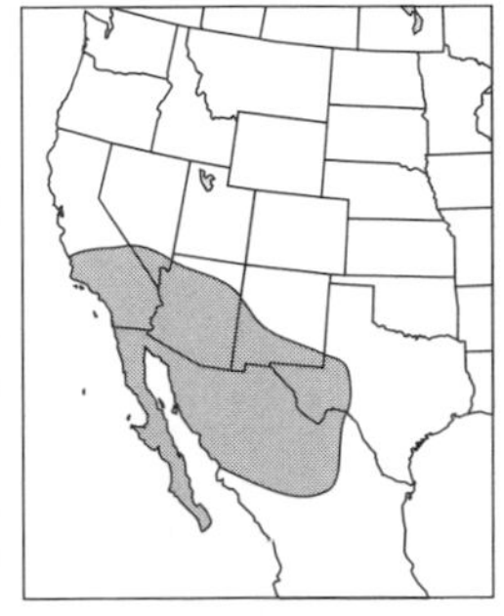

arrow-weed spurthroated grasshopper

Identification: A thin-bodied, pale-green species with long forewings that lack spots. Like most *Melanoplus* species, it is marked with a black stripe extending from behind the eye across the lateral lobes of the pronotum. Hind femora lack bands or stripes. Hind tibiae are greenish blue. Males are 24–26 mm long, females 26–28 mm.

In males, the extensions of the furcula are very broad in the basal and central regions, abruptly tapering to narrow points that point inward. The cerci taper gradually, twisting inward and bearing a slight depression at the tips.

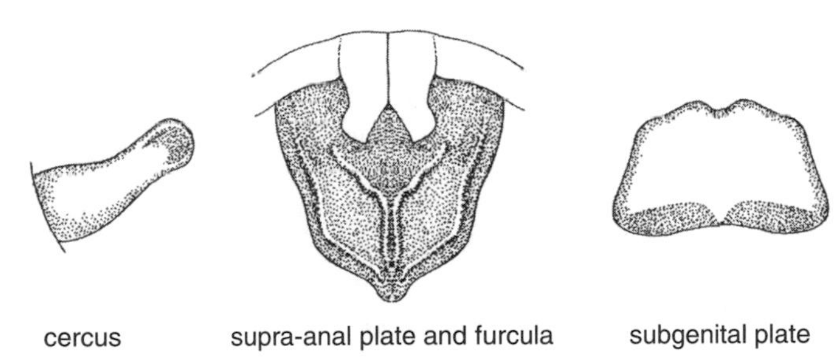

Figure 40. Diagnostic features of male arrow-weed spurthroated grasshopper

Ecology: Found along streams or other areas where water may accumulate, and often associated with arrow-weed. It sometimes disperses from its favored habitat and food plant, however, and feeds on many other plants.

SMALL SPURTHROATED GRASSHOPPER

Melanoplus infantilis (Plate 25, Fig. 41)

Distribution: Found in the northern Rocky Mountains region and northern Great Plains, and south to Colorado and Nebraska.

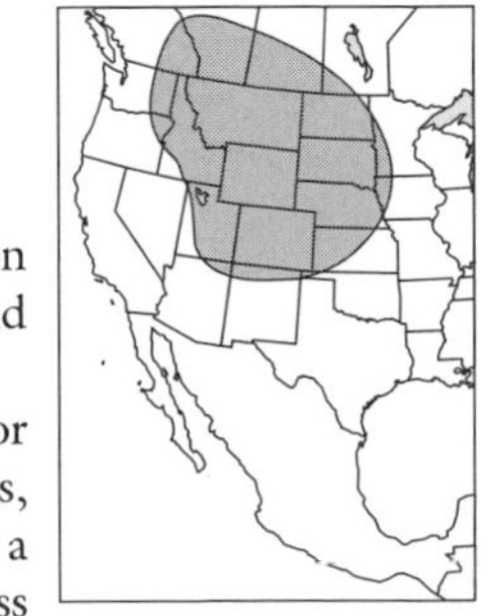

small spurthroated grasshopper

Identification: As the name suggests, this is a small species for this genus of mostly moderate- to large-sized grasshoppers, though it is not the smallest. It is brownish or gray, with a broad black stripe extending from the back of the eye across the lateral lobe of the pronotum. Forewings are long, extending beyond the tip of the abdomen, and marked centrally with a row of small alternating light and dark spots. Hind femora bear two oblique pale bands. Hind tibiae are pale blue. Males are about 15 mm in length, females 20 mm.

In males, the cerci are swollen at the base, narrowing at the middle, and forked distally. The furcula is small and triangular. The subgenital plate is expanded and notched at the tip.

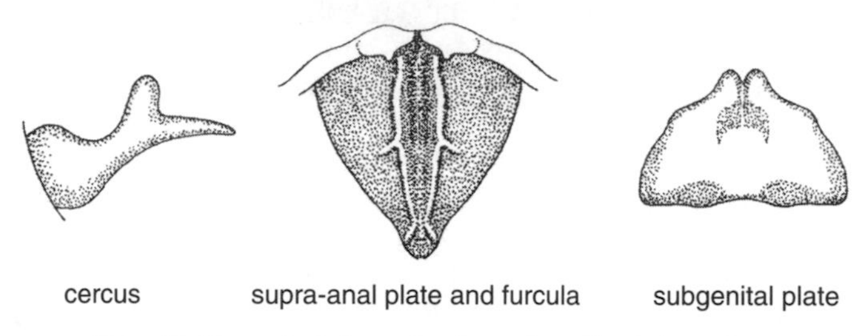

Figure 41. Diagnostic features of male small spurthroated grasshopper

Ecology: Inhabits open grassy areas including areas of moderate elevation. Feeds principally on grasses. Eggs hatch rather early, in midspring. Adults are found from August to October.

KEELER'S SPURTHROATED GRASSHOPPER

Melanoplus keeleri (Plate 27, Fig. 42)

Distribution: Found throughout the United States and southern Canada east of the Rocky Mountains, but largely absent from the Gulf Coast.

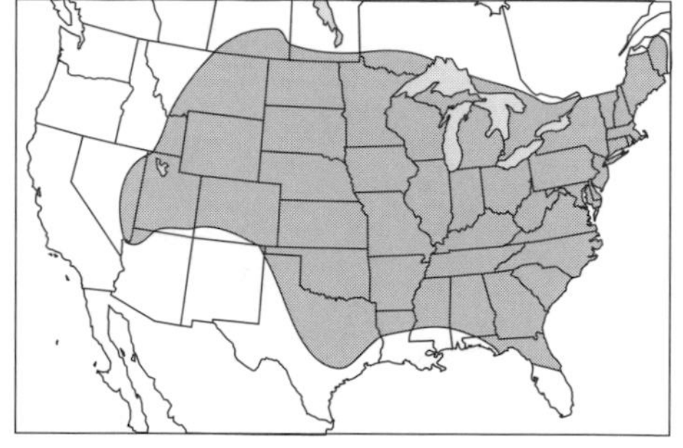

Keeler's spurthroated grasshopper

Identification: Moderately large for the genus *Melanoplus.* Grayish brown or reddish brown above and yellowish below. A dark stripe extends from the back of the eye onto the lateral lobe of the pronotum. Forewings bear a series of small brown spots centrally, and extend to the tip of the abdomen or beyond. Hind femora are marked with diffuse dark bands. Hind tibiae are coral red. Males are 17–29 mm long, females 19–38 mm.

In males, cerci are distinctively shaped, resembling a boot with a rounded toe directed upward and a pointed heel directed downward. The furcula is reduced to very small lobes.

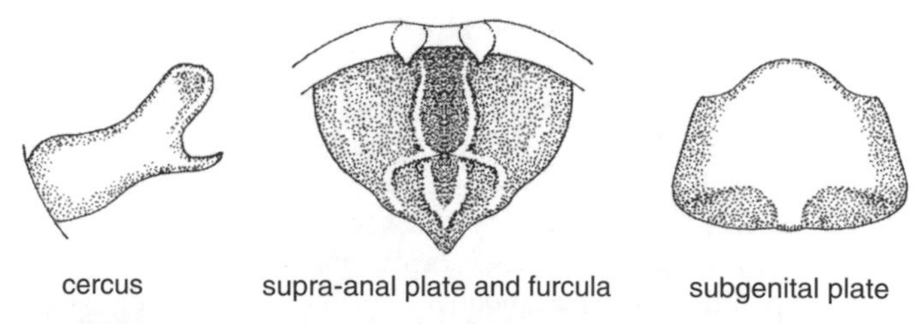

Figure 42. Diagnostic features of male Keeler's spurthroated grasshopper

Ecology: Inhabits dry prairie, pasture, and open woods. In many eastern woodlands it is the most common *Melanoplus.* Normally feeds on broadleaf shrubs, but sometimes on grasses and crop plants. Eggs commence hatching relatively late, in early summer. In Florida, adults are found from July to December.

LAKIN'S GRASSHOPPER

Melanoplus lakinus (Plate 25, Fig. 43)

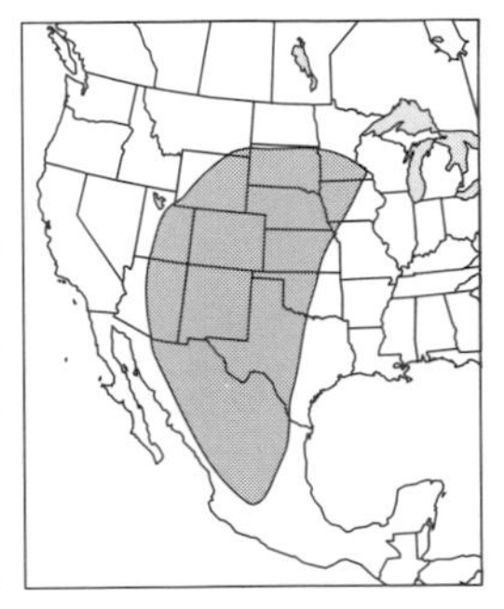

Lakin's grasshopper

Distribution: Found over most of the Great Plains and southern Rocky Mountains, south to central Mexico; absent from Canada.

Identification: A grayish brown, grasshopper with a dark stripe that extends from the top of the head onto the pronotum. A black stripe extends from the back of the eye onto the lateral lobe of the pronotum. Forewings are moderately short, about the length of the head and pronotum, and taper to a point. They are dark but marked centrally with a line of light spots. Hind femora are yellowish brown and marked with two oblique purplish bands. Hind tibiae are blue. Males are about 22 mm long, females 30 mm.

In males, the cerci are swollen basally and marked with a ridge. They taper markedly in the basal half and develop into a finger-like projection that extends upward and inward. The extensions of the furcula are small, triangular, and widely separated.

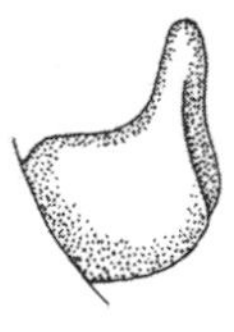

cercus

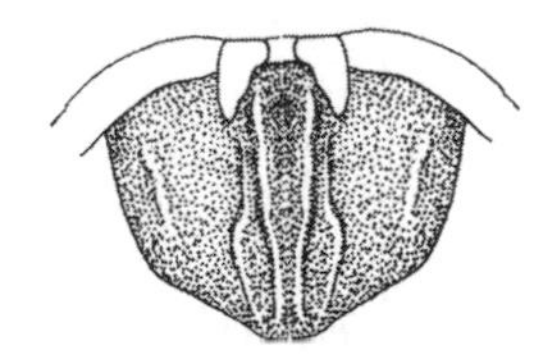

supra-anal plate and furcula

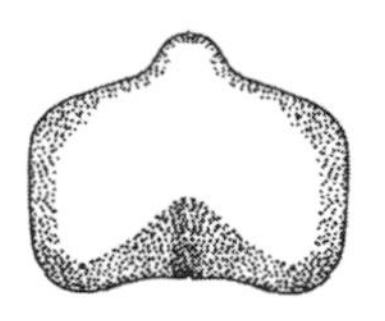

subgenital plate

Figure 43. Diagnostic features of male Lakin's grasshopper

Ecology: Commonly inhabits rangelands but is not considered damaging. Its diet is restricted to the family Chenopodiaceae, particularly Russian thistle. It develops late in the season, with eggs hatching in early summer and adults found from July or August to October.

MARGINED SPURTHROATED GRASSHOPPER

Melanoplus marginatus (Plate 27, Fig. 44)

margined spurthroated grasshopper

Distribution: Found only in the central and coastal areas of California.

Identification: A brownish or greenish grasshopper, sometimes with forewings and pronotum marked with red. It occurs in two forms, short-winged and long-winged. In the short-winged form, the length of the forewings is about equal to the length of the pronotum, and they taper to a point. The sides of the abdominal segments bear dark spots. Hind femora lack the bands or large spots commonly found on *Melanoplus* species. Hind tibiae are brown or olive green. Both males and females are 14–23 mm long.

In males, the cerci are bluntly rounded and constricted slightly near the middle. The furcula is small. The subgenital plate bears a strong tubercle.

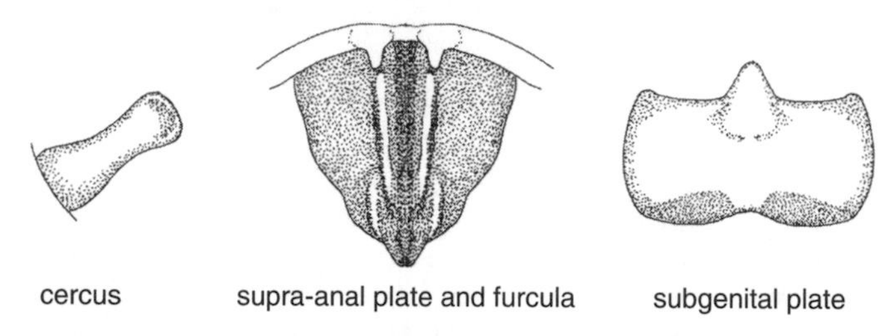

Figure 44. Diagnostic features of male margined spurthroated grasshopper

Ecology: Inhabits grasslands and weedy areas, but normally is found only at low elevations. It occurs in a variety of agricultural environments, where it can be abundant enough to cause injury to vegetable and field crops and to fruit. It is most abundant, and damaging, in fields of alfalfa.

FLABELLATE GRASSHOPPER

Melanoplus occidentalis (Plate 27, Fig. 45)

Distribution: Found widely in the western Great Plains and Rocky Mountains region from southern Canada to northern Mexico.

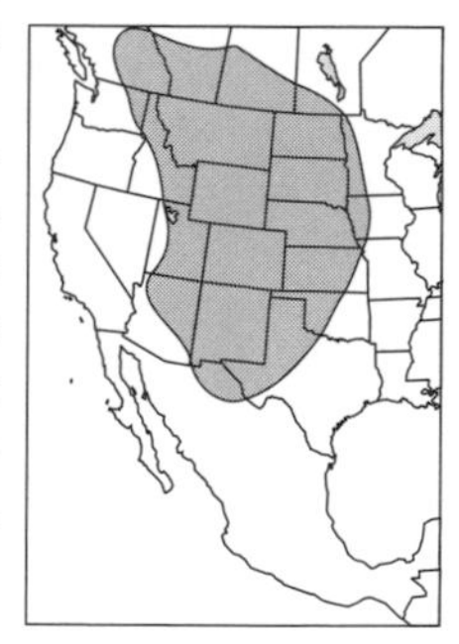

flabellate grasshopper

Identification: A reddish-brown species, dark above and yellowish below. Top of the head is marked with a dark stripe that sometimes continues onto the pronotum. Lateral lobes of the pronotum bear a dark stripe or patch. Forewings extend to, or beyond, the tip of the abdomen. They are dark but bear alternating light and dark square-shaped spots centrally. Hind femora are dark, with oblique pale bands. Hind tibiae are light blue. Males are 18 mm long, females 25 mm.

In males, the cerci are unusually large and triangular or ear-like in shape, wide basally and rounded at the tip. The extensions of the furcula are small and triangular.

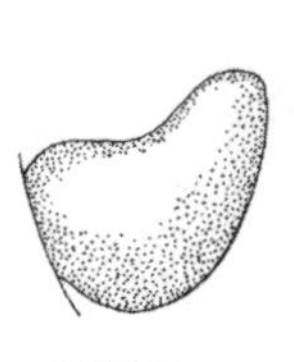

cercus

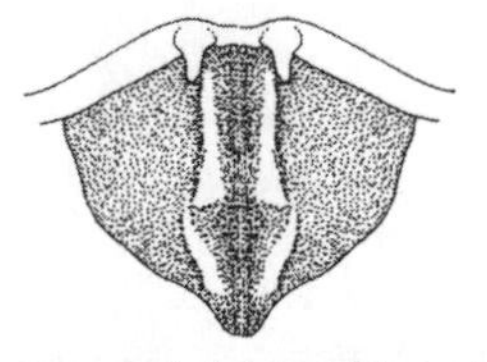

supra-anal plate and furcula

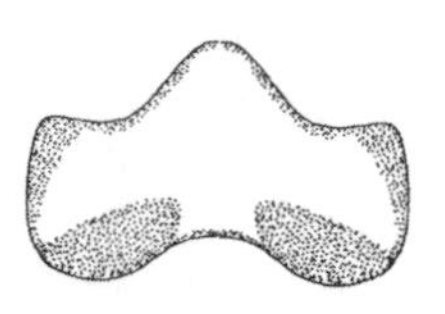

subgenital plate

Figure 45. Diagnostic features of male flabellate grasshopper

Ecology: Inhabits open grasslands, including high-altitude pastures. Feeds both on grasses and on broadleaf plants but prefers the latter. Eggs commence hatching relatively early, in midspring. Adults are found from June to October.

Similar species: A related species, the Nevada sage grasshopper, *Melanoplus rugglesi*, is found in California, the northwestern states, and adjacent areas of Canada. It can be distinguished by the shape of the male cerci. Instead of the triangular shape found in the flabellate grasshopper, the cerci of the Nevada sage grasshopper are slightly quadrate, the tip not tapering to a single point.

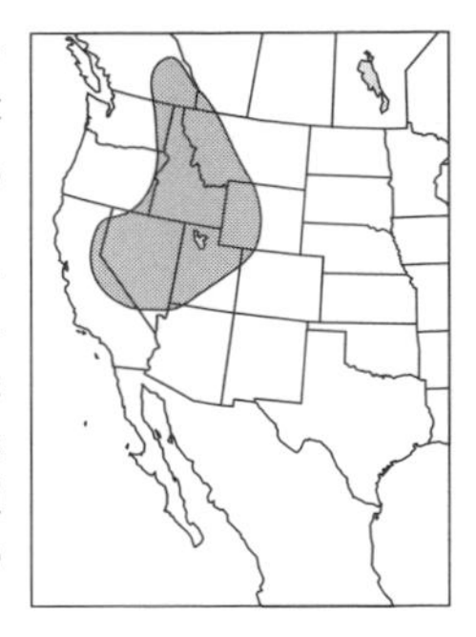

Nevada sage grasshopper

PACKARD'S GRASSHOPPER

Melanoplus packardii (Plate 28, see Fig. 38)

Distribution: Found widely in the West, although absent from the Pacific Coast area, and east to Illinois.

Identification: Grayish, brownish, or yellowish-brown above and yellowish below. A diffuse dark stripe extends from the top of the head over the top of the pronotum. A black stripe begins behind the eye and extends onto the lateral lobes of the pronotum. Forewings are grayish brown, usually with a few small spots, and extend beyond the tip of the abdomen. Hind femora are yellowish with a dark stripe along the upper edge. Hind tibiae are red or blue. Males are 22–33 mm long, females 26–37 mm.

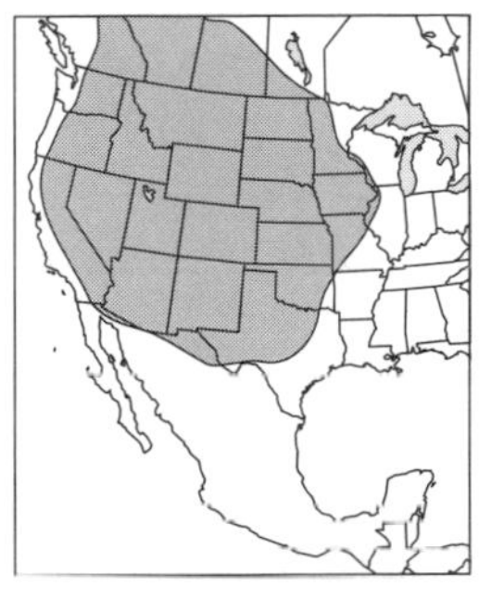

Packard's grasshopper

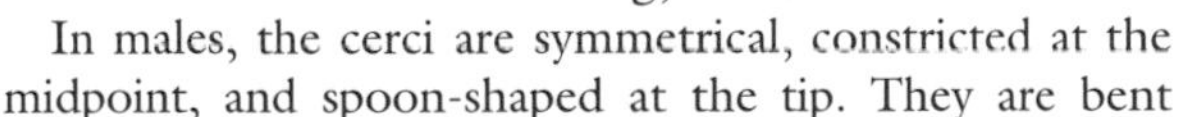

In males, the cerci are symmetrical, constricted at the midpoint, and spoon-shaped at the tip. They are bent inward at the tip. The extensions of the furcula are broad basally but short, extending not even half the length of the supra-anal plate. (See Fig. 38.)

Ecology: Common on western prairie. Feeds on numerous plants, including both broadleaf plants and grasses, and is reported to prefer legumes. Sometimes it becomes numerous enough to damage rangeland and grain crops, particularly in the northern areas of its range. Eggs commence hatching in midspring. Adults are found from July to September.

Similar species: Easily confused with the foedus grasshopper, *Melanoplus foedus* (Plate 26), Packard's grasshopper often can be distinguished by its darker body coloration, particularly the presence of a dark but diffuse stripe extending from the top of the head across the top of the pronotum.

PONDEROUS SPURTHROATED GRASSHOPPER

Melanoplus ponderosus (Plate 28, Fig. 46)

Distribution: Found from the midwestern states of Iowa and Kentucky south to Mississippi and west to southern New Mexico. It also occurs in northeastern Mexico.

Identification: A moderately large species for the genus, it is olive, gray, or brownish above and yellowish below. In males, a black stripe

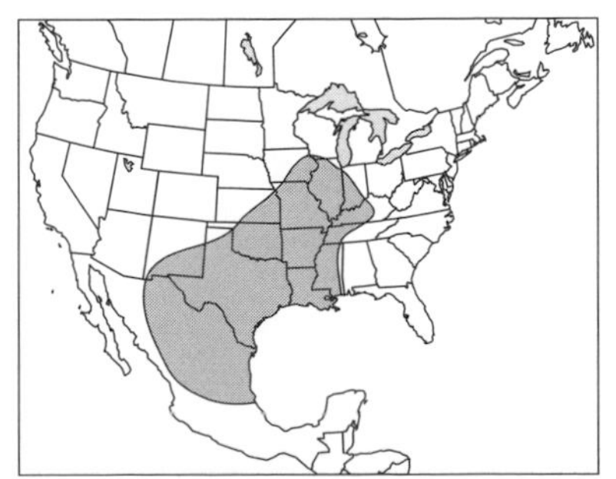

ponderous spurthroated grasshopper

extends from the back of the eye onto the lateral lobes of the pronotum, but this is weak or absent in females. Wing length is variable, ranging from about the length of the head and pronotum, to past the tip of the abdomen. Hind femora bear 2 to 3 dark bands. Hind tibiae normally are red, though sometimes yellow tinged with red. Males are 25–32 mm long, females 27–38 mm.

In males, the cerci are large and flared distally, with both dorsal and ventral extensions, but broadly rounded at the tip. The furcula is not visible.

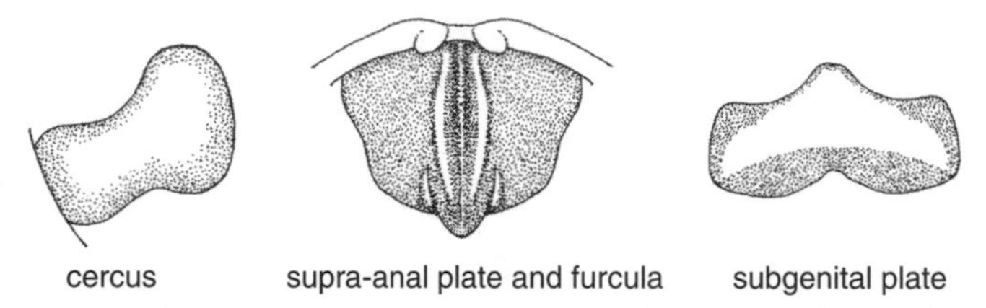

Figure 46. Diagnostic features of male ponderous spurthroated grasshopper

Ecology: This little-known species is most often associated with underbrush in wooded areas. It is reported to inhabit both dry areas and vegetation along streams. Males are active, but females are sluggish. They have been observed to deposit their eggs within small holes in stone walls, but it is not certain that this unusual oviposition site is normal.

FLORIDA LEAST SPURTHROATED GRASSHOPPER

Melanoplus puer (Plate 27, Fig. 47)

Distribution: Found through most of the Florida peninsula but not elsewhere.

Florida least spurthroated grasshopper

Identification: A very small, short-winged species. General color is reddish brown or purplish gray above and yellowish below. Forewings are shorter than the length of the pronotum, elongate oval in shape, and widely separated above. Males display a shiny black stripe behind the eye that extends across the pronotum to the first abdominal segments, but this pattern is indistinct in females. This black stripe is very wide on the anterior portion of the pronotum, narrowing markedly on the posterior region of the pronotum. Hind femora are dull yellow, sometimes with black spots but not complete bands. Hind tibiae are purplish green. Males are 10–17 mm long, females 16–22 mm.

In males, the furcula is a very short structure. The slender cerci taper gradually to a blunt tip but are not completely symmetrical; the upper edge is slightly concave. The tip of the cercus is not flattened. The subgenital plate is only weakly elongate.

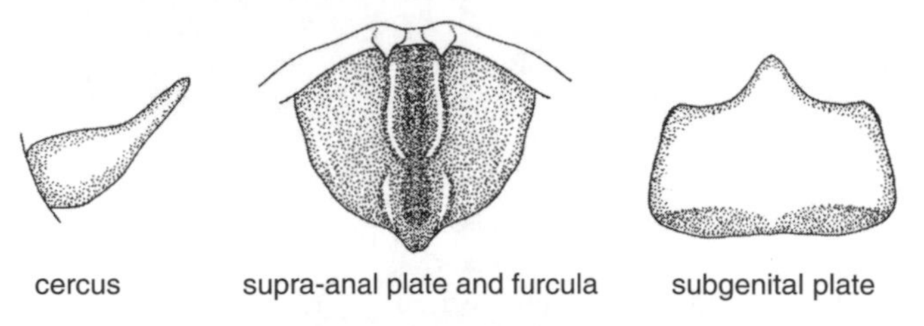

Figure 47. Diagnostic features of male Florida least spurthroated grasshopper

Ecology: Inhabits wiregrass patches in open woods, particularly scrub and high-pine habitats.

Similar species: The narrowing of the black stripe on the back of the lateral lobe of the pronotum is an important character in distinguishing this species from several other uncommon small *Melanoplus* species found in Florida: the Apalachicola spurthroated grasshopper, *M. apalachicolae*, Gurney's spurthroated grasshopper, *M. gurneyi*, the Trail Ridge scrub grasshopper, *M. ordwayae*, and the Tequesta spurthroated grasshopper, *M. tequestae*. The tip of the cercus is not flattened, as it is in St. John's spurthroated grasshopper, *M. adelogyrus*.

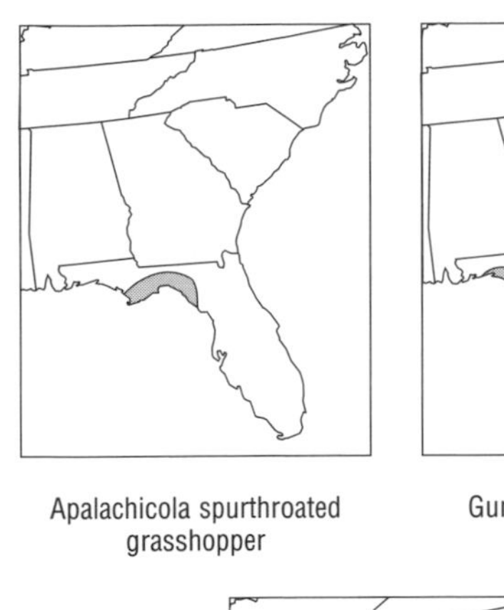

Apalachicola spurthroated grasshopper

Gurney's spurthroated grasshopper

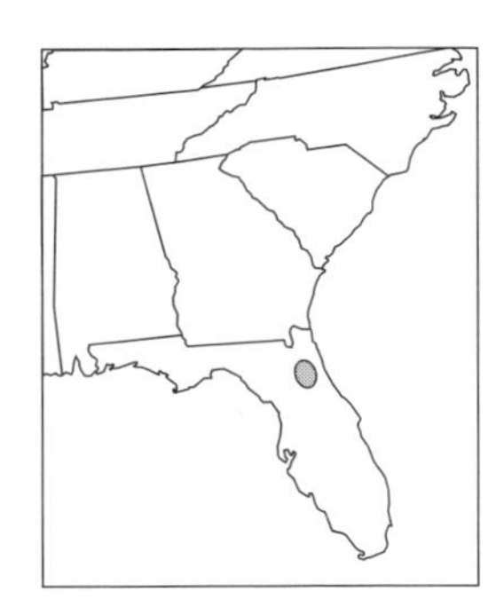

Trail Ridge scrub grasshopper

Tequesta spurthroated grasshopper

St. John's spurthroated grasshopper

PINETREE SPURTHROATED GRASSHOPPER

Melanoplus punctulatus (Plate 28, Fig. 48)

Distribution: Found throughout the eastern states and west to North Dakota and New Mexico.

Identification: This large, grayish species is unusual among the *Melanoplus* species in that its body and forewings bear numerous dark brown or black spots of moderate size. The underside may be reddish or yellowish, but it usually is light gray. Wings are long. There is

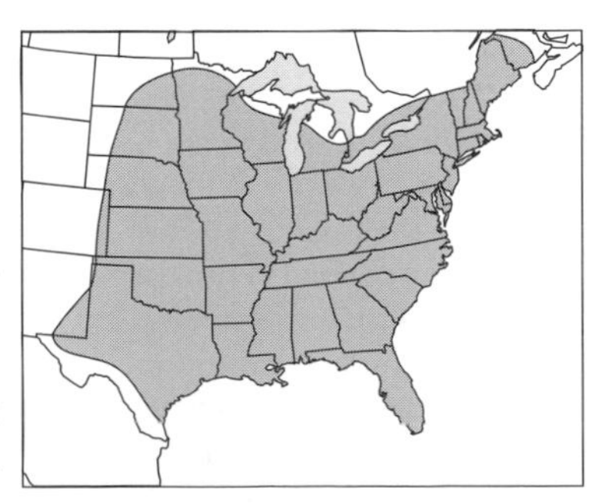

pinetree spurthroated grasshopper

no distinct dark stripe behind the eye. Hind femora are marked with alternating blackish and grayish bands on the outer face. Hind tibiae are reddish or gray in color. Males are 19–31 mm long, females 27–45 mm.

In males, the furcula is barely visible. The cerci are large and markedly expanded beyond the middle, often somewhat boot shaped. The subgenital plate ends with an upward extension.

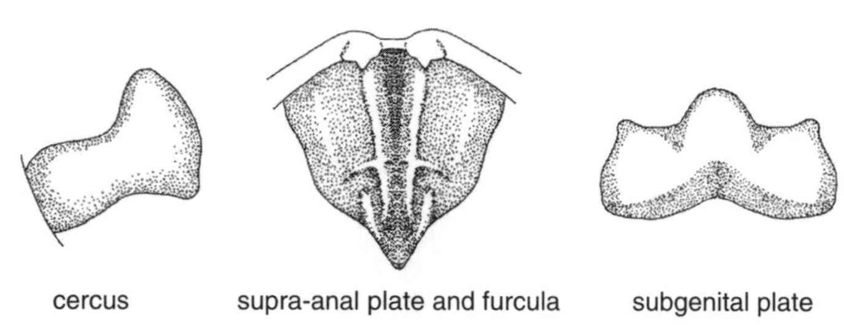

Figure 48. Diagnostic features of male pinetree spurthroated grasshopper

Ecology: This poorly known species is associated with pine and tamarack trees, though it is also reported from broadleaf trees, and it may be primarily nocturnal. It is most often observed resting on the trunks of trees, where it blends in well with mottled bark and the moss and lichens growing there. The female reportedly deposits her eggs within holes or crevices of dead tree trunks, a relatively uncommon habit among grasshoppers. Adults are found from July to September in Michigan but can be seen at least until January in Florida.

PYGMY SPURTHROATED GRASSHOPPER

Melanoplus pygmaeus (Plate 28, Fig. 49)

Distribution: Found only in western Florida.

Identification: This small, short-winged species is poorly known. It is reddish brown above and yellowish below. The abbreviated forewings are elongate oval and widely separated where they meet. As is the case with most *Melanoplus* species, a dark stripe is found behind the eye, extending onto the pronotum. The stripe is about equal in width from the front to the back margin of the lateral lobe of the pronotum, but it sometimes expands slightly posteriorly. Hind femora are yellowish brown with three dark blotches dorsally. Hind tibiae are purplish blue. Males are 14–20 mm long, females 23–30 mm.

pygmy spurthroated grasshopper

In males, the furcula is not visible. The cerci are constricted near the middle and elbowed, with the tip turned upward. The tip is broadly rounded and flattened, and slightly concave or grooved.

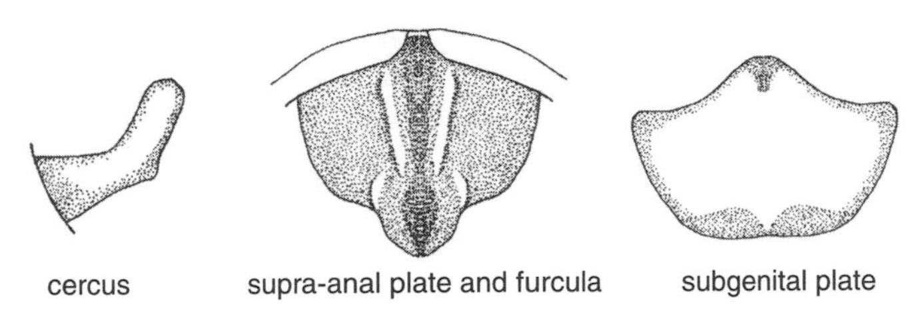

Figure 49. Diagnostic features of male pygmy spurthroated grasshopper

Ecology: Inhabits high pine and scrub habitats.

Similar species: The presence of a recessed or grooved area on the tip of the cerci cause the pygmy spurthroated grasshopper to resemble the round-winged spurthroated grasshopper, *Melanoplus rotundipennis* (Plate 29), a much more abundant and widespread species. The absence of a furcula and a pallium (an erect conical structure near the tip of the abdomen) serve to distinguish the former from the latter.

OAK SPURTHROATED GRASSHOPPER

Melanoplus querneus (Plate 28, Fig. 50)

Distribution: Found only in northern Florida and adjacent areas of Alabama and Georgia.

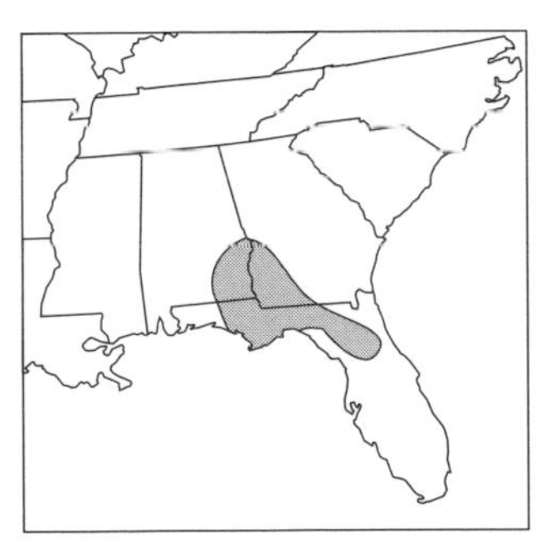

oak spurthroated grasshopper

Identification: This heavy-bodied *Melanoplus* with forewings of intermediate length is flightless. Forewings normally extend two-thirds to three-fourths the length of the abdomen. Body and forewings are brown with yellowish or grayish markings above, yellowish green below. A dark stripe extends from the eye onto the pronotum but sometimes is relatively indistinct. Outer face of the hind femora is marked with two dark bands. Hind tibiae are reddish. Males are 22–29 mm long, females 28–40 mm.

In males, the furcula is greatly reduced or not apparent. The cerci are large and expanded beyond the middle, especially upward.

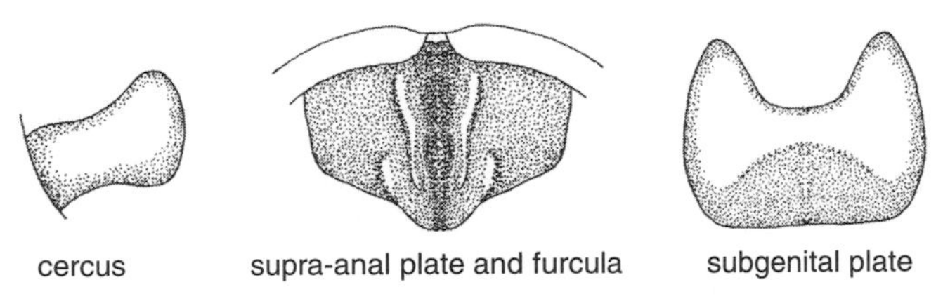

Figure 50. Diagnostic features of male oak spurthroated grasshopper

Ecology: Inhabits the undergrowth and margins of woodlands. It seems to prefer moist areas or taller vegetation.

REGAL SPURTHROATED GRASSHOPPER

Melanoplus regalis (Plate 29, Fig. 51)

Distribution: Found in the southern Great Plains and southwestern states, and in northern Mexico.

Identification: A heavy-bodied, light-brown species, often with contrasting red, green, and white markings. Wings are about as long as the abdomen. Hind femora often bear chevron-like markings. Hind tibiae are blue. Males are 20–26 mm long, females 27–34 mm.

In males, the cerci are flat and taper to a sharp point. The upper edge is concave, the lower edge convex. The furcula is about one-third the length of the supra-anal plate, with the extensions narrow and pointed.

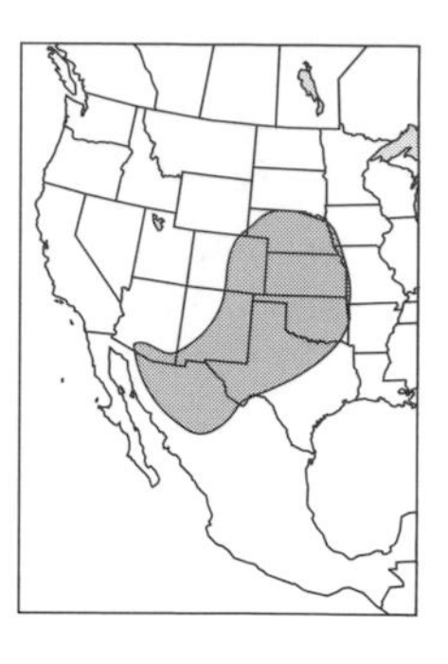

regal spurthroated grasshopper

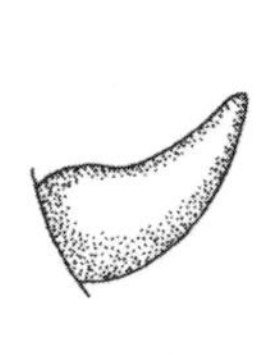

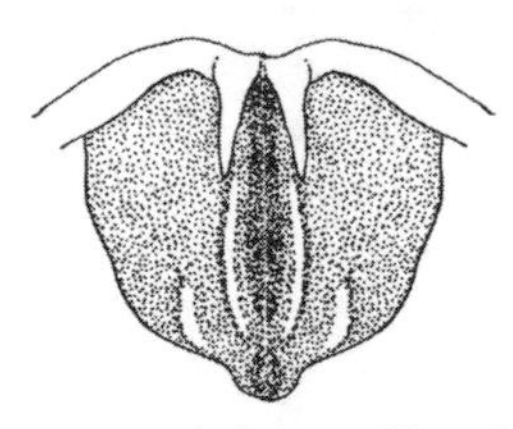

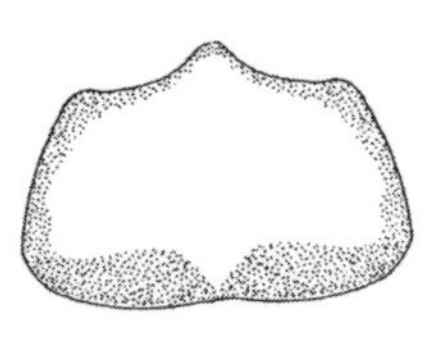

Figure 51. Diagnostic features of male regal spurthroated grasshopper

Ecology: Commonly associated with shortgrass prairie, but also found in association with mixed grass and shrubs.

ROUND-WINGED SPURTHROATED GRASSHOPPER

Melanoplus rotundipennis (Plate 29, Fig. 52)

Distribution: Found only in northern Florida and southern Georgia.

Identification: A small grasshopper, but about average in size for the short-winged species. It is reddish brown above and yellowish below. A dark stripe behind the eye extends over the length of the pronotum and onto the abdomen in males, but only to about the middle of the pronotum in females. Hind femora are yellowish or brownish, with two bars often evident. Forewings are not really round, despite the common name, but they are only slightly elongate oval. Forewings are widely separated where they meet. Hind tibiae are bluish. Males are 13.5–19 mm long, females 17–25 mm.

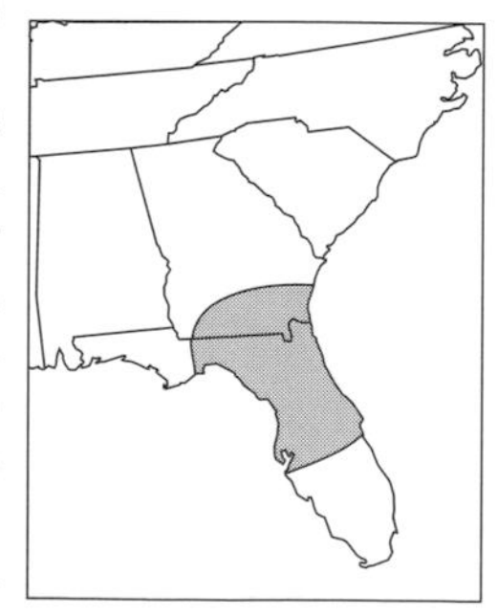

round-winged spurthroated grasshopper

In males, the cerci are constricted near the middle, and slightly widened and flattened or shallowly grooved at the tip. The furcula is very short, consisting of rounded lobes. The most striking feature of this species is the male's enlarged pallium, an erect conical structure near the tip of the abdomen. This structure protrudes dorsally and is an important character for distinguishing this species from most other short-winged *Melanoplus.*

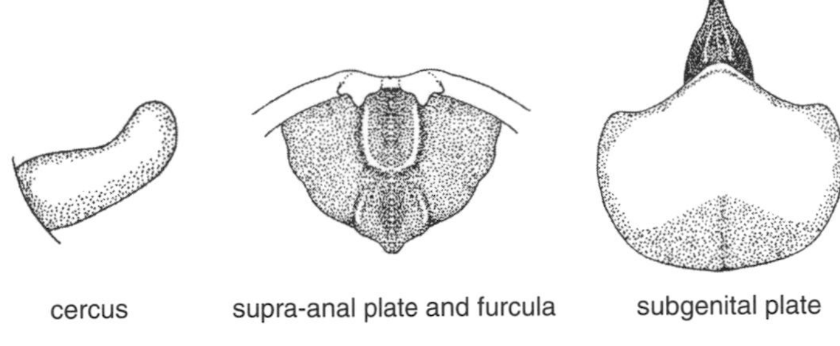

Figure 52. Diagnostic features of male round-winged spurthroated grasshopper

Ecology: Common in a number of southeastern xeric and mesic habitats, including deciduous and pine woodlands and scrub oak. It is particularly common along edges of woods and sometimes ventures out into old fields. In Florida, eggs commence hatching in April or May, and adults are found from May to December.

Similar species: The presence of a pallium, distinguishes this species from most other short-winged *Melanoplus* species. The Withlacoochee grasshopper, *M. withlacoocheensis*, also possesses the enlarged pallium, but it is easily distinguished because the tips of the cerci have a small ventral tooth and are swollen, appearing bulbous when viewed from above. The Withlacoochee grasshopper occurs only in a small area on the west coast of peninsular Florida.

Withlacoochee grasshopper

MIGRATORY GRASSHOPPER

Melanoplus sanguinipes (Plate 29, Fig. 53)

Distribution: Found throughout the United States and southern Canada, though it is more abundant in western states.

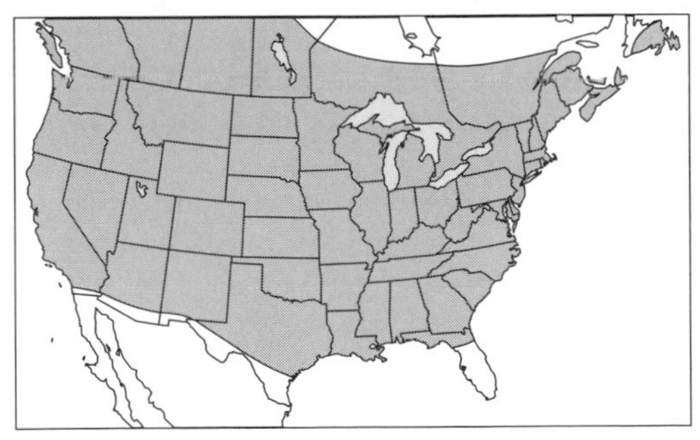

migratory grasshopper

Identification: A grayish-brown species with a black stripe that usually extends from the eye onto the lateral lobe of the pronotum. Forewings are long, brownish, and bear a row of dark-brown spots centrally. They extend at least to the tip of the abdomen and sometimes considerably beyond. Hind femora usually have two oblique dark bands. Hind tibiae are normally red but sometimes blue. Males are 19–24 mm long, females 18–32 mm.

In males, the furcula is slender with diverging extensions, and measures about one-fourth to one-third the length of the supra-anal plate. The cerci are compact, about twice as long as broad, extending slightly upward and rounded at the tip. The tip of the subgenital plate is extended; when viewed from above it is clearly notched in the middle.

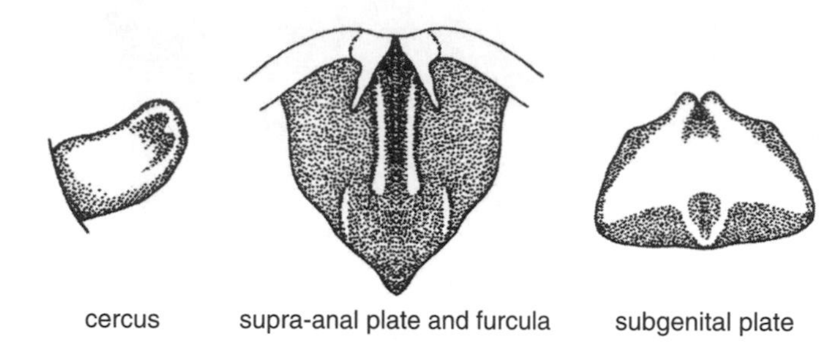

Figure 53. Diagnostic features of male migratory grasshopper

Ecology: Favored habitats are weedy pastures, crops, and similar disturbed areas where annual weeds are abundant, though it can also be found in open pastures and rangelands. Feeds both on grasses and on broadleaf plants. Among grasshoppers, it is one of the most significant crop pests. In western states, it sometimes attains very high and damaging densities. At high densities, a behavioral change occurs wherein the grasshoppers become gregarious, moving as a group. During such times the grasshoppers may disperse long distances; thus the species' common name, "migratory grasshopper." Eggs commence hatching relatively early, in midspring. Adults are found from August to October in most areas. In warm-weather areas, including regions as far north as Kansas, two or three generations may occur.

SCUDDER'S SHORT-WINGED GRASSHOPPER

Melanoplus scudderi (Plate 29, Fig. 54)

Distribution: Widely distributed in the eastern United States, west to Nebraska and Texas. In many areas, it is the most abundant of the short-winged species.

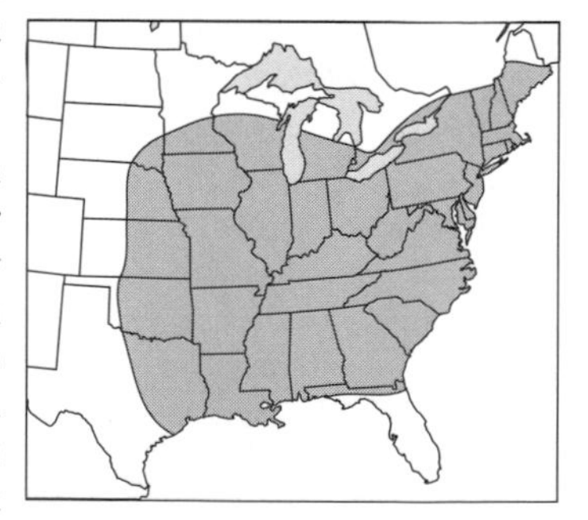

Scudder's short-winged grasshopper

Identification: A short-winged species of medium size, brownish or reddish brown. The oval or elongate-oval forewings overlap, or are only slightly separated above. Forewings vary in length, from shorter than the pronotum to slightly longer. The dark stripe normally found behind the eye in *Melanoplus* species may be present or weak in both sexes. Hind femora lack bands on the outer face, but two dark spots may be present on the upper edge. Hind tibiae are red. Males are 14–20 mm long, females 22–26 mm.

In males, the furcula consists of minute triangular structures that sometimes are not apparent. The cerci taper from the base to a broadly rounded point, usually curving upward. The tip of each cercus is slightly concave or grooved.

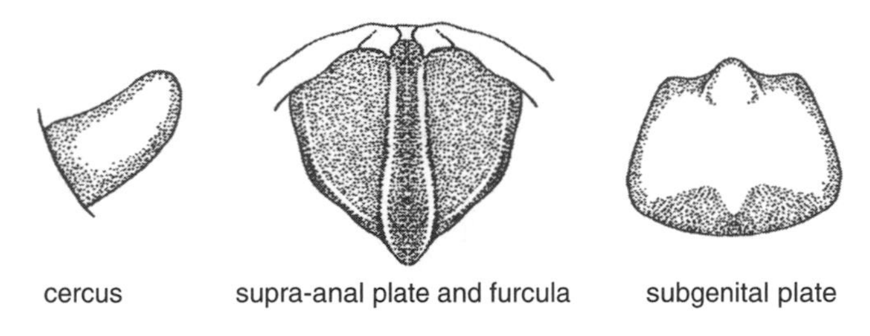

Figure 54. Diagnostic features of male Scudder's short-winged grasshopper

Ecology: Inhabits open woods, brushy areas, rocky slopes, and edges of wooded areas. Feeds principally on broadleaf plants. Over most of its range it is a late-season species.

VALLEY GRASSHOPPER

Oedaleonotus enigma (Plate 29)

Distribution: Most abundant in California, but occurs east to Utah and Arizona, and north to Oregon and Washington.

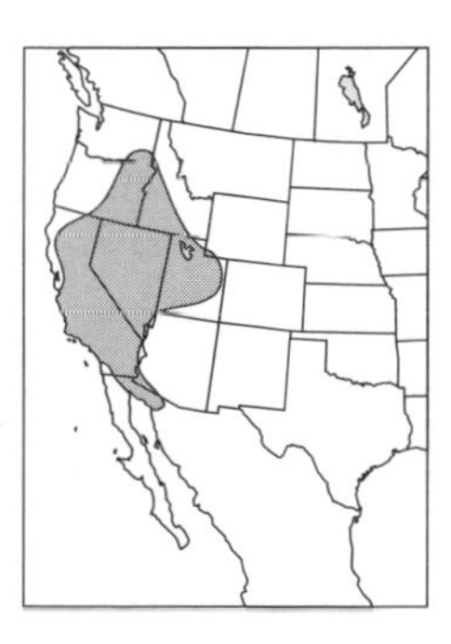

valley grasshopper

Identification: A small but robust species. It is yellow, but orange or black markings may be present on the pronotum, forewings, and femora. The front edge of the pronotum has a white margin, and the lateral lobe of the pronotum may bear a black stripe, but this latter character is variable. Forewing length is variable, often measuring only about the length of the pronotum but sometimes exceeding the tips of the hind femora. Only long-winged individuals can fly. Hind femora are yellow, sometimes with three black transverse markings. Hind tibiae are blue. Males are 20–23 mm long, females 22–25 mm.

Male cerci are broad basally and do not taper much until the tip, which quickly reaches a rounded point.

Ecology: Inhabits sagebrush grasslands and other arid habitats, and feeds both on grasses and on broadleaf plants, including shrubs. Eggs hatch in April to May, and adults are found as early as late May. At high densities, nymphs form bands and disperse in groups. Long-winged adults disperse by flight from high-density populations.

Similar species: This genus contains other species of uncertain validity. They are largely restricted to California. *Oedaleonotus* resembles *Melanoplus* in body form but can be distinguished by its "hump-backed" or thick-bodied appearance: the front portion of the pronotum is slightly elevated.

ATLANTIC GRASSHOPPER

Paroxya atlantica (Plate 30)

Distribution: Found in most of the eastern United States, west to about the Mississippi River and north to Pennsylvania.

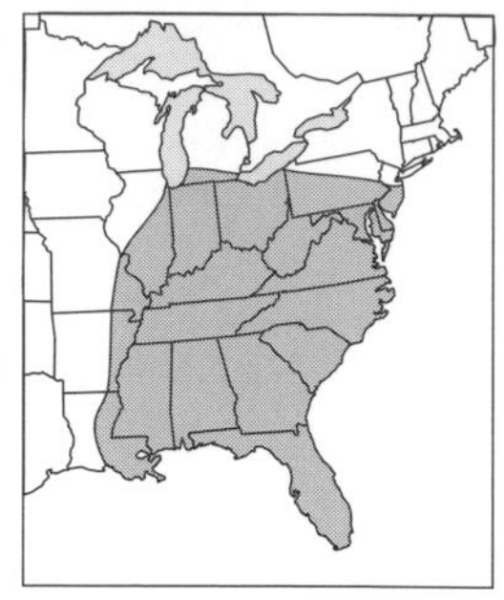
Atlantic grasshopper

Identification: Usually yellowish brown, but sometimes tends toward maroon or red. Face is moderately slanted. A black stripe behind the eye fades near the middle of the pronotum. Antennae and pronotum are unusually long. In males, the antennae are about twice the length of the pronotum, and the dorsal surface of the pronotum is elongate—about twice as long as wide. Hind tibiae are bright blue or greenish blue. Males are 16–26 mm long, females 22–30 mm.

In males, the cerci are long, slender, constricted at the middle, strongly incurved, and with the tip flattened and broadly rounded. The furcula is short or barely visible.

Ecology: Inhabits wet areas and is particularly common on the vegetation around ponds and swamps, and in coastal salt marshes. In the north, adults are found from July to October. In northern Florida, nymphs are found in the winter, spring, and summer months, and adults can be found from April to December. In southern Florida, adults are found year-round.

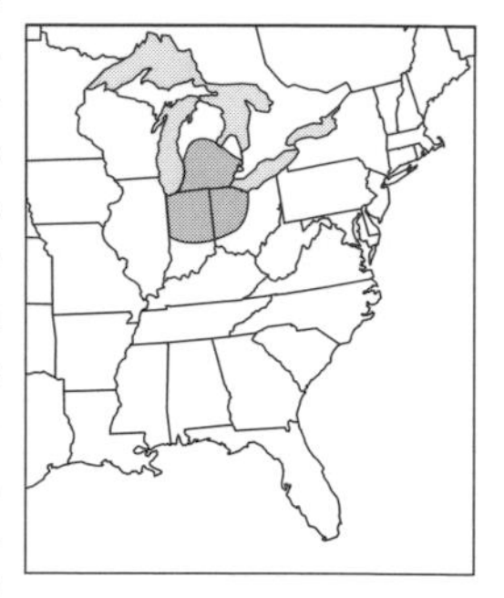
Indiana swamp grasshopper

Similar species: The length of the antennae and pronotum serves to distinguish this species from *Melanoplus* species. The moderate size of the Atlantic grasshopper distinguishes it from the similar, but larger, olive-green swamp grasshopper, *Paroxya clavuliger* (Plate 30). Also, the antennae of the male Atlantic grasshopper are shorter than the hind femora, whereas in the male olive-green swamp grasshopper the antennae are longer than the hind femora. The Indiana swamp grasshopper, *Paroxya hoosieri*, is a related species found only in the Midwest, and is distinguished from the Atlantic and the olive-green swamp grasshoppers by the relatively short wings. The forewings extend only about two-thirds the length of the abdomen in the Indiana swamp grasshopper but reach the tip of the abdomen or extend beyond the tip in the other *Paroxya* species.

OLIVE-GREEN SWAMP GRASSHOPPER

Paroxya clavuliger (Plate 30)

olive-green swamp grasshopper

Distribution: Widespread in the southeastern United States, occurring as far north as Massachusetts and as far west as eastern Texas. Unlike the similar Atlantic grasshopper, it is not abundant in the midwestern states.

Identification: A greenish to greenish black species, closely resembling the Atlantic grasshopper in most respects. A dark stripe extends from the eye onto the pronotum, usually continuing to the hind margin of the pronotum. In males, the upper surface of the pronotum is elongate, about twice as long as wide. Antennae are strikingly long, in males measuring longer than the hind femora. Hind tibiae are bluish green. Males are 20–33 mm long, females 29–46 mm.

In males, the cerci are long, slender, constricted at the middle, strongly incurved, and with the tip flattened and broadly rounded. The furcula is evident, measuring one-fourth to one-third the length of the supra-anal plate.

Ecology: Inhabits wet areas and is normally associated with the edges of ponds, freshwater marshes, and coastal salt marshes. In the north, adults are found from July to October. In northern Florida, nymphs are found in the spring and summer, and adults can be found from April to December. In southern Florida, adults are found year-round, though they are not usually numerous.

Similar species: The length of the antennae and pronotum serve to distinguish this species from *Melanoplus* species. The larger size of the olive-green swamp grasshopper distinguishes it from the similar, but smaller, Atlantic grasshopper, *Paroxya atlantica* (Plate 30). The antennae of the male Atlantic grasshopper are shorter than the hind femora, whereas in the male olive-green swamp grasshopper the antennae are longer than the hind femora. The length of the pronotal stripe, which extends to the hind margin of the pronotum, also helps to separate the olive-green swamp grasshopper from the Altantic grasshopper; in this latter species the stripe fades before reaching the hind margin. The Indiana swamp grasshopper, *P. hoosieri*, is a related species found only in the Midwest, and is distinguished from the other *Paroxya* species by the relatively short wings. The forewings extend only about two-thirds the length of the abdomen in the Indiana swamp grasshopper but reach the tip of the abdomen or extend beyond the tip in the other *Paroxya* species.

LARGE-HEADED GRASSHOPPER

Phoetaliotes nebrascensis (Plate 30, Fig. 55)

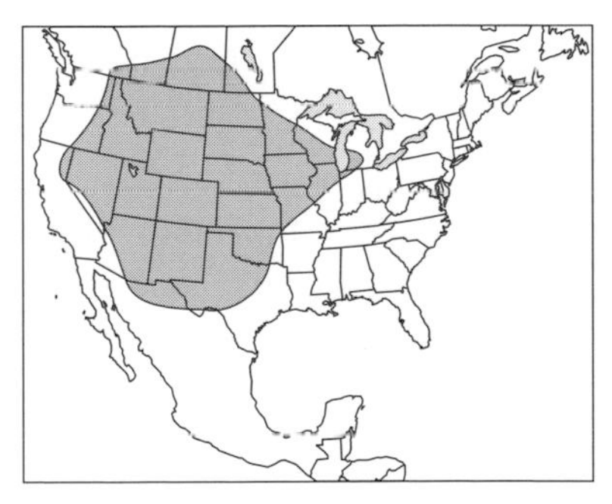

large-headed grasshopper

Distribution: Found widely in the western states, from Indiana to California and from southern Canada to northern Mexico.

Identification: Olive green with brown markings, the abdomen lighter. As suggested by the name, the head is exceptionally large as compared with the thorax. Viewed from above, the prothorax looks unusually slender relative to the head. Forewings usually are pointed and short, only about the length of the pronotum, but occasional specimens have long wings. A black stripe extends from the back of the eye onto the lateral lobe of the pronotum. Hind femora are reddish brown but yellowish below. Hind tibiae are blue, greenish, or purplish. Males are 21–25 mm long, females 23–33 mm.

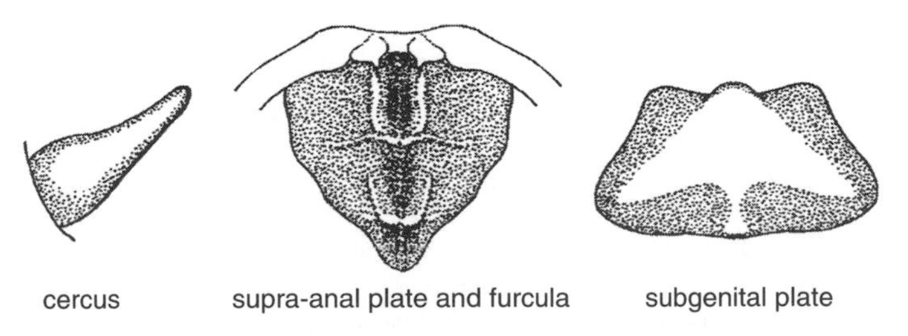

Figure 55. Diagnostic features of male large-headed grasshopper

In males, the cerci are broad basally, tapering to a rounded point. The upper edge of the cercus is slightly concave, the lower surface straight. The furcula is barely visible and the lobes widely separated.

Ecology: This species is found in numerous habitats, sometimes accounting for a substantial portion of the grasshopper assemblage in an area, especially in tallgrass prairie. Feeds both on grasses and on broadleaf plants but prefers the former. Eggs hatch relatively late, in May and June. Adults are found from July or August to October.

Similar species: This species greatly resembles *Melanoplus* species but can be distinguished by the enlarged head.

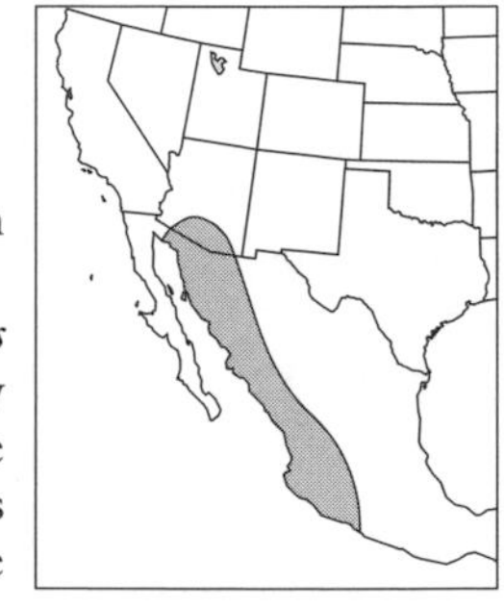

panther-spotted grasshopper

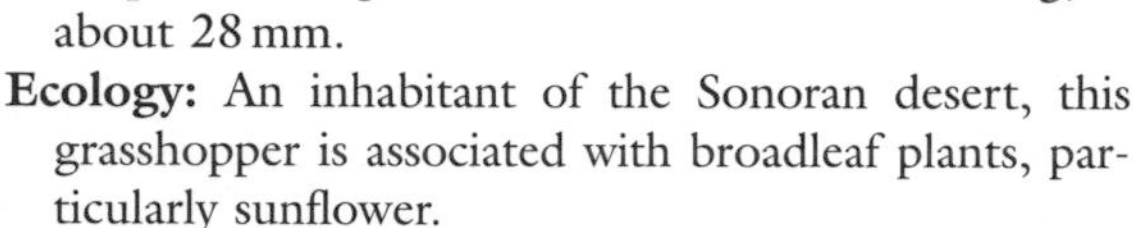

PANTHER-SPOTTED GRASSHOPPER

Poecilotettix pantherinus (Plate 30)

Distribution: Found in southern Arizona and northwestern Mexico.

Identification: Resembles a typical long-winged *Melanoplus* or *Hesperotettix* in body form but is quite distinctly marked. The body is yellowish and the forewings are greenish yellow. The head, antennae, pronotum, and legs bear rows of alternating black and white spots. Hind tibiae are pale bluish green. Males are about 25 mm long, females about 28 mm.

Ecology: An inhabitant of the Sonoran desert, this grasshopper is associated with broadleaf plants, particularly sunflower.

Similar species: This distinctively marked species is not easily confused with other grasshoppers. However, another brightly colored *Poecilotettix*, *P. sanguineus*, occurs in some of the same southwestern desert areas. *P. sanguineus* is known as the red-lined grasshopper because it bears a red line on the top of the head and pronotum, and also on the side of the pronotum.

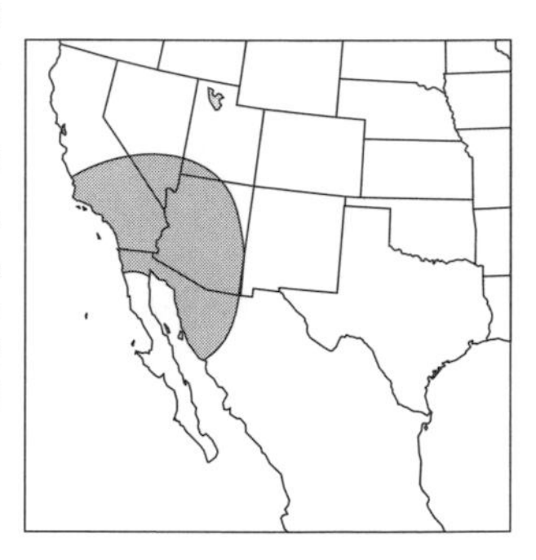

red-lined grasshopper

LEATHER-COLORED BIRD GRASSHOPPER

Schistocerca alutacea (Plate 31, Figs. 56, 57)

Distribution: Found throughout the eastern United States except in the northernmost regions.

Identification: Highly variable in appearance. Body is often olive green or golden brown but may also be reddish brown. A dorsal yellowish stripe usually runs the length of the body. Occasionally, the stripe is lacking or indistinct brownish spots mark the forewings. Hind tibiae are brownish, with yellow spines bearing dark tips. The male supra-anal plate bears a pair of small tubercles near the center. Males are 30–35 mm long, females 43–55 mm.

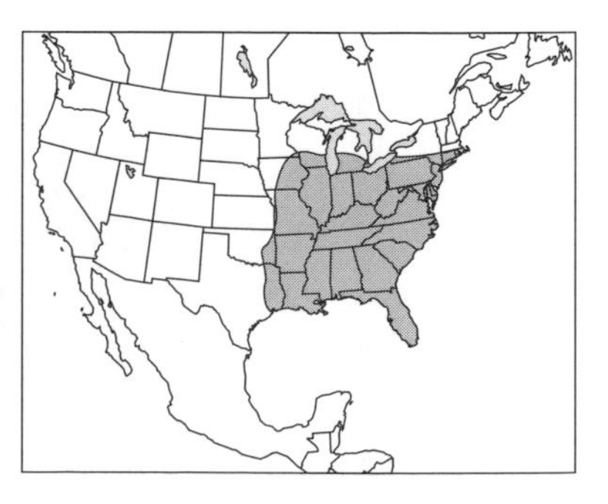

leather-colored bird grasshopper

Ecology: Commonly inhabits open woods. It may occur in pastures and the margin of wooded areas. Moist areas are favored. Adults are found from July to October.

Similar species: The *Schistocerca* species have confused orthopterists because of their variability and lack of distinguishing structural characters. Recent studies by H. Song of Ohio State University (unpublished) have helped to clarify the species relationships, and the results are included herein.

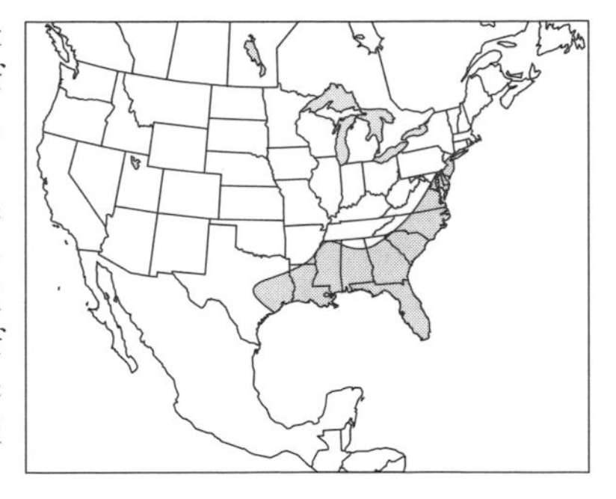

rusty bird grasshopper

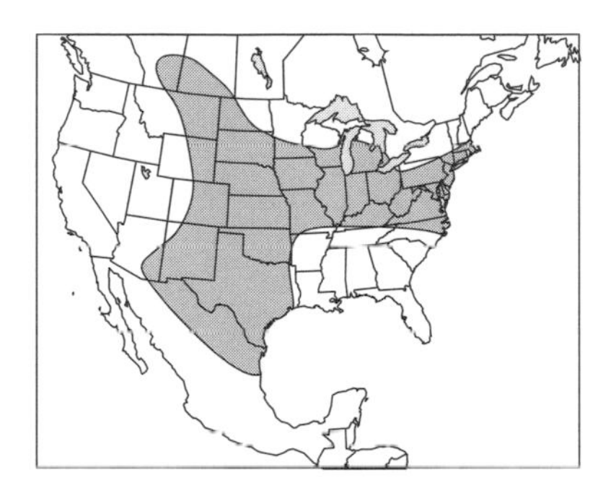

spotted bird grasshopper

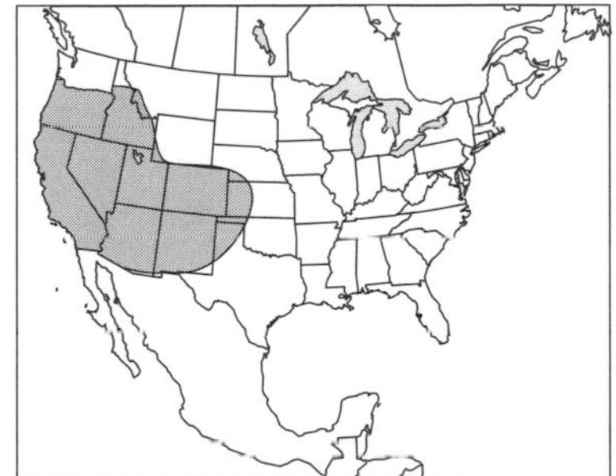

green bird grasshopper

The olive-green forms are easily confused with the obscure bird grasshopper, *Schistocerca obscura* (Plate 31), and related species. The best approach to distinguish between them is to examine the tip of the male abdomen. In the obscure bird grasshopper, the notch of the male's subgenital plate is V-shaped, whereas in the leather-colored bird grasshopper it is U-shaped (see Fig. 56). The obscure bird grasshopper also tends to be a larger grasshopper, and the base color of the hind tibiae is purplish or blackish, rather than the brown color of the leather-colored bird grasshopper.

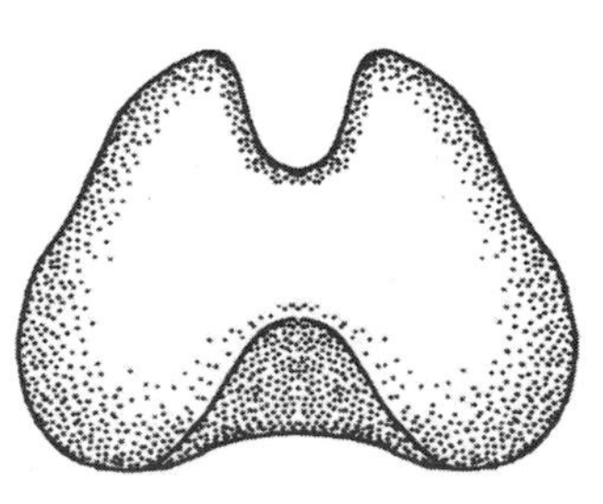

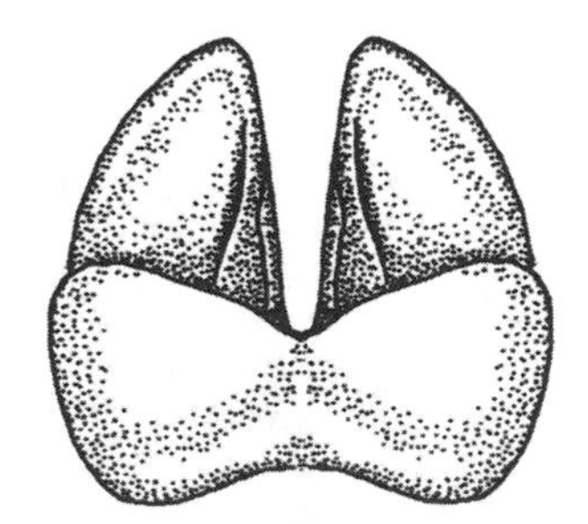

Figure 56. The male subgenital plate of two similar *Schistocerca* species, the leather-colored bird grasshopper and the obscure bird grasshopper, showing a U-shaped (left) and V-shaped (right) notch.

In the southeastern states, particularly the coastal plain, the similar rusty bird grasshopper, *Schistocerca rubiginosa*, is found. It has usually been treated as a subspecies of *S. alutacea* but likely warrants species-level status. It more commonly frequents dry habitats and normally lacks the dorsal stripe. Also, whereas the tip of the male cercus in the rusty bird grasshopper is blunt or only slightly indented, in the leather-colored it usually is bilobed and the ventral lobe is considerably longer than the dorsal lobe (see Fig. 57).

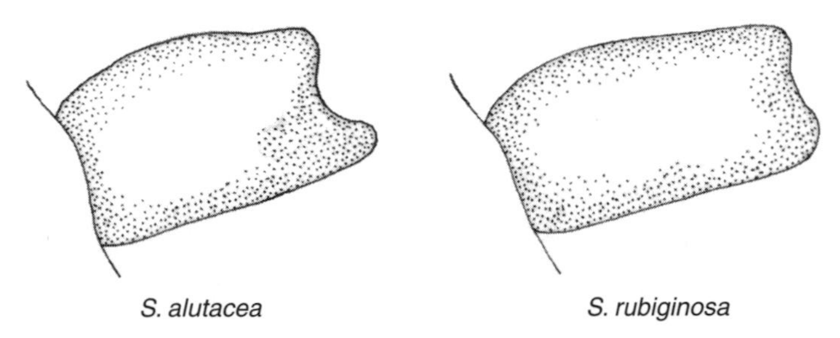

Figure 57. Cerci of two similar *Schistocerca* species, the leather-colored bird grasshopper and the rusty bird grasshopper, showing a strongly bilobed condition (left) and weakly bilobed condition (right).

Other similar (and variable) species are the spotted bird grasshopper, *Schistocerca lineata* (also known as *S. emarginata*), and the green bird grasshopper, *S. shoshone*. They are distinguished from the rusty and the leather-colored bird grasshoppers by their inflated front and middle femora (about twice as thick as the tibiae; in the spotted and the green bird grasshoppers the front and middle femura are not much thicker than the tibiae). The spotted bird grasshoppers occurs widely in the United States and extends to the southern portion of Canada's Prairie Provinces. It is perhaps the most confusing of the *Schistocerca* species because it can resemble most other species. In western states, from California to Texas, the similar-appearing green bird grasshopper occurs, but it tends to be a brighter, grass-green color, and its hind tibiae tend to be red, pink, or orange; those of the spotted bird grasshopper are brown or black.

AMERICAN BIRD GRASSHOPPER

Schistocerca americana (Plate 31)

Distribution: Found throughout the eastern United States west to the Rocky Mountains.

Identification: A large, strong-flying species, normally brownish or yellowish brown, with lighter and darker areas. Immediately after molting to the adult stage it has a pinkish or reddish color, but after a week or so the typical brown or yellow-brown color is acquired. A dorsal creamy-white stripe normally extends along the front of the head to the tip of the forewings. Forewings extend well beyond the tip of the abdomen and bear large, dark-brown spots. Hind tibiae are reddish brown. Males are 39–52 mm long, females 48–68 mm.

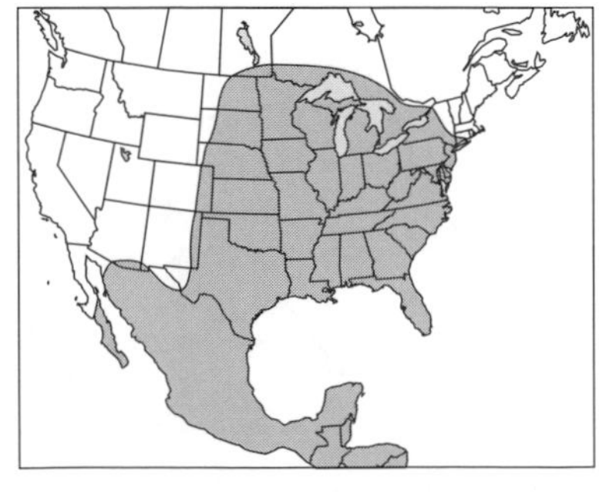

American bird grasshopper

Ecology: When weather and food conditions allow this species to become abundant, behavioral changes occur. Nymphs and adults become gregarious, moving in unison and dispersing in swarms. These swarms can be very damaging to crops. Inhabits open fields, and open oak and pine woodlands. Feeds on a wide variety of grasses, broadleaf plants, shrubs, and trees. When disturbed, it often flies into trees, or a considerable distance from the source of disturbance, a characteristic of most bird grasshoppers (genus *Schistocerca*). Two generations annually occur in southern states, and even on occasion in the cooler midwestern states. In the South, one generation occurs in April to June and another begins in August or September. Adults are long-lived and eggs do not hatch synchronously, however, so the generations may be indistinct. The adults overwinter and are active in the winter whenever it is warm and sunny.

ROSEMARY BIRD GRASSHOPPER

Schistocerca ceratiola (Plate 31)

Distribution: Found only in central and northcentral Florida.

rosemary bird grasshopper

Identification: This species is quite slim in general appearance. It is mottled gray and brown, with green on the abdomen. A faint pale stripe occurs on the upper surface of the head and pronotum, extending weakly along the forewings. The underside is markedly paler. Hind tibiae are red or brown. Males are 28–32 mm long, females 36–40 mm.

Ecology: Despite its moderately large size, this grasshopper is not often noticed. It escapes detection because of its restricted range—it occurs only on Florida's sandy ridges; its restricted diet—it feeds only on rosemary (*Ceratiola ericoides*); and its restricted period of activity—it is active only at night. It hides deep within rosemary bushes, where it is effectively camouflaged during the daylight hours, moving to the surface of the bushes at night or during periods of heavy cloud cover. The adults resemble the stems of rosemary, and the green and yellow nymphs blend perfectly with the foliage. The degree of host specificity displayed by the rosemary bird grasshopper, wherein a single plant species serves as the host plant, is extremely unusual in grasshoppers. Adults are found in late summer to autumn.

MISCHIEVOUS BIRD GRASSHOPPER

Schistocerca damnifica (Plate 31)

Distribution: Found throughout the eastern United States, east to the Great Plains, except for New England and the Great Lakes region.

mischievous bird grasshopper

Identification: Reddish brown in color, with an elevated medial ridge on the pronotum, and sometimes with a narrow pale or yellow line along the head and pronotum. Lacks the pronounced yellowish dorsal line commonly appearing along the head, pronotum, and forewings of some of the other *Schistocerca* species. Forewings extend beyond the tip of the abdomen but to a lesser degree than in most other *Schistocerca* species. Hind tibiae are brown or reddish brown. A relatively small member of the genus. Males are 25–35 mm long, females 37–47 mm.

Ecology: Inhabits old fields and open woodlands; in the latter environment it can be quite common at times. Adults are found from June to September in the North, but throughout the winter in Florida.

Similar species: The elevated ridge on the pronotum, lack of a wide dorsal stripe, uniform brown color, and small size serve to distinguish the mischievous bird grasshopper from other *Schistocerca* species.

GRAY BIRD GRASSHOPPER

Schistocerca nitens (Plate 31)

Distribution: Found in the southwestern United States, and south through Mexico to northern South America.

gray bird grasshopper

Identification: Quite variable both in color and size. Neither particularly heavy-bodied nor especially slender, but somewhat intermediate in bulk. Generally gray and mottled with small, irregular dark spots, but sometimes is brownish or greenish, or uniform in color. Except in uniformly colored individuals, a light dorsal stripe extends from the top of the head across the thorax. Hind tibiae are brown or blue. Males are 24–46 mm long, females 40–66 mm.

Ecology: Often associated with mesquite and weeds; these grasshoppers sometimes disperse to irrigated crops and suburban yards, where they damage plants. Though not often abundant, these large insects attract notice.

Similar species: The gray body color, large size, and limited geographic distribution serve to distinguish typical specimens from related common species, though, as in all *Schistocerca* species, atypical individuals can be difficult to identify.

OBSCURE BIRD GRASSHOPPER

Schistocerca obscura (Plate 31, see Fig. 56)

Distribution: Found throughout the eastern United States, west to the Rocky Mountains, but absent from the northernmost states.

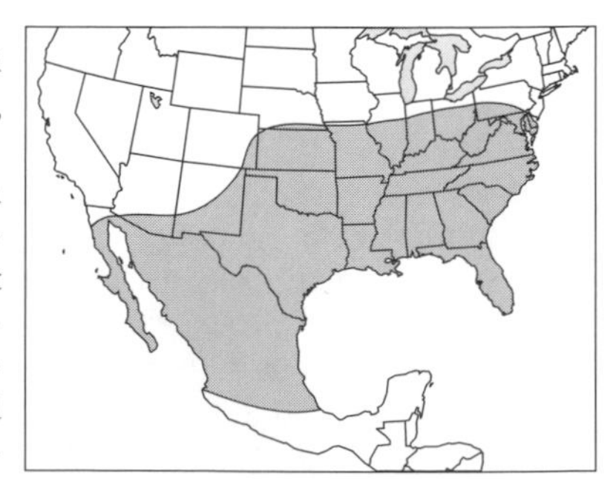

obscure bird grasshopper

Identification: A large, green species, with olive-green forewings and usually a pale-yellow dorsal stripe extending from the front of the head to the tip of the forewings. Occasionally females lack the stripe or bear indistinct brownish spots on the forewings. Hind tibiae are blackish purple with yellow, black-tipped spines. Males are 36–45 mm long, females 50–65 mm.

In males, the supra-anal plate lacks a pair of small tubercles near the center. The tips of the subgenital plate are flared outward.

The cerci are highly bilobed, with the lower lobe protruding more than the upper lobe (similar to those of *S. alutacea*; see Fig. 57).

Ecology: Preferred habitats are fields and open woodlands. Sometimes feeds on flowers and shrubs of economic value, so it may become a pest on occasion. Adults are found only during the summer.

Similar species: The obscure bird grasshopper can be confused with the leather-colored bird grasshopper, *Schistocerca alutacea* (Plate 31), but it usually is considerably larger than this latter species. To distinguish between the two species, examine the tip of the male abdomen. In the obscure bird grasshopper, the notch of the male's subgenital plate is V-shaped, whereas in the leather-colored bird grasshopper, it is U-shaped (see Fig. 56).

white-lined bird grasshopper

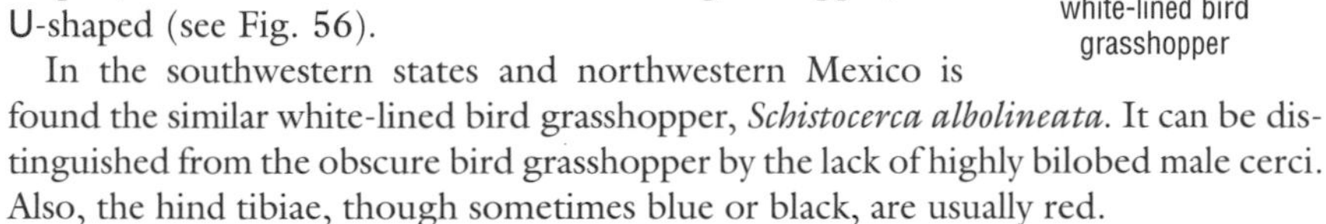

In the southwestern states and northwestern Mexico is found the similar white-lined bird grasshopper, *Schistocerca albolineata*. It can be distinguished from the obscure bird grasshopper by the lack of highly bilobed male cerci. Also, the hind tibiae, though sometimes blue or black, are usually red.

GLASSY-WINGED TOOTHPICK GRASSHOPPER

Stenacris vitreipennis (Plate 30)

Distribution: Found only in the coastal plain of the southeastern states, from North Carolina to Mississippi.

Identification: A very slender, elongate grasshopper with a distinctly pointed head and flattened, sword-shaped antennae. It superficially resembles slantfaced grasshoppers but can be distinguished by the presence of a spine beneath the head, in front of the forelegs. Green to brownish green, usually with a dark or pale lateral line extending from the eye to the base of the front legs. Head is shorter than the pronotum. Males are 24–30 mm long, females 27–40 mm.

glassy-winged toothpick grasshopper

Ecology: Inhabits semiaquatic vegetation, such as cattails and pickerelweed. This species flies readily if disturbed, tending to alight on emergent vegetation where it dodges to the side of the plant opposite the source of disturbance. Thus, it is similar in habits to the cattail toothpick grasshopper, *Leptysma marginicollis*. In Florida, adults are found from March to November.

Similar species: The glassy-winged toothpick grasshopper is easily confused with the cattail toothpick grasshopper, *Leptysma marginicollis* (Plate 30), but in the latter the head is as long as, or longer than, the pronotum, whereas in the former the head is shorter than the pronotum. The antennal segments, although flattened, are not nearly as wide as in the cattail toothpick grasshopper. Other thin-bodied grasshoppers, such as the long-headed toothpick grasshopper, *Achurum carinatum* (Plate 1), and the Wyoming toothpick grasshopper, *Paropomala wyomingensis* (Plate 3), are similar in appearance, but lack a spine between the front legs and have shorter wings. The stubby toothpick grasshopper, *Achurum minimipenne*, and Sumichrast's toothpick grasshopper, *Achurum sumichrasti*, bear long wings and resemble the glassy-winged toothpick grasshopper but lack the spine and occur in western states.

Lubber Grasshoppers

Family Acrididae, subfamily Romaleinae

Some entomologists consider this group to be a family, Romaleidae. A distinctive feature of lubber grasshoppers is the presence of a spine on both the inner and outer surface at the tip of the hind tibiae. Some other grasshoppers have a moveable spur, which resembles a spine, but only the lubbers have immovable spines at this position. Lubber grasshoppers also bear a spine ventrally, between the front legs, as do the spurthroated grasshoppers (subfamily Cyrtacanthacridinae). Lubber grasshoppers are large, robust, sometimes colorful, and usually bear short wings. For most people, it is this set of characters that serve to set these grasshoppers apart from others. Indeed, the terms "lubber" and "lubberly" refer to clumsy, stout individuals, and this is an appropriate term for these large insects. The shape of the head, though variable, is usually broadly rounded. The hind femora are enlarged. When disturbed, lubber grasshoppers may hiss and spread their wings. Males may stridulate. The forewings and hind wings sometimes are brightly colored. Only a few species occur in North America, although many are known from South America.

PLAINS LUBBER GRASSHOPPER

Brachystola magna (Plate 32)

Distribution: Found throughout the Great Plains region, west to Arizona and south into Mexico.

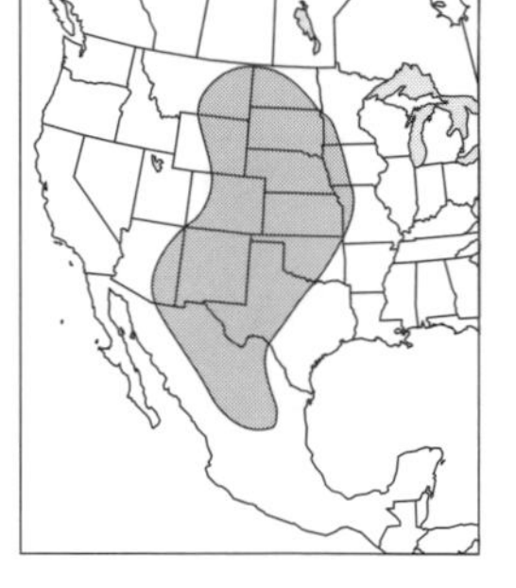

plains lubber grasshopper

Identification: A large, heavy-bodied, flightless species. There are two basic color forms: one is reddish brown and marked with green; the other is predominantly green marked with brown. Pronotum is rough and often edged with white. Short, oval forewings are pinkish with black spots. Hind femora are robust. Hind tibiae bear large and numerous spines. Males are 40 mm long, females 50–60 mm.

Ecology: This species is most abundant in areas with poor soil, or in weedy areas adjacent to fields and roadways. Feeds on weedy broadleaf plants, such as sunflower and ragweed, but also accepts grasses. Sometimes it moves to crops such as cotton and bean, where it causes injury. Though not predatory, it feeds readily on insect cadavers. Eggs usually hatch in May to June, and they are reported to require two winters of incubation before hatching. Adults are found from June or July to November.

GRAY DRAGON LUBBER GRASSHOPPER

Dracotettix monstrosus (Plate 32)

Distribution: Found only in California and Baja California, Mexico.

gray dragon lubber grasshopper

Identification: An unusual species, distinguished by the greatly elevated but deeply cut crest on the pronotum and by the protuberance on the top of the head. It is fairly large, heavy bodied, and short winged. Colors are variable—

normally brown or gray but often marked with white on the pronotum. Hind tibiae are black at the base and reddish at the tip. Males are 18–26 mm long, females 31–47 mm.

Ecology: Found at moderate to high altitudes in dry, but not desert, regions. It typically inhabits sandy and gravelly soils, often in conjunction with sage, but also in open forests. The body color blends in well with the rocky soil where this species is found.

Similar species: The dusky dragon lubber, *Dracotettix plutonius*, is a related but less common species that is found in California and Nevada. Although similar in appearance, the dusky dragon lubber bears a less-elevated crest on the pronotum, and the protuberance on the top of the head is less pronounced.

dusky dragon lubber grasshopper

ROBUST TOAD LUBBER GRASSHOPPER

Phrynotettix robustus (Plate 32)

Distribution: Distribution is quite limited in the United States, where it is found only in southeastern New Mexico and western Texas. It also occurs through much of central Mexico.

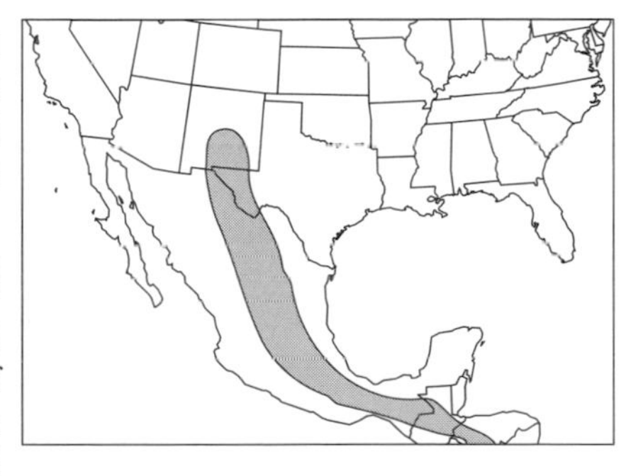

robust toad lubber grasshopper

Identification: A large, heavy-bodied, flightless species. The bumpy "toad-like" texture on much of the body surface and the heavy body account for the name of this unusual grasshopper. Color pattern is variegated white, creamy white, and various tones of brown. Pronotum is extended and relatively flat. Forewings are intermediate in length, and do not attain the tip of the abdomen. Hind femora are whitish, often tinged with yellow or pink. Males are 29–35 mm long, females 38–56 mm.

Ecology: Inhabits pebbly ground in arid areas. It blends in well with gravelly soil and is hard to detect.

EASTERN LUBBER GRASSHOPPER

Romalea microptera (Plate 32)

Distribution: Found throughout the southeastern states. Locally, it is called the Georgia thumper. In the scientific literature, it also is known as *Romalea guttata*. Despite the confusion concerning the correct name, no one should have trouble recognizing this insect. It is probably the best-known species of grasshopper in the Southeast.

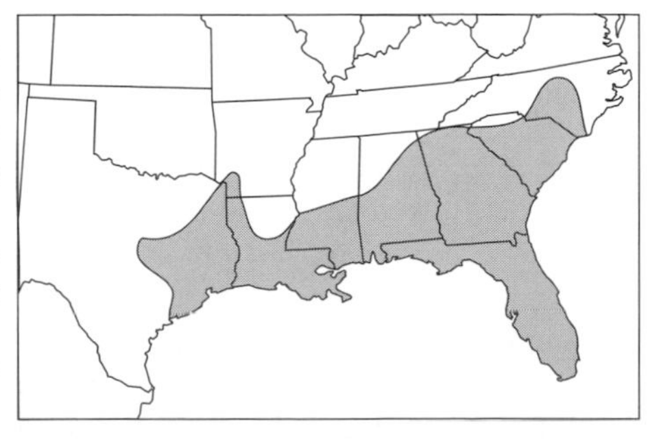

eastern lubber grasshopper

Identification: Adults are large and colorful, but the color pattern varies. They may be mostly black, or mostly yellow with red and black markings and rose or red on the forewings. Intermediate forms also exist. Adults have small wings, measuring no more than two-thirds the length of the abdomen, and are flightless. Males are 43–55 mm long, females 50–70 mm.

The nymphs are mostly black with a narrow yellow or red stripe running from the head to the tip of the abdomen, and with red on the head and front legs. Their color pattern is distinctly different from the adult stage, and so they commonly are mistaken for a different species by the general public.

Ecology: Preferred habitat is low, moist areas of dense undergrowth including wet wooded areas with moderately dense overstory, but as they mature they disperse widely and can be found in nearly all habitats. Young lubbers tend to be gregarious and dispersive. This commonly brings them into contact with people and gardens, accounting for their familiarity. On occasion, they are abundant enough to damage citrus or vegetables. They commonly seek out and defoliate amaryllis and related plants in flower gardens, but they also consume numerous other plants. Both sexes stridulate by rubbing the forewing against the hind wing. When alarmed, the eastern lubber will spread its wings, hiss, and secrete foul-smelling froth from the spiracles. In Florida, nymphs occur from February to May, and adults are most commonly found from June to September but sometimes to November.

HORSE LUBBER GRASSHOPPER

Taeniopoda eques (Plate 32)

Distribution: Found in the southwestern states from Arizona to Texas. It also occurs south through Mexico to Costa Rica.

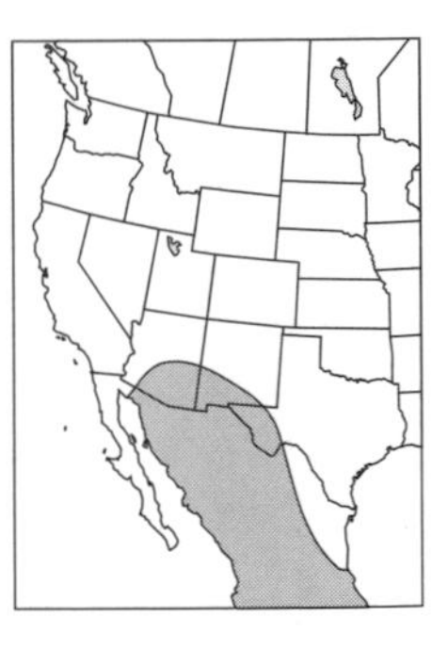
horse lubber grasshopper

Identification: A large, shiny grasshopper that is easily recognizable. Mostly black, but the hind margin of the pronotum and veins of the forewings are yellow. Forewings normally extend beyond the tip of the abdomen but may be shorter. Hind wings are rose red with black borders.

Ecology: Inhabits desert shrub and oak communities, feeding on various shrubs and broadleaf weeds. In the United States, it is often associated with mesquite. Horse lubbers sometimes are sufficiently abundant to cause injury to vegetation. They also feed readily on insect cadavers. They rarely have been observed to fly. When disturbed, they usually drop to the ground, display their wings, and hiss. Males make a clacking sound with their wings.

Pygmy Grasshoppers

Family Tetrigidae

This peculiar group of insects is also known as groundhoppers, reflecting their propensity to remain in association with leaf litter, soil, and leaf mold. Sometimes they are also called grouse grasshoppers, because of their tendency to burst from cover when disturbed. They are distinguished by their small size, 6–16 mm long; their dull coloration, usually brownish gray, gray, black, or mottled but never green; their pronounced eyes; and especially their greatly elongated pronotum, which often extends back to the tip of the abdomen and

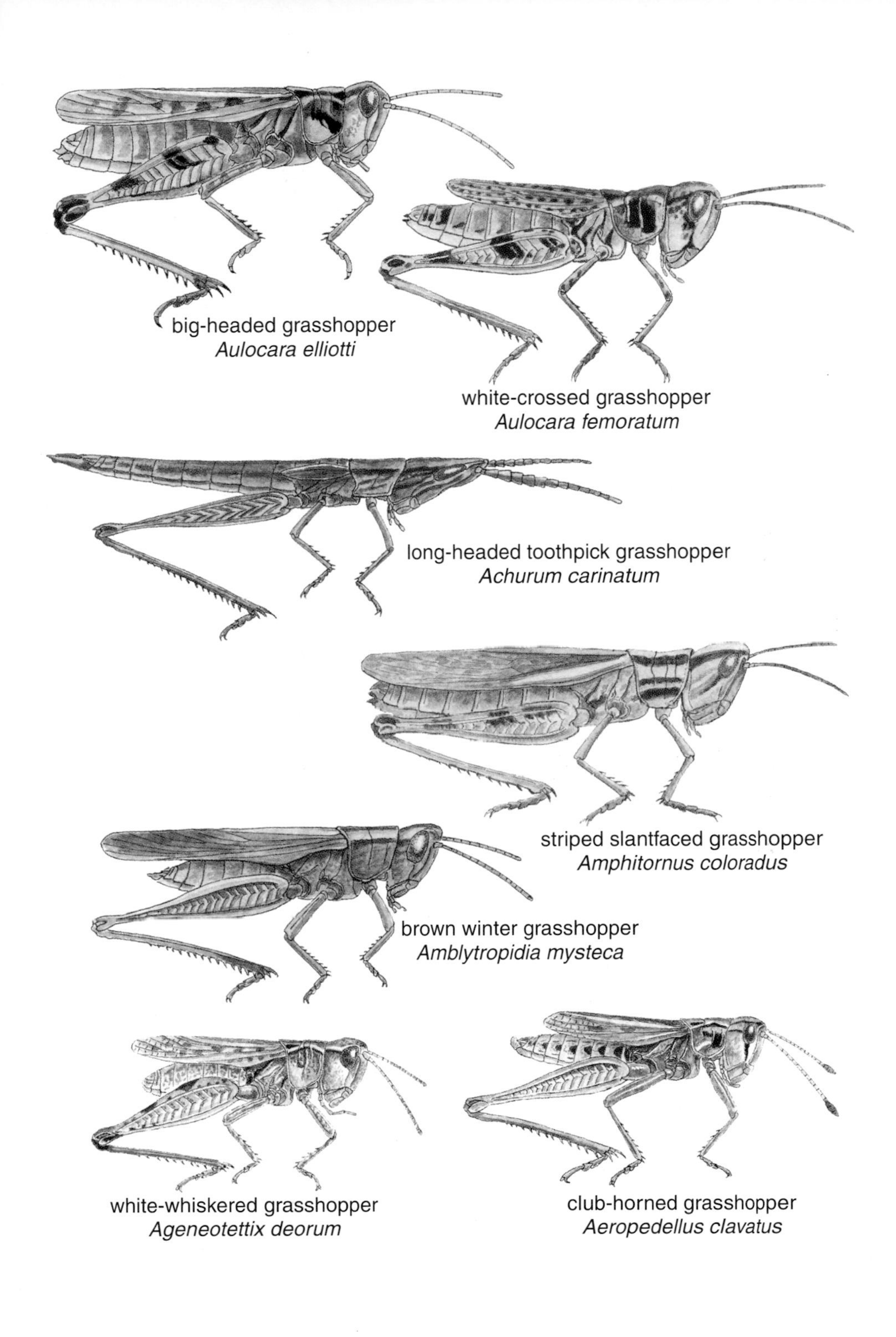

Plate 1: Stridulating Slantfaced Grasshoppers
subfamily Gomphocerinae

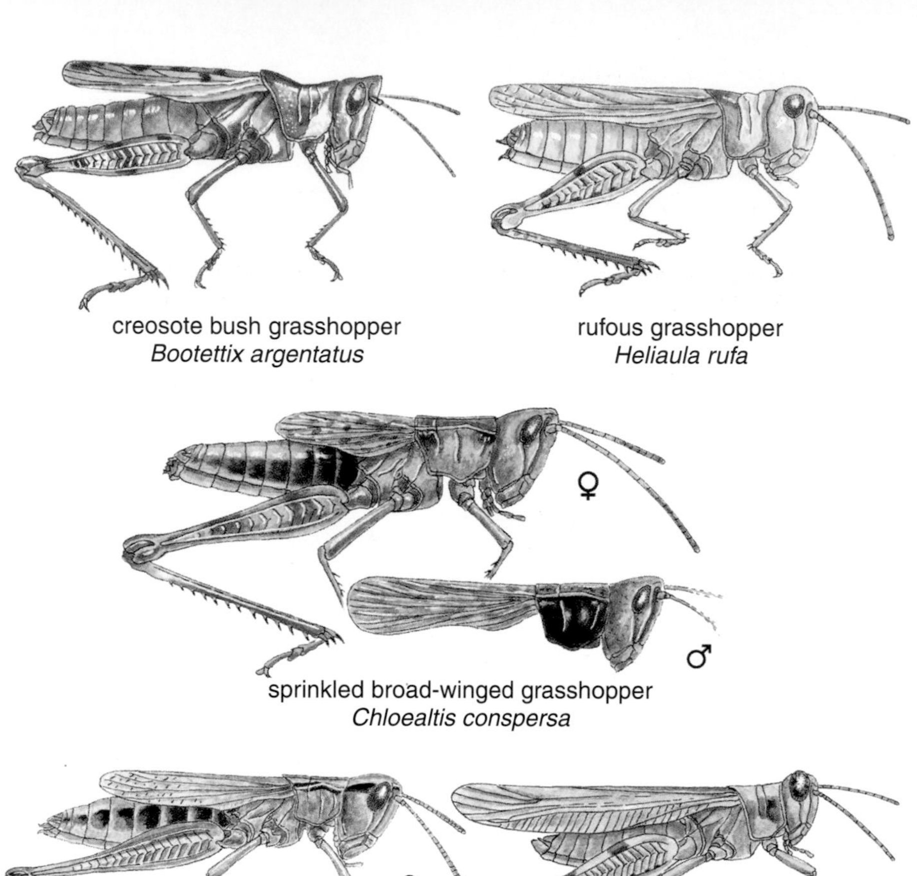

creosote bush grasshopper
Bootettix argentatus

rufous grasshopper
Heliaula rufa

sprinkled broad-winged grasshopper
Chloealtis conspersa

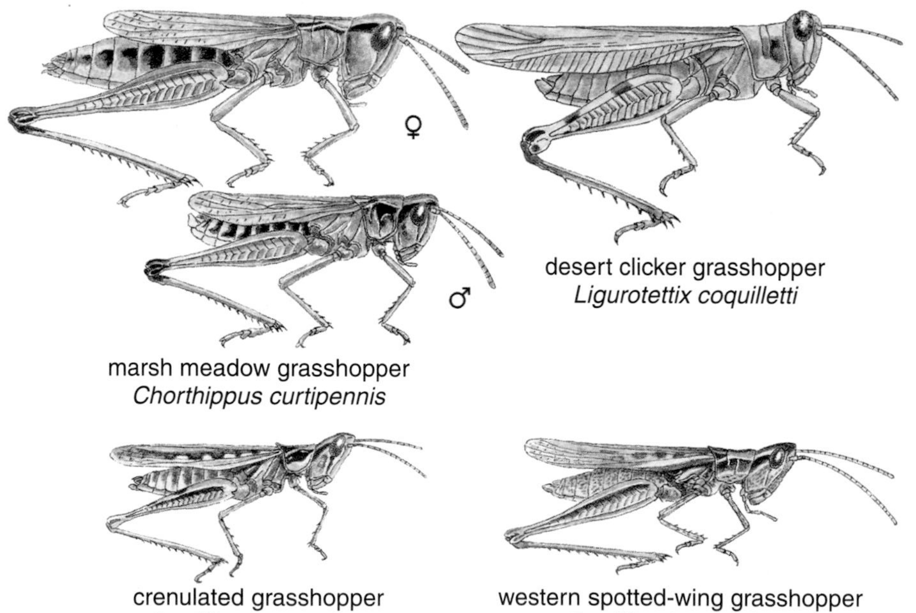

marsh meadow grasshopper
Chorthippus curtipennis

desert clicker grasshopper
Ligurotettix coquilletti

crenulated grasshopper
Cordillacris crenulata

western spotted-wing grasshopper
Cordillacris occipitalis

Plate 2: Stridulating Slantfaced Grasshoppers
subfamily Gomphocerinae

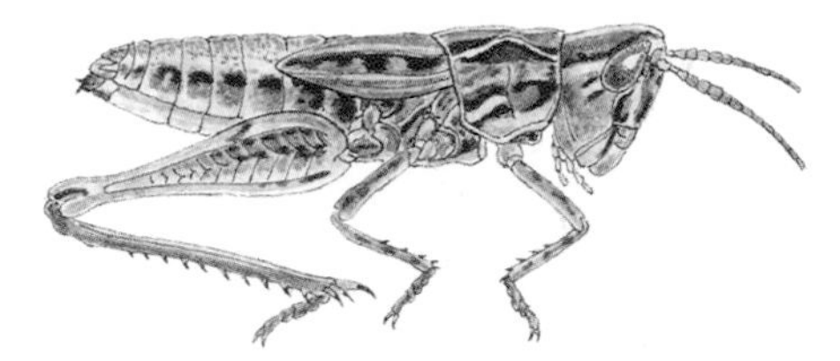

Texas short-wing slantfaced grasshopper
Eritettix arbortivus

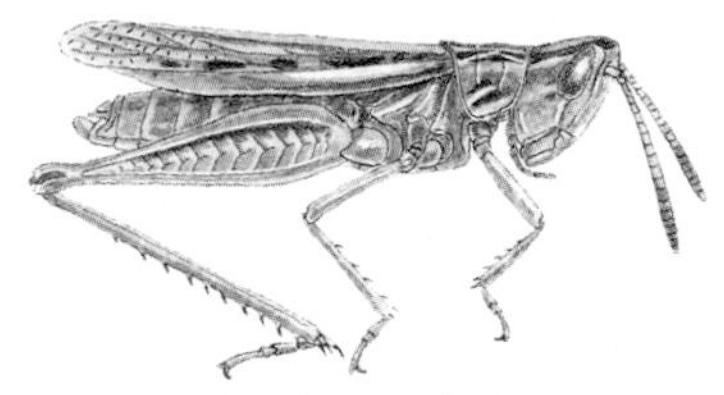

velvet-striped grasshopper
Eritettix simplex

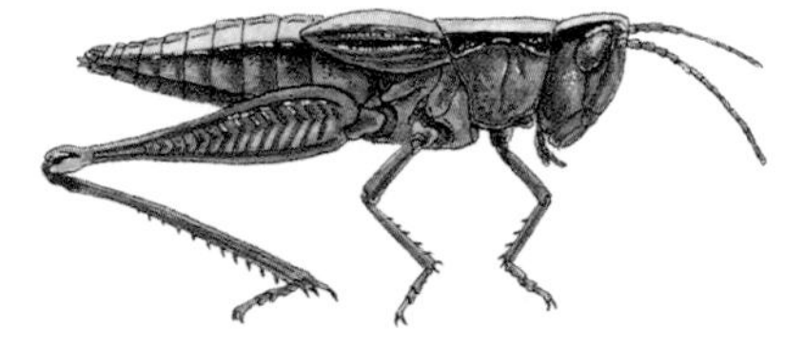

obscure slantfaced grasshopper
Eritettix obscurus

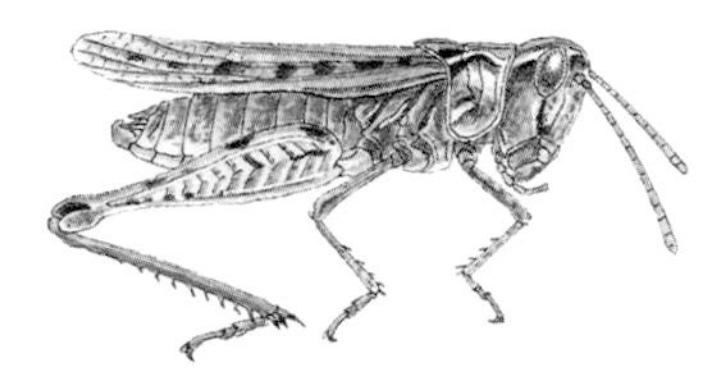

brown-spotted range grasshopper
Psoloessa delicatula

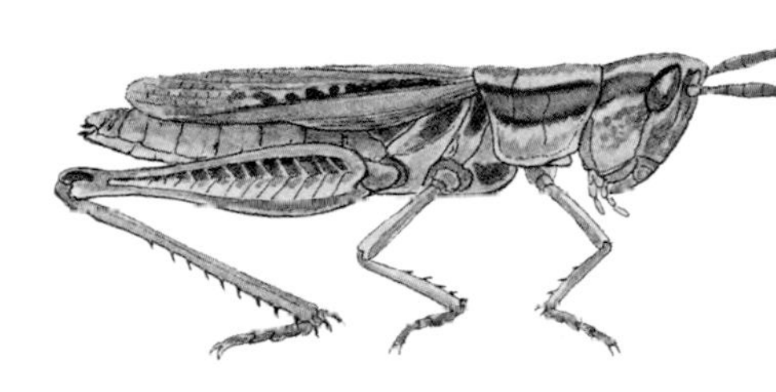

obscure grasshopper
Opeia obscura

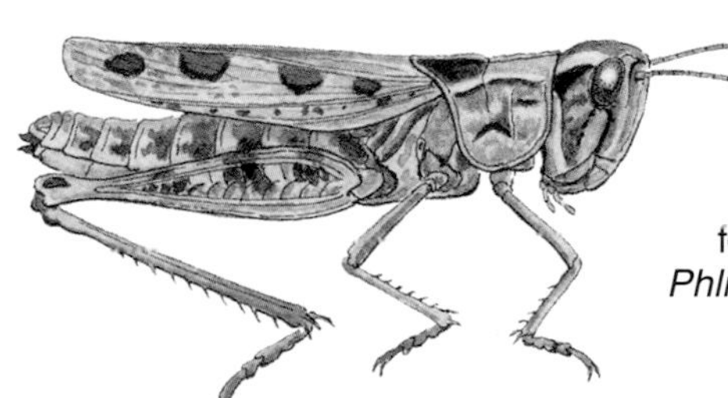

four-spotted grasshopper
Phlibostroma quadrimaculatum

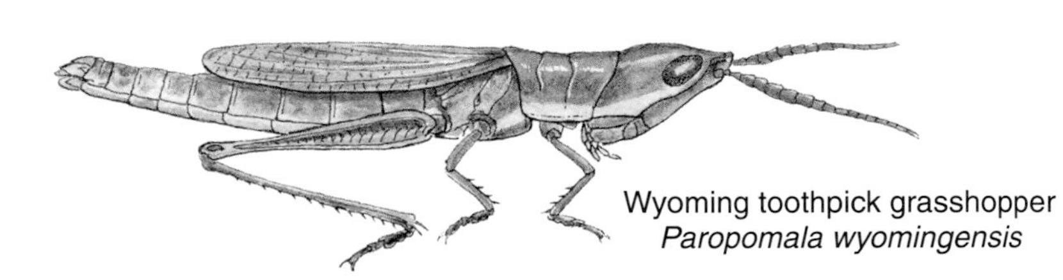

Wyoming toothpick grasshopper
Paropomala wyomingensis

Plate 3: Stridulating Slantfaced Grasshoppers
subfamily Gomphocerinae

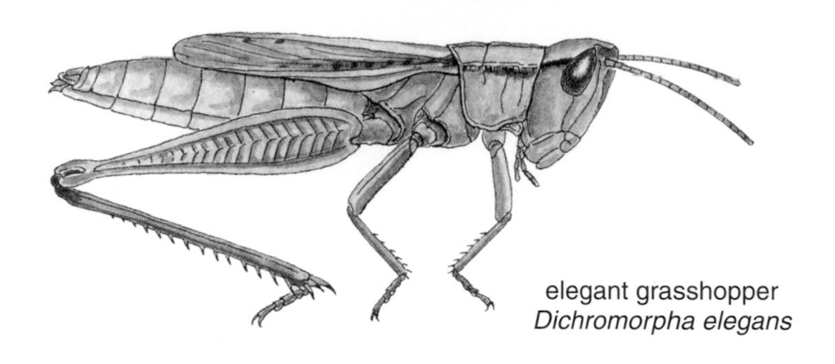

elegant grasshopper
Dichromorpha elegans

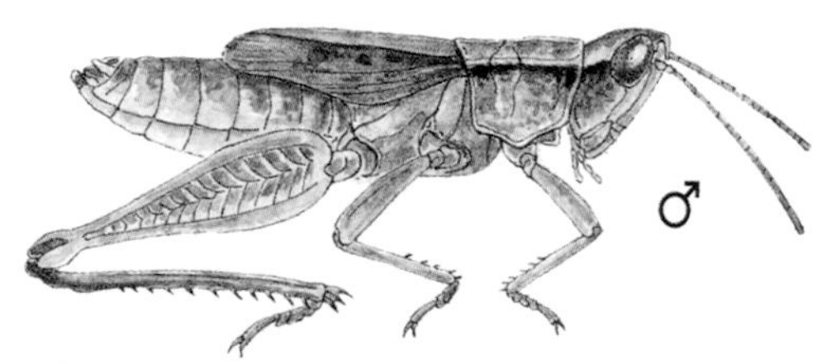

short-winged green grasshopper (green phase)
Dichromorpha viridis

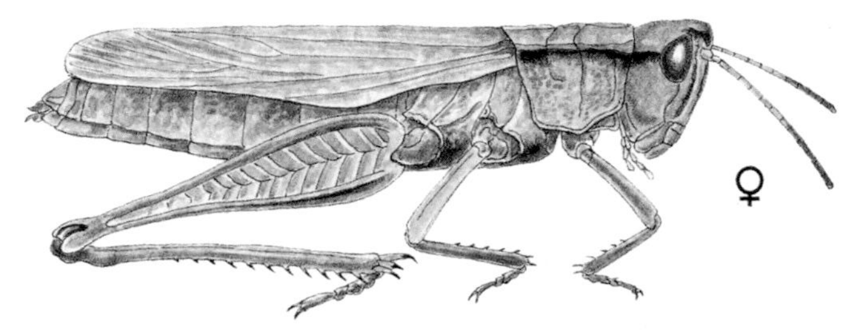

short-winged green grasshopper (green phase, long-winged form)
Dichromorpha viridis

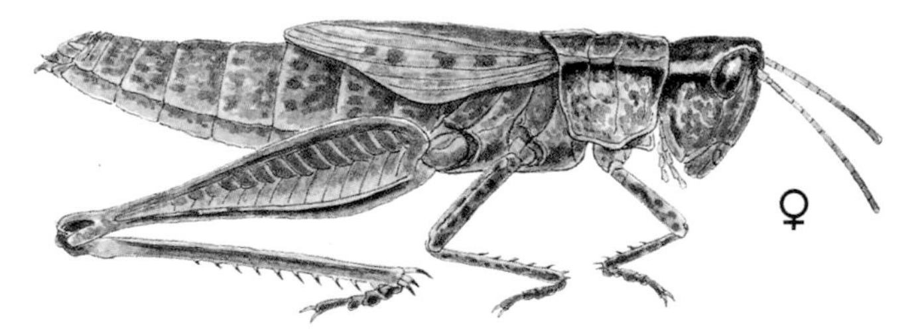

short-winged green grasshopper (brown phase, short-winged form)
Dichromorpha viridis

Plate 4: Stridulating Slantfaced Grasshoppers
subfamily Gomphocerinae

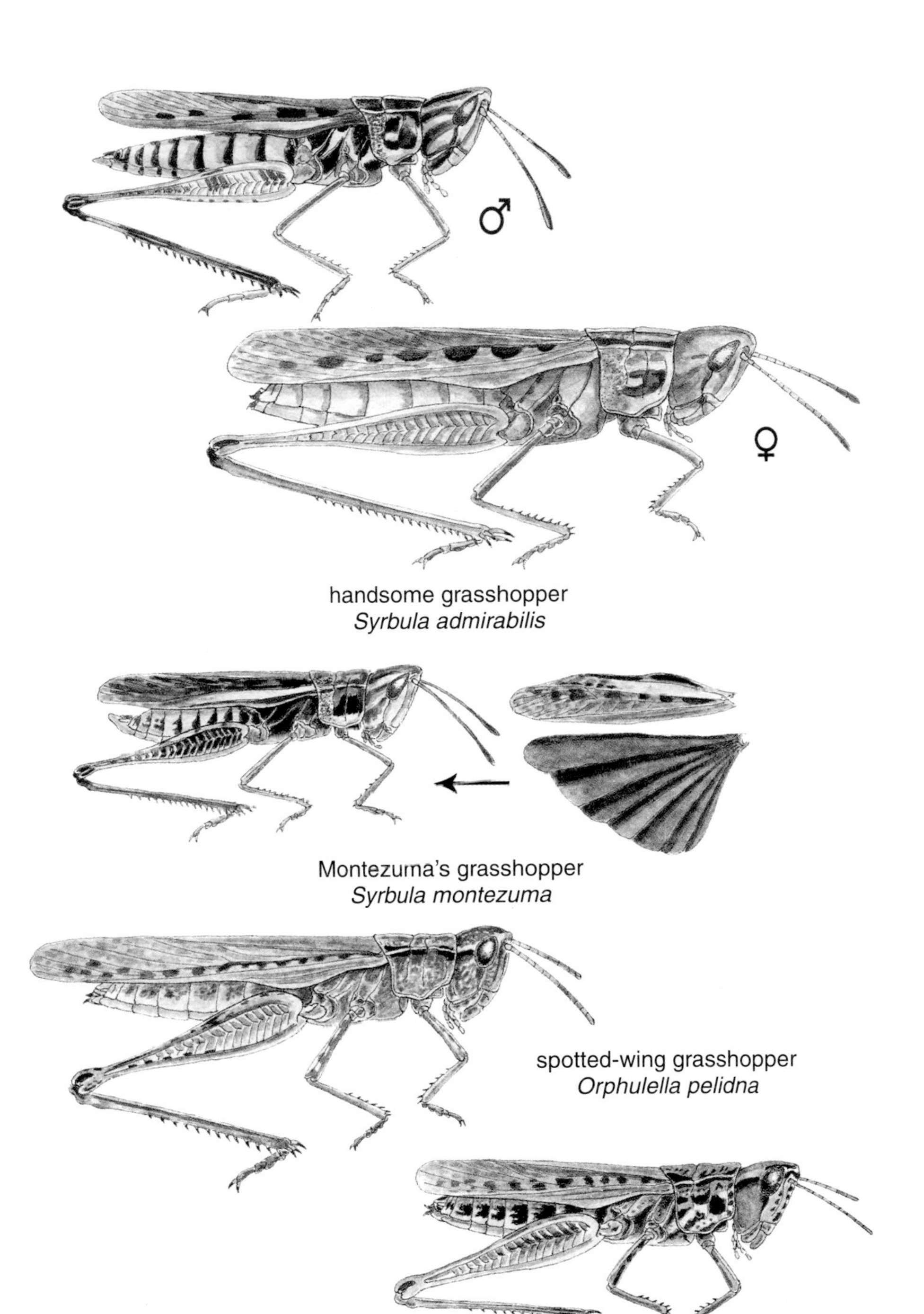

Plate 5: Stridulating Slantfaced Grasshoppers
subfamily Gomphocerinae

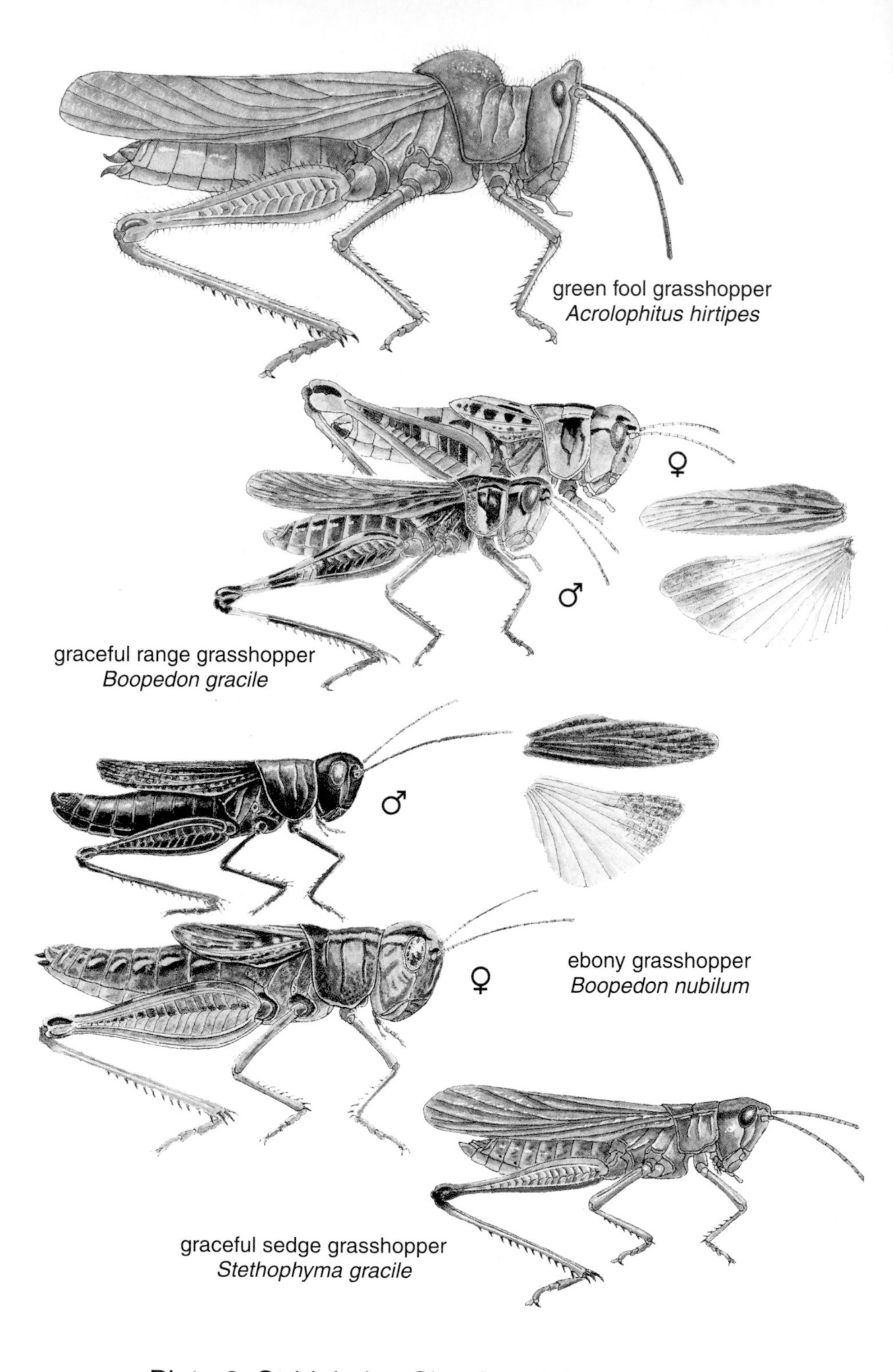

Plate 6: Stridulating Slantfaced Grasshoppers
subfamily Gomphocerinae

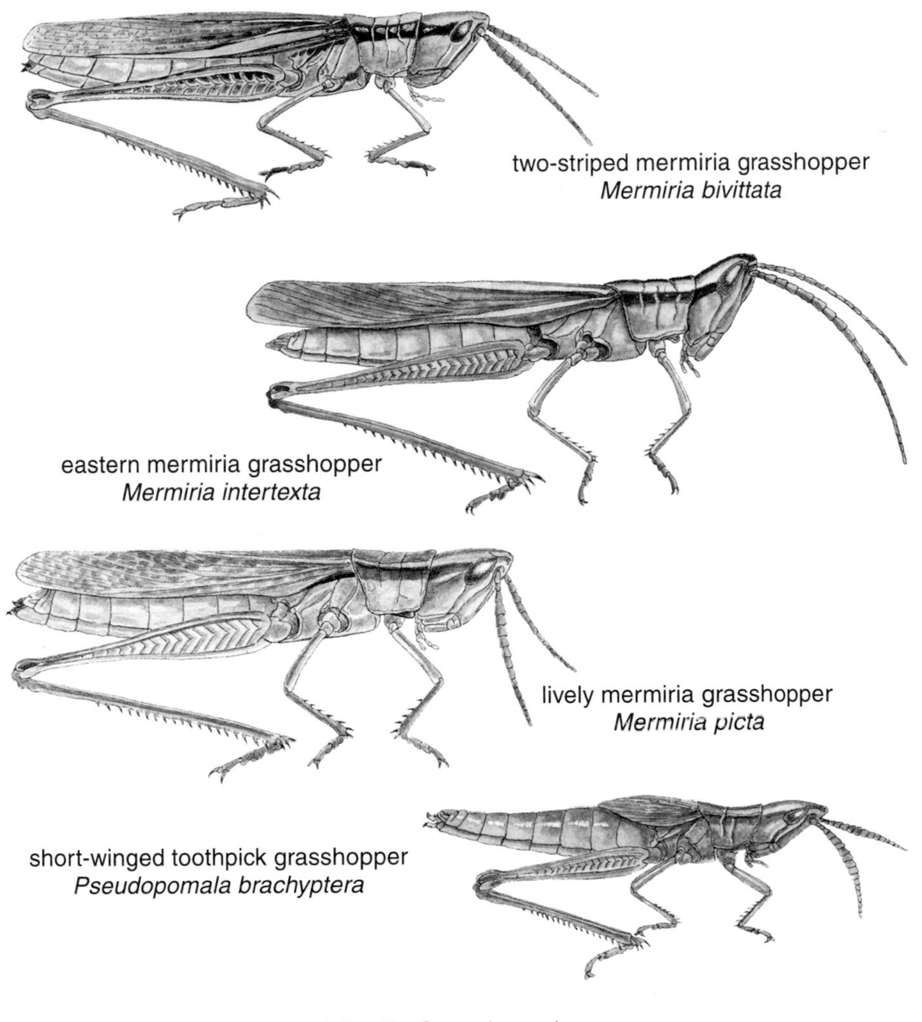

two-striped mermiria grasshopper
Mermiria bivittata

eastern mermiria grasshopper
Mermiria intertexta

lively mermiria grasshopper
Mermiria picta

short-winged toothpick grasshopper
Pseudopomala brachyptera

subfamily Gomphocerinae

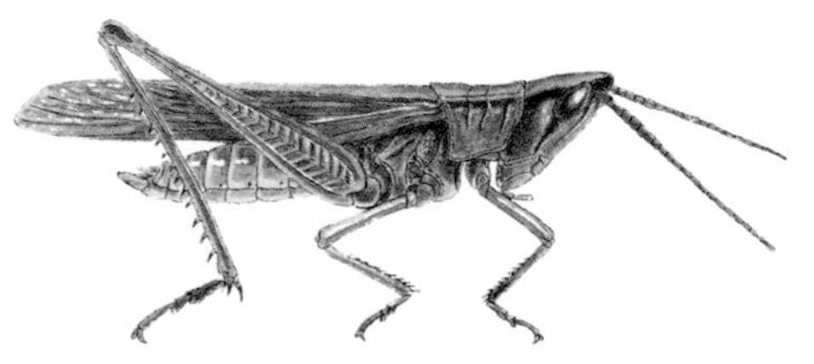

clipped-wing grasshopper
Metaleptea brevicornis

subfamily Acridinae

Plate 7: Stridulating Slantfaced Grasshoppers and Silent Slantfaced Grasshoppers

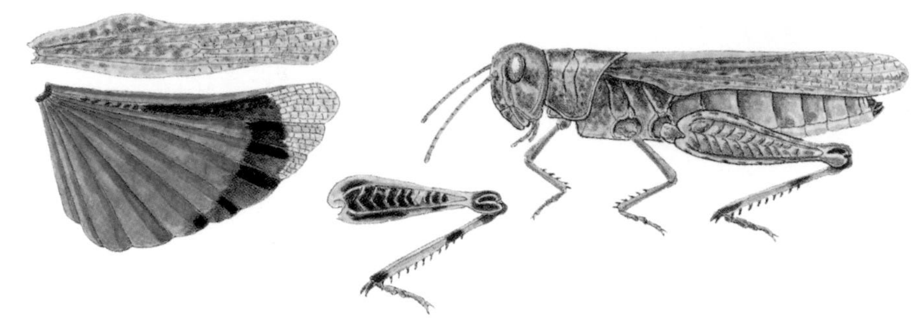

speckle-winged rangeland grasshopper
Arphia conspersa

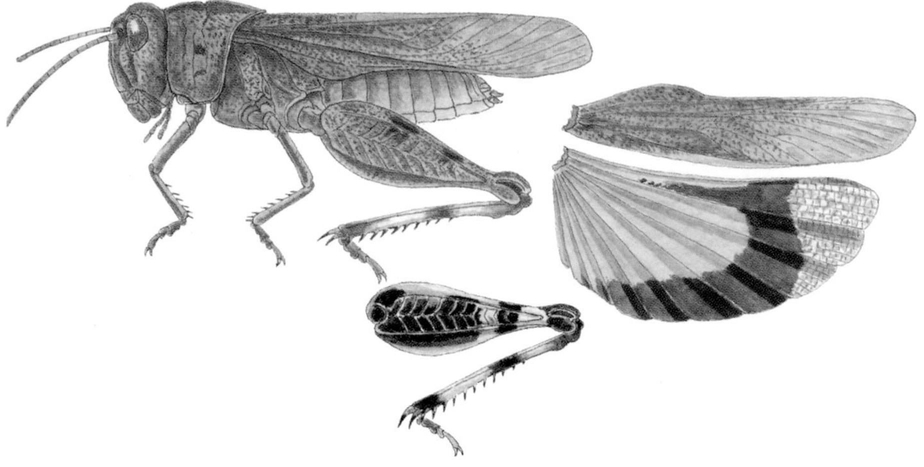

southern yellow-winged grasshopper
Arphia granulata

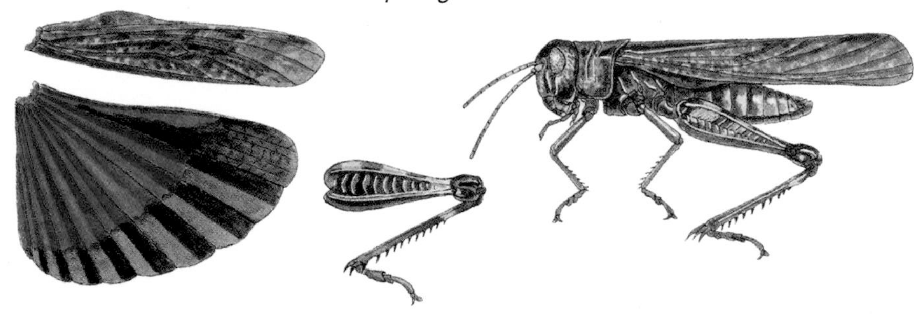

northwestern red-winged grasshopper
Arphia pseudonietana

Plate 8: Band-winged Grasshoppers
subfamily Oedipodinae

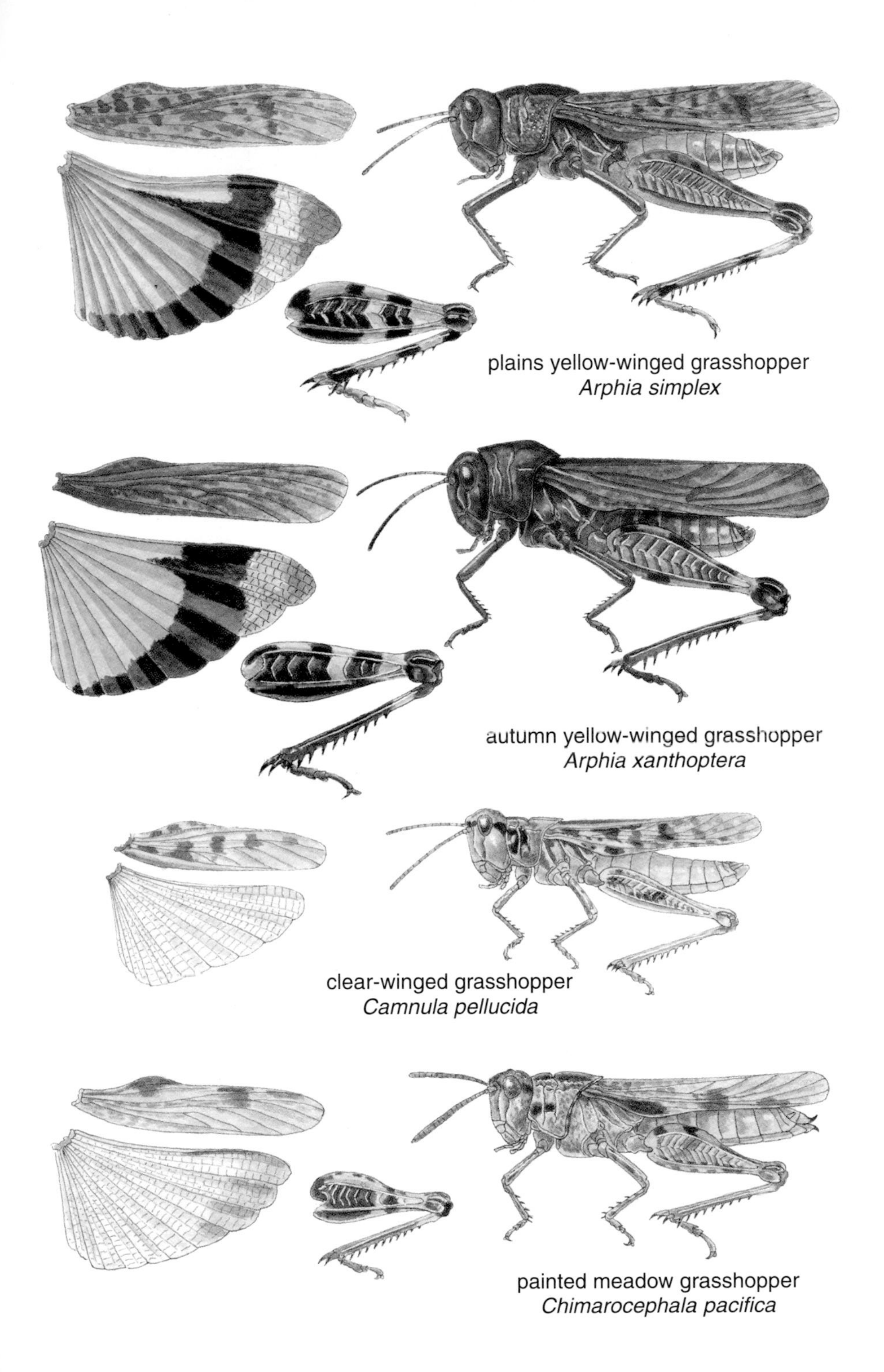

Plate 9: Band-winged Grasshoppers
subfamily Oedipodinae

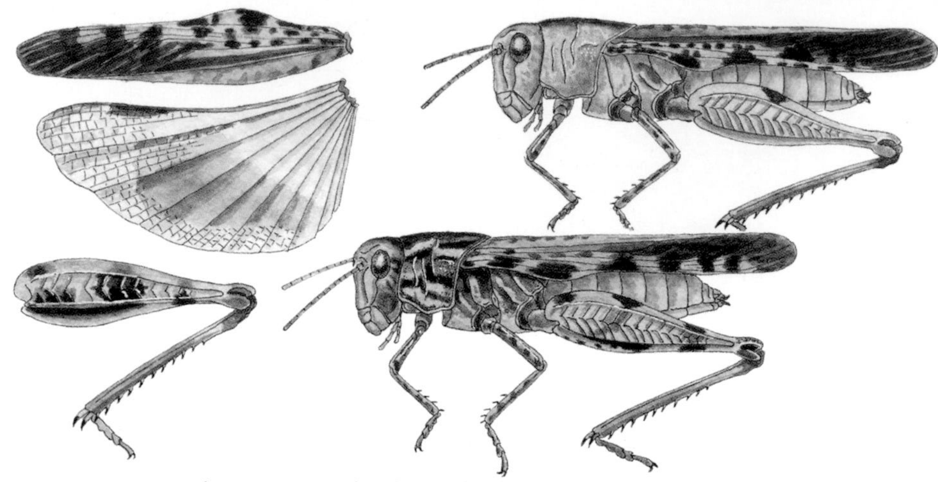

southern green-striped grasshopper (green & brown forms)
Chortophaga australior

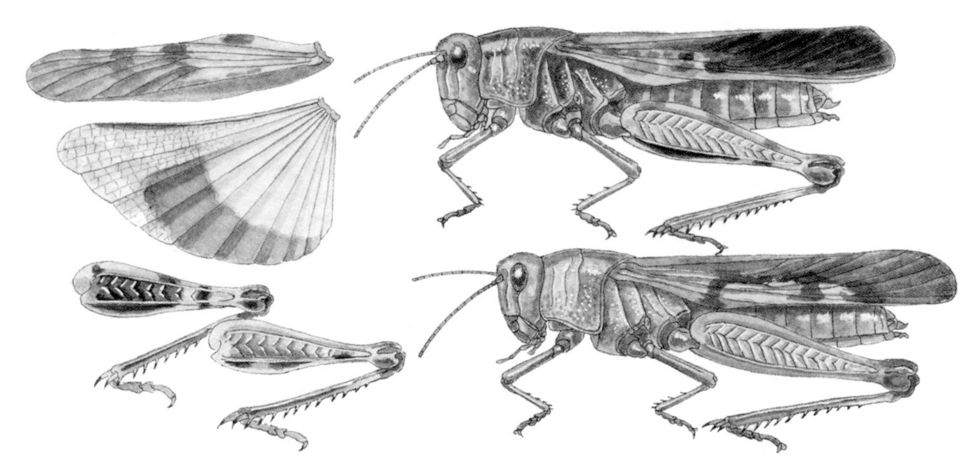

northern green-striped grasshopper (green & brown forms)
Chortophaga viridifasciata

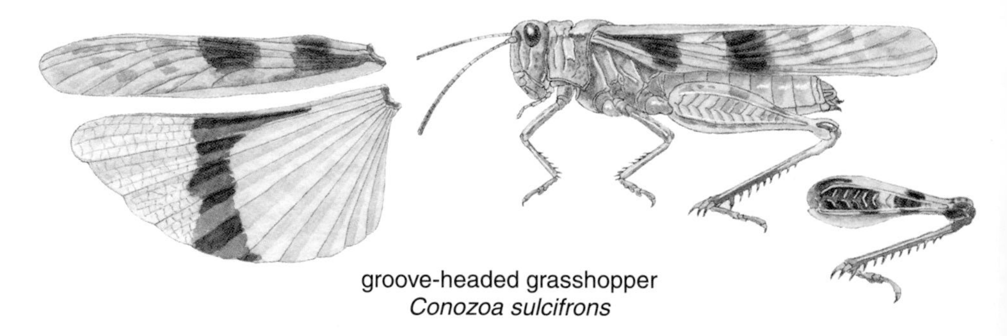

groove-headed grasshopper
Conozoa sulcifrons

Plate 10: Band-winged Grasshoppers

subfamily Oedipodinae

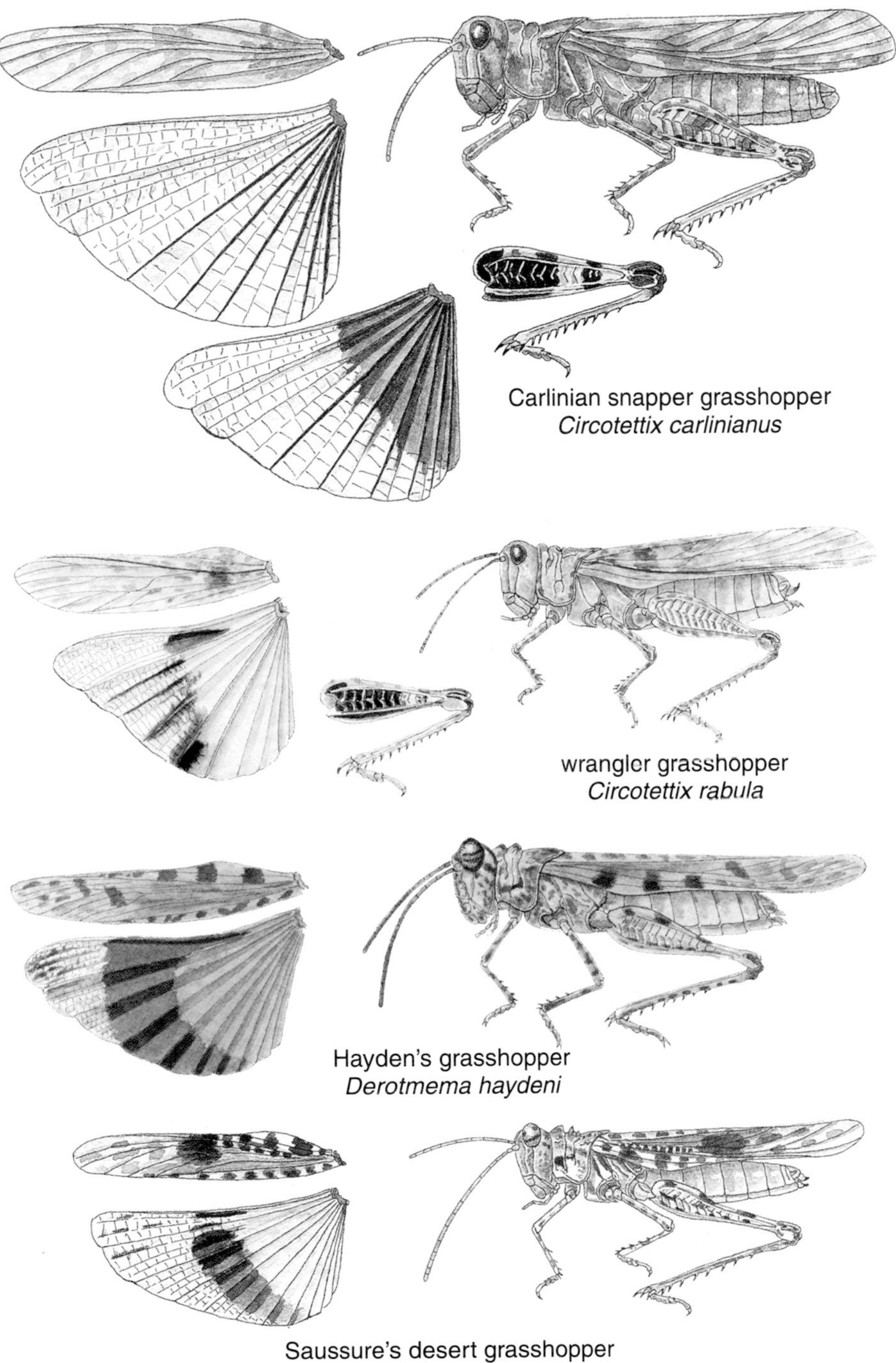

Carlinian snapper grasshopper
Circotettix carlinianus

wrangler grasshopper
Circotettix rabula

Hayden's grasshopper
Derotmema haydeni

Saussure's desert grasshopper
Derotmema saussureanum

Plate 11: Band-winged Grasshoppers
subfamily Oedipodinae

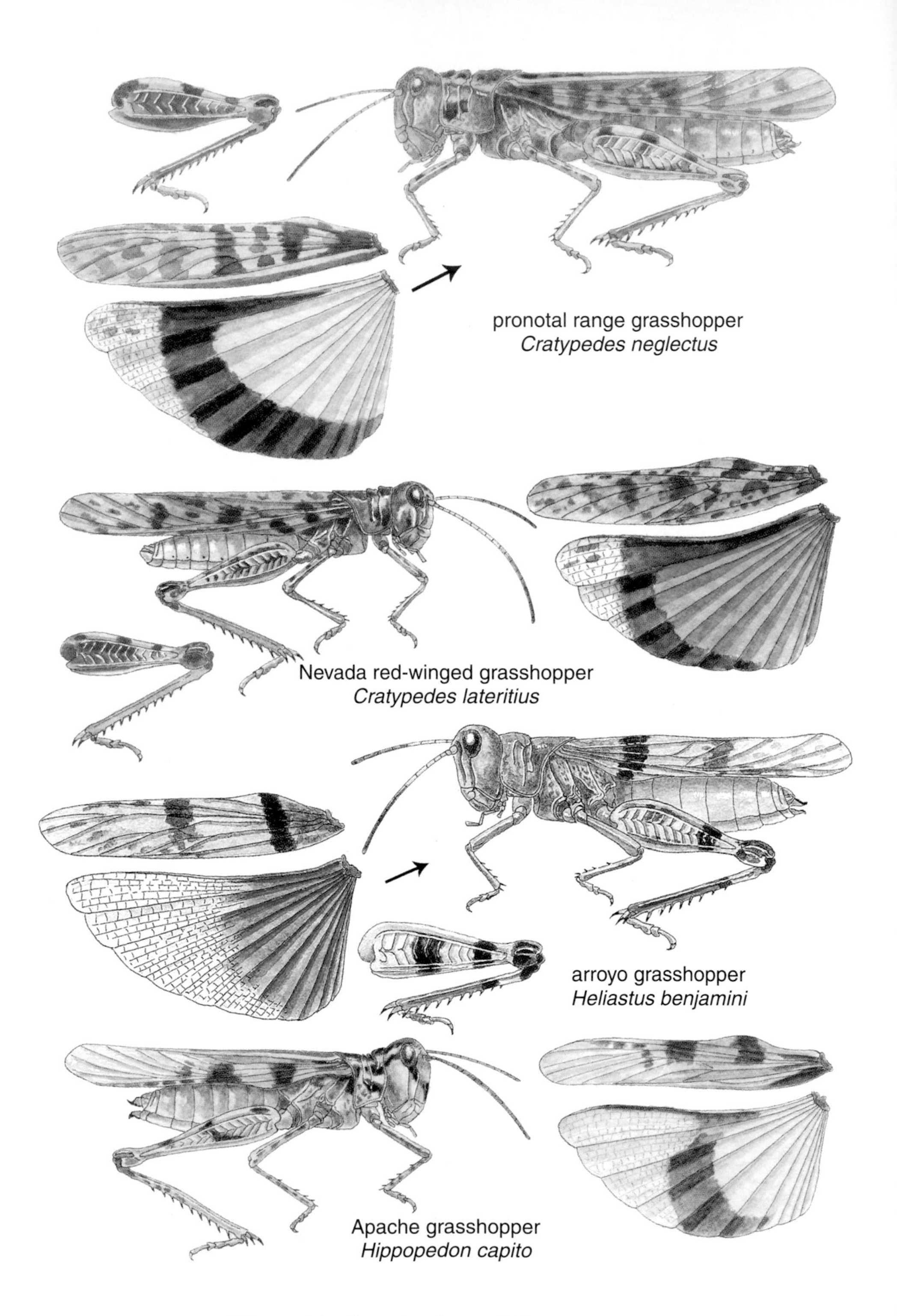

Plate 12: Band-winged Grasshoppers
subfamily Oedipodinae

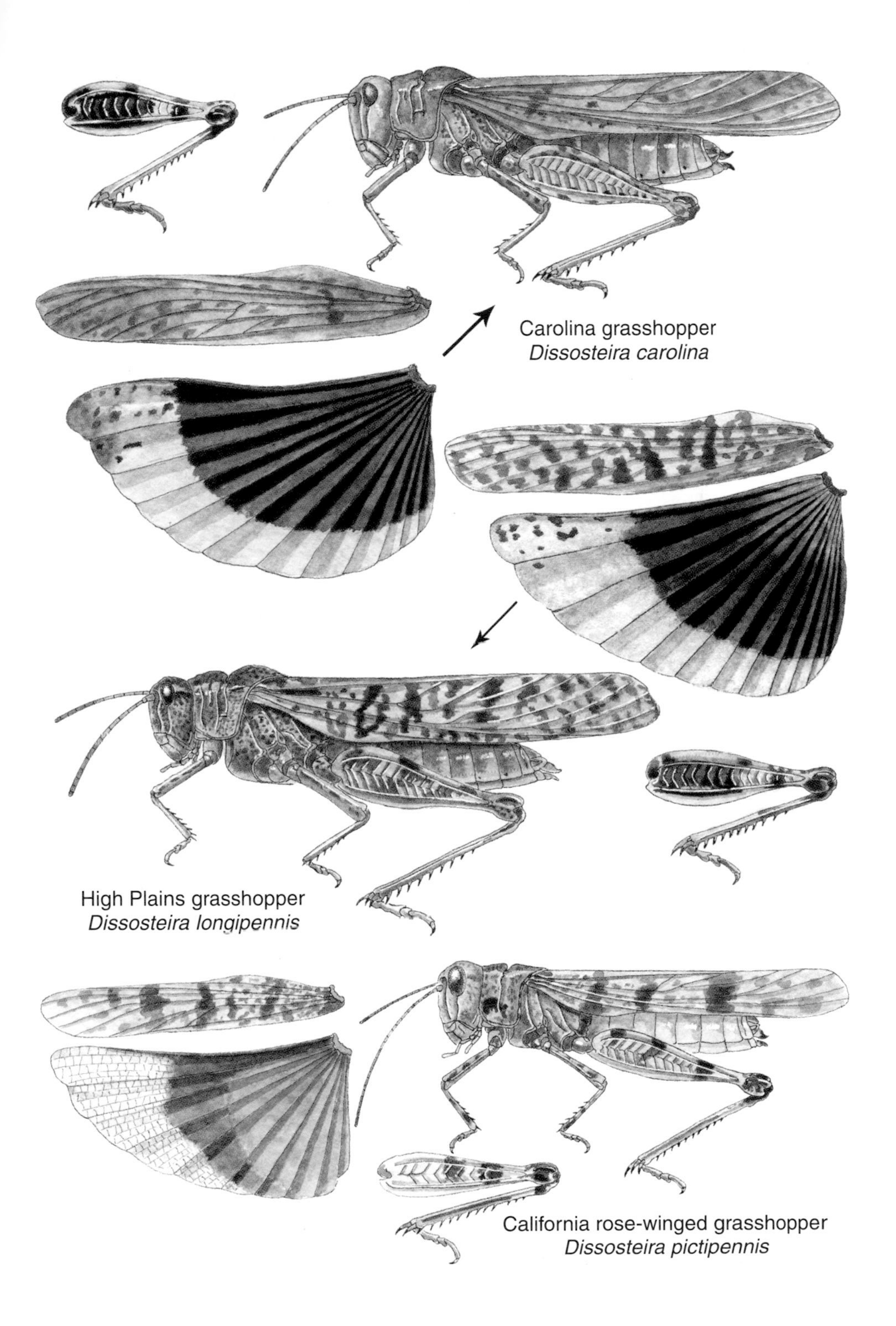

Plate 13: Band-winged Grasshoppers
subfamily Oedipodinae

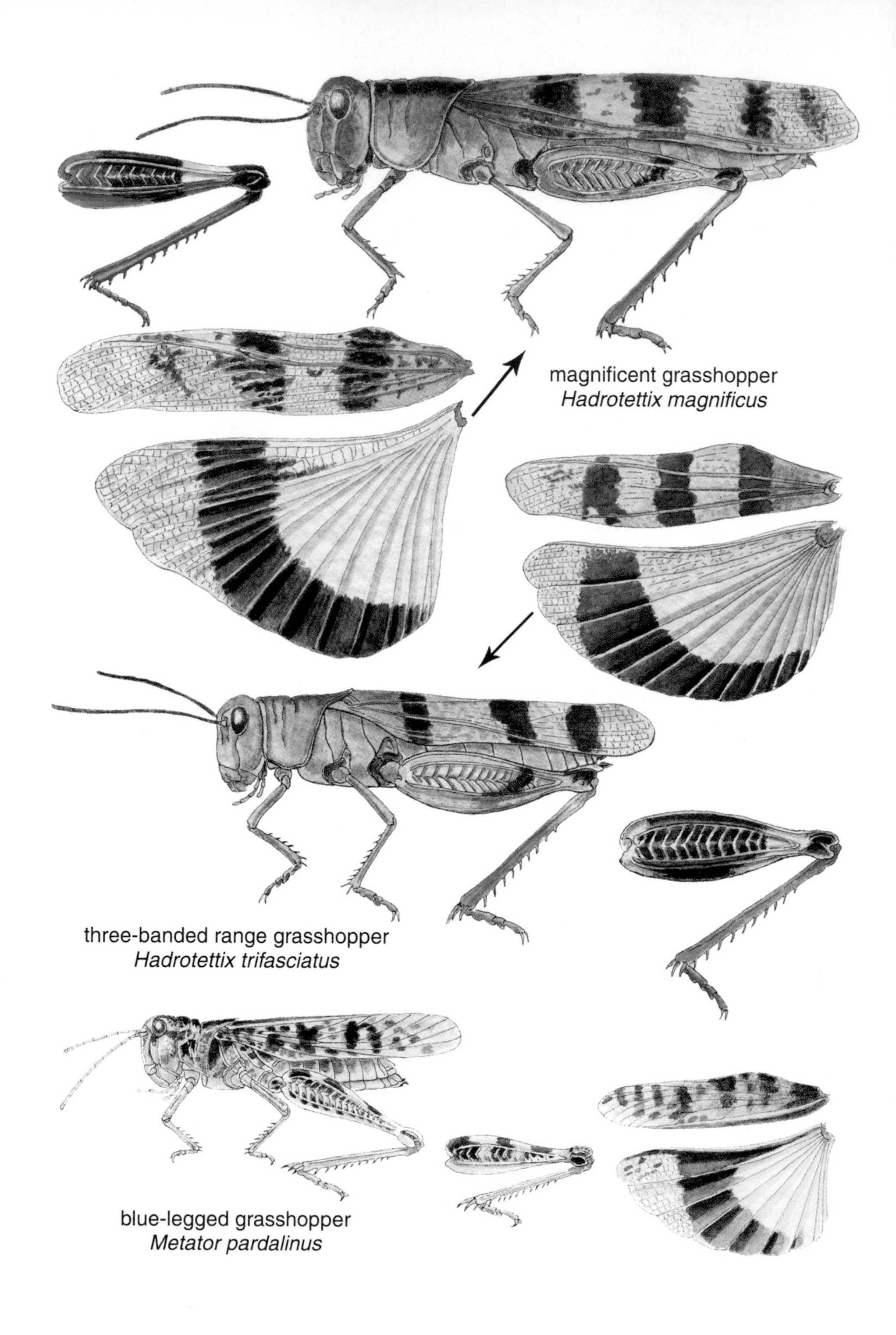

Plate 14: Band-winged Grasshoppers
subfamily Oedipodinae

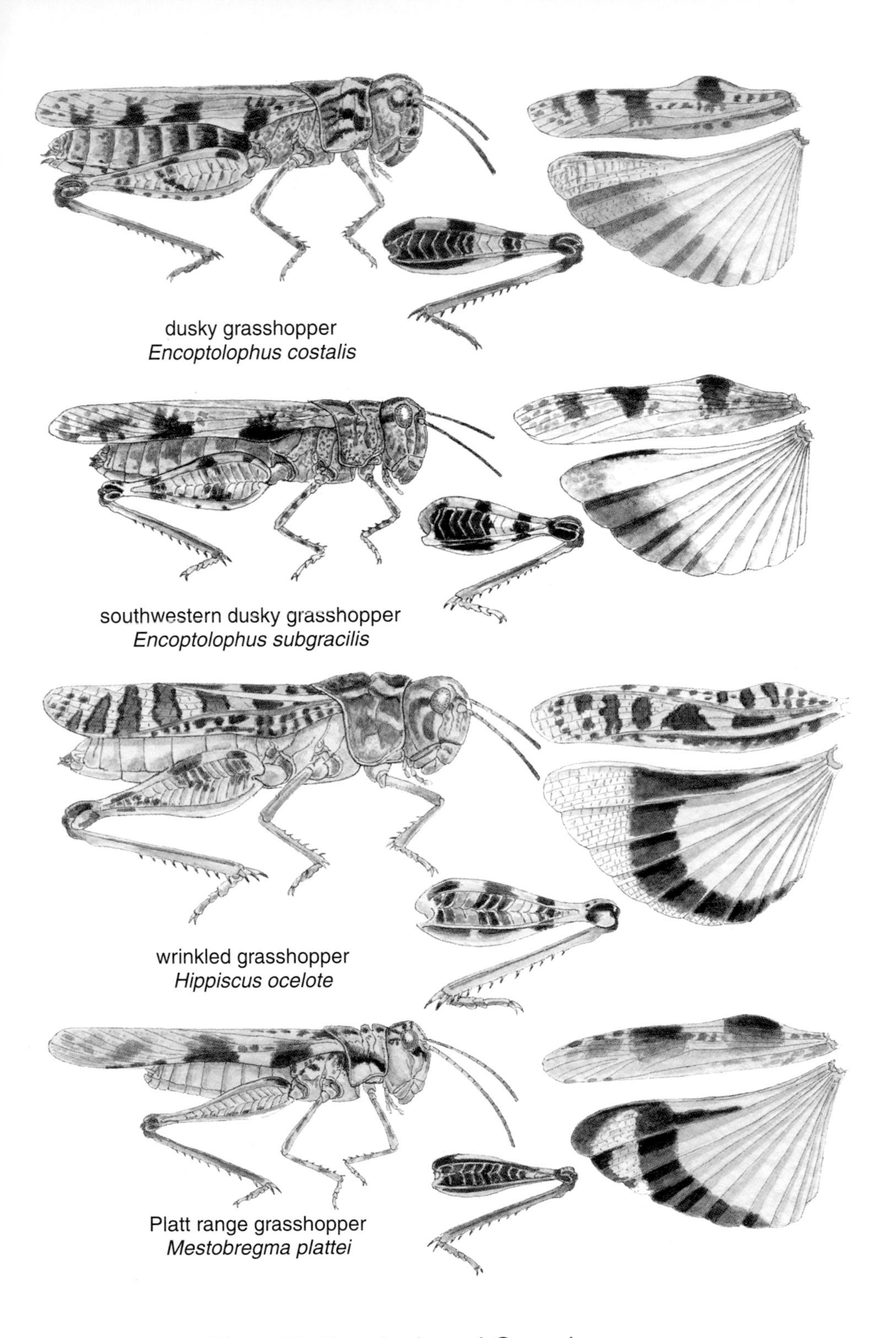

Plate 15: Band-winged Grasshoppers
subfamily Oedipodinae

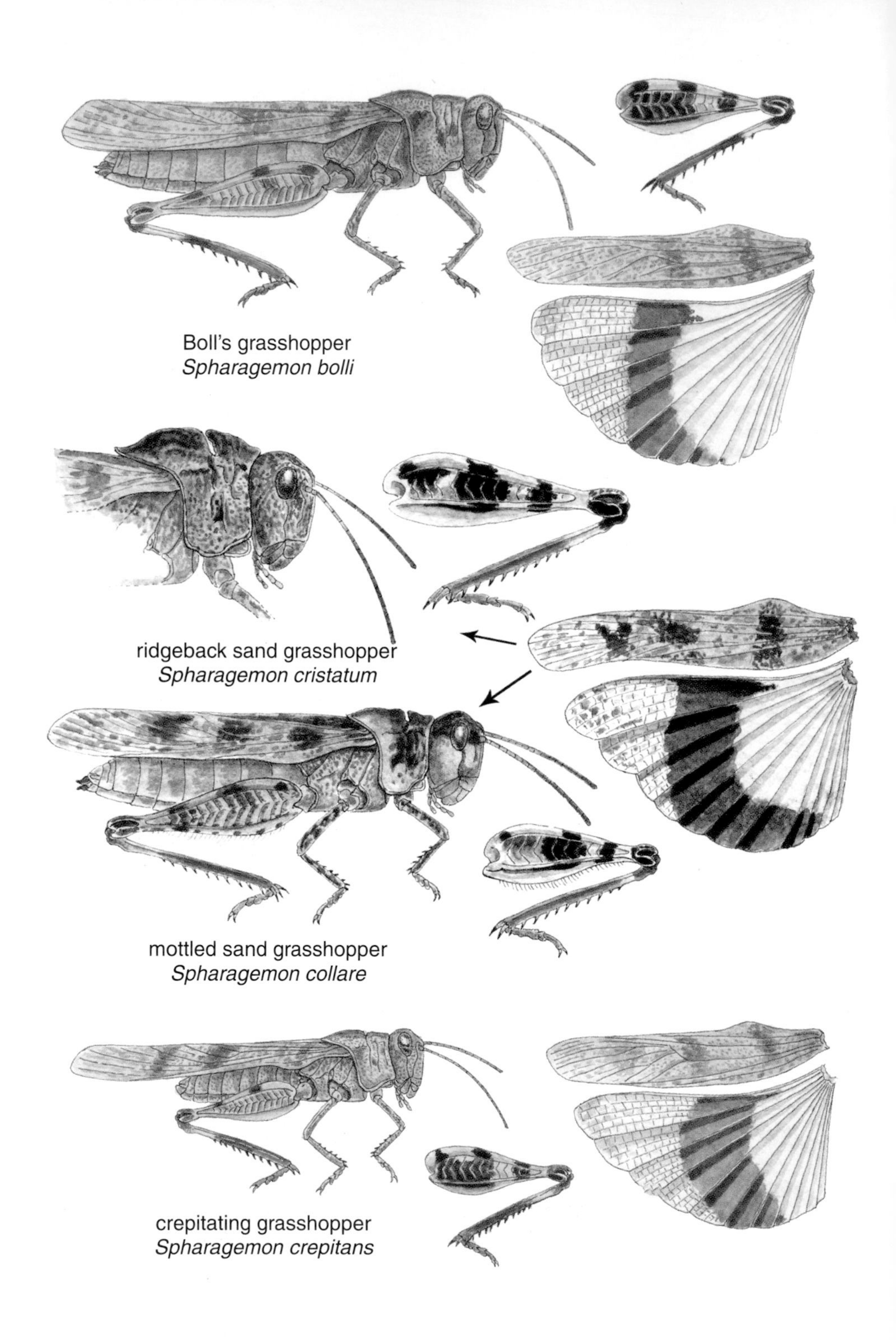

Plate 16: Band-winged Grasshoppers
subfamily Oedipodinae

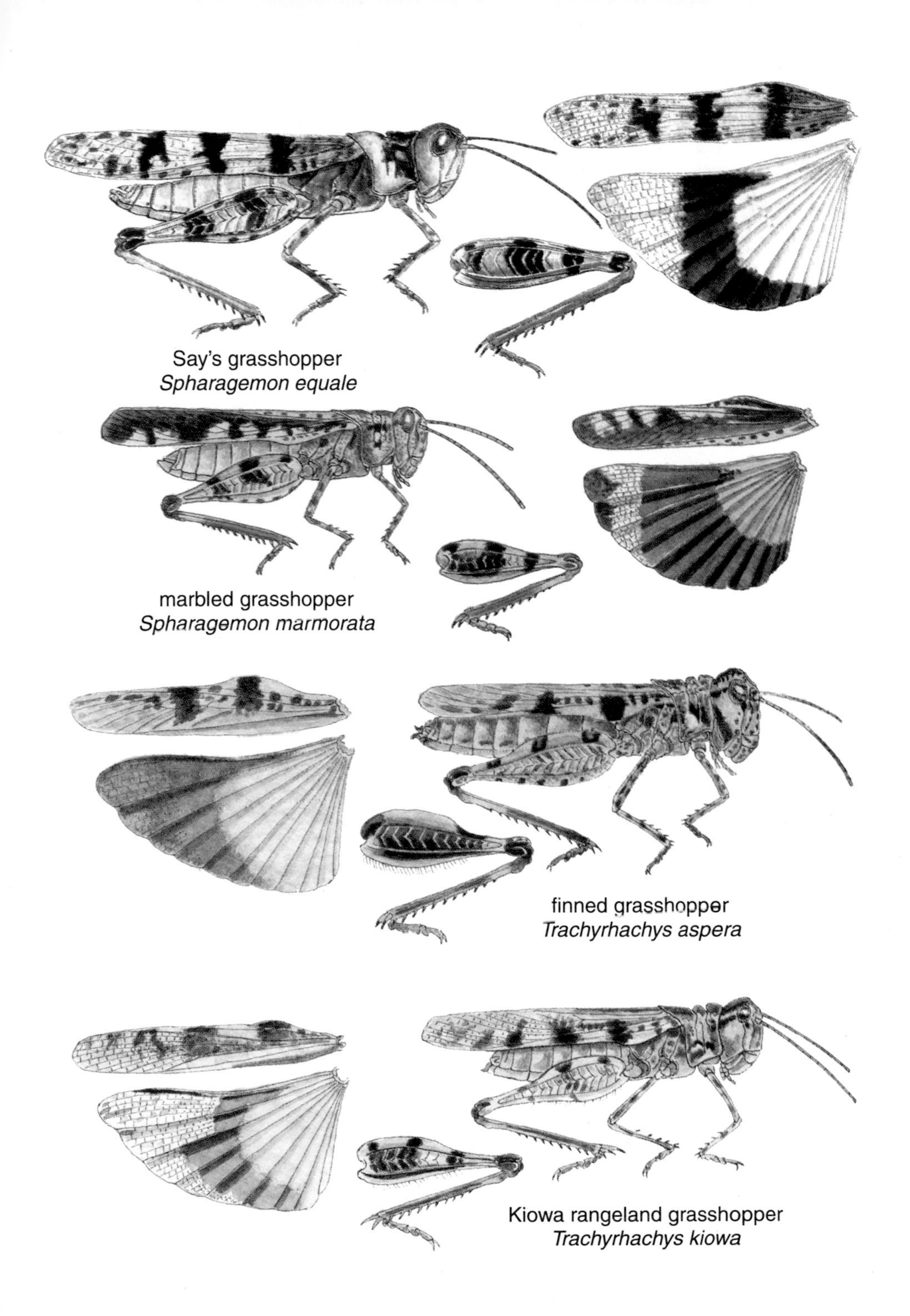

Plate 17: Band-winged Grasshoppers
subfamily Oedipodinae

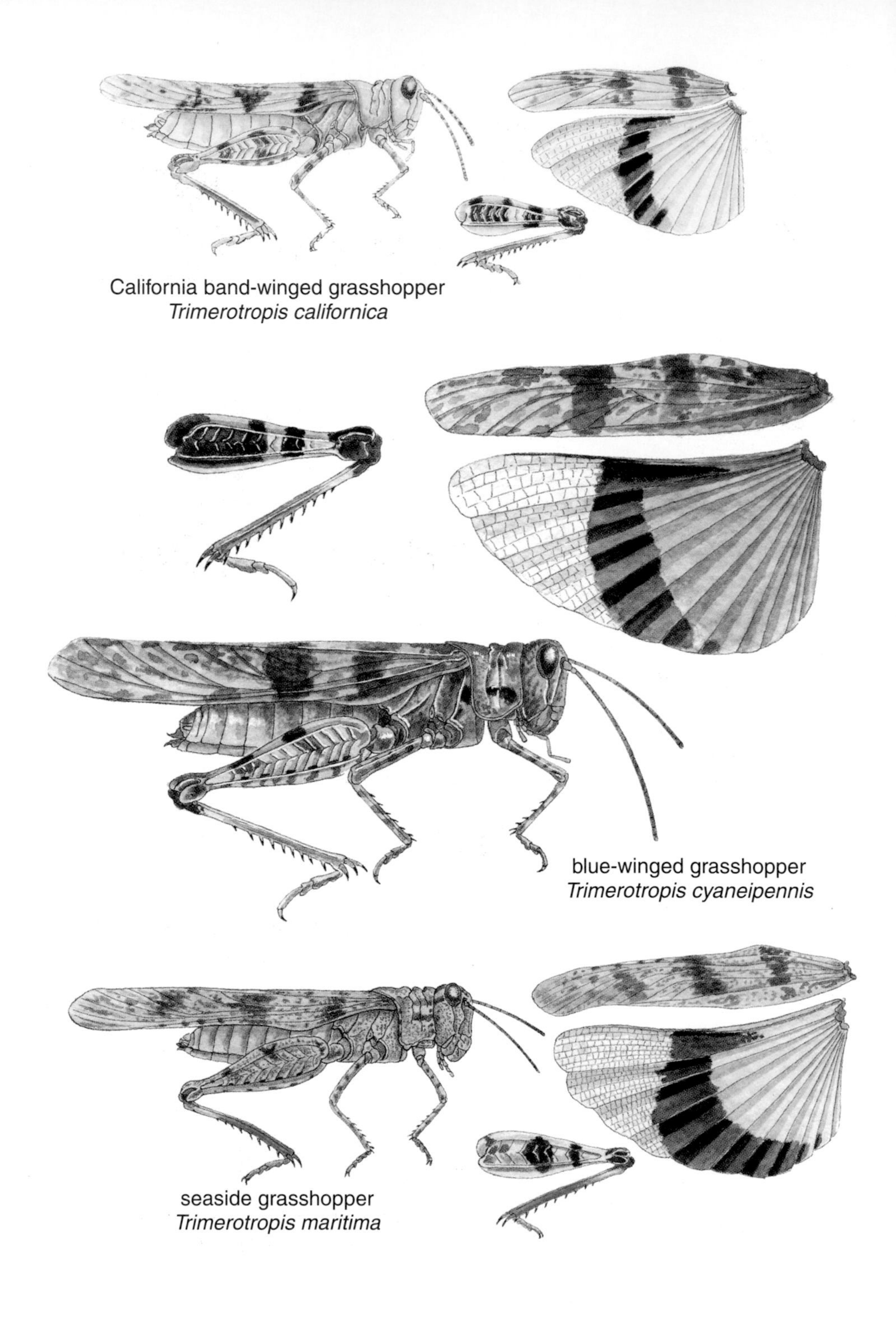

Plate 18: Band-winged Grasshoppers
subfamily Oedipodinae

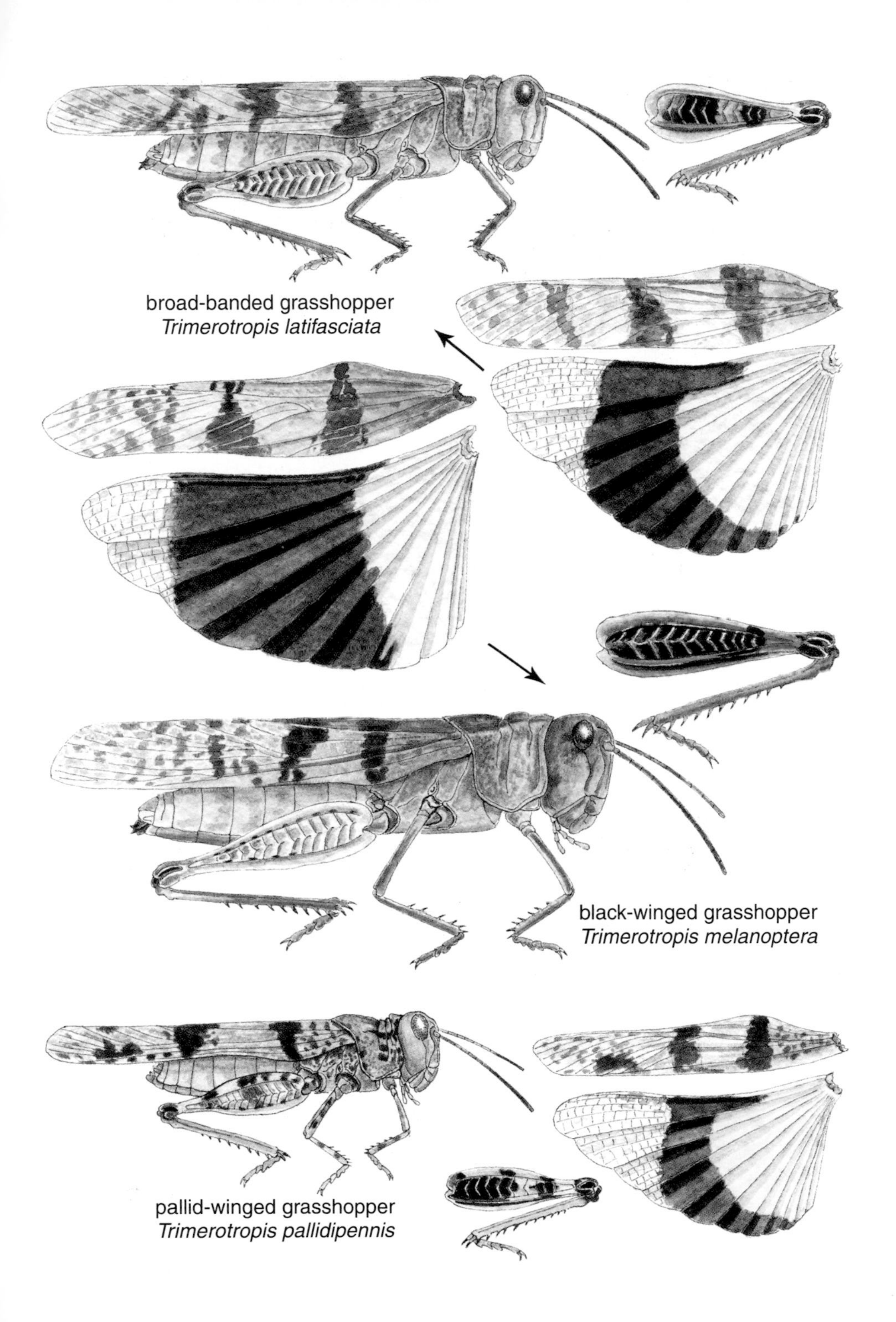

Plate 19: Band-winged Grasshoppers
subfamily Oedipodinae

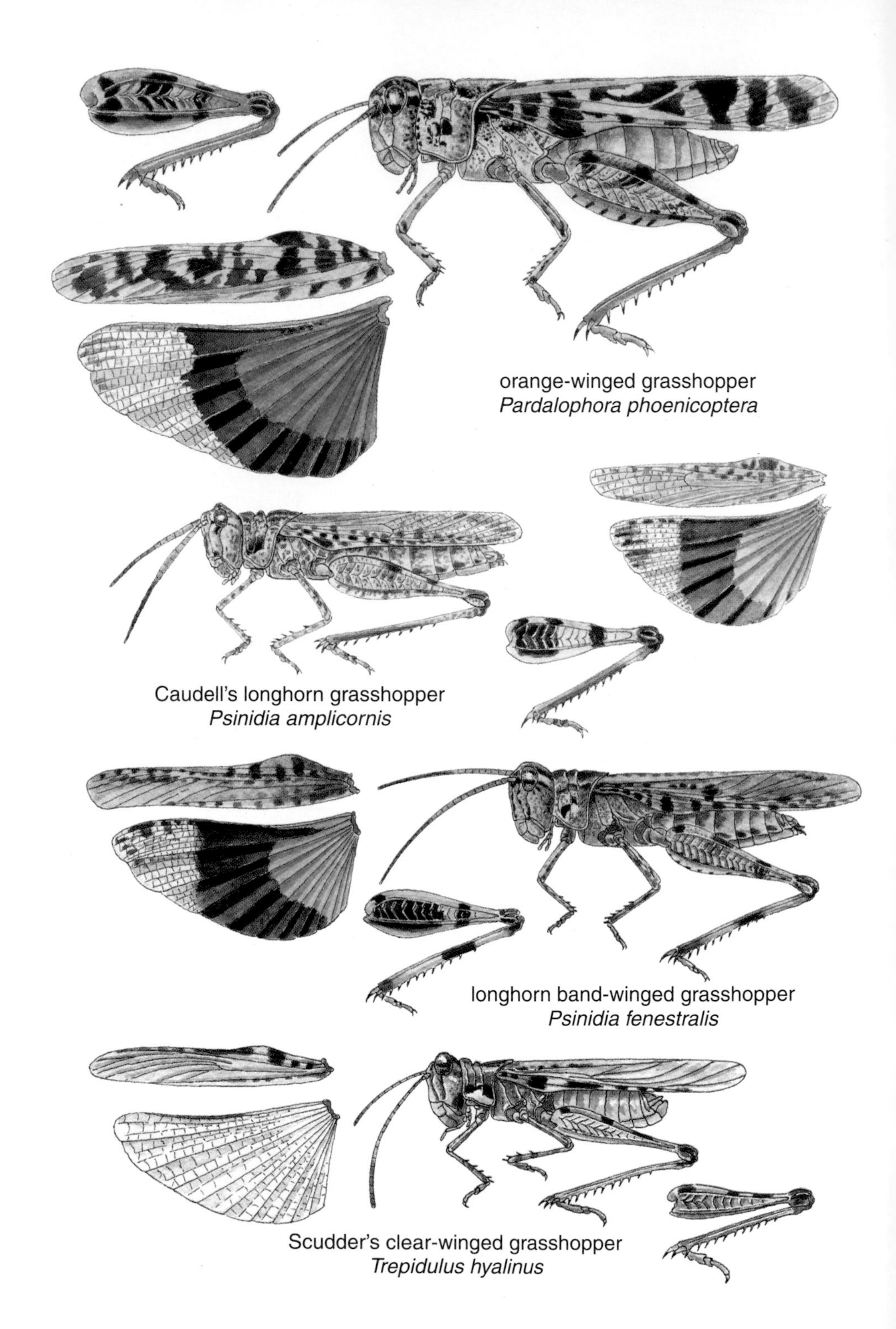

Plate 20: Band-winged Grasshoppers
subfamily Oedipodinae

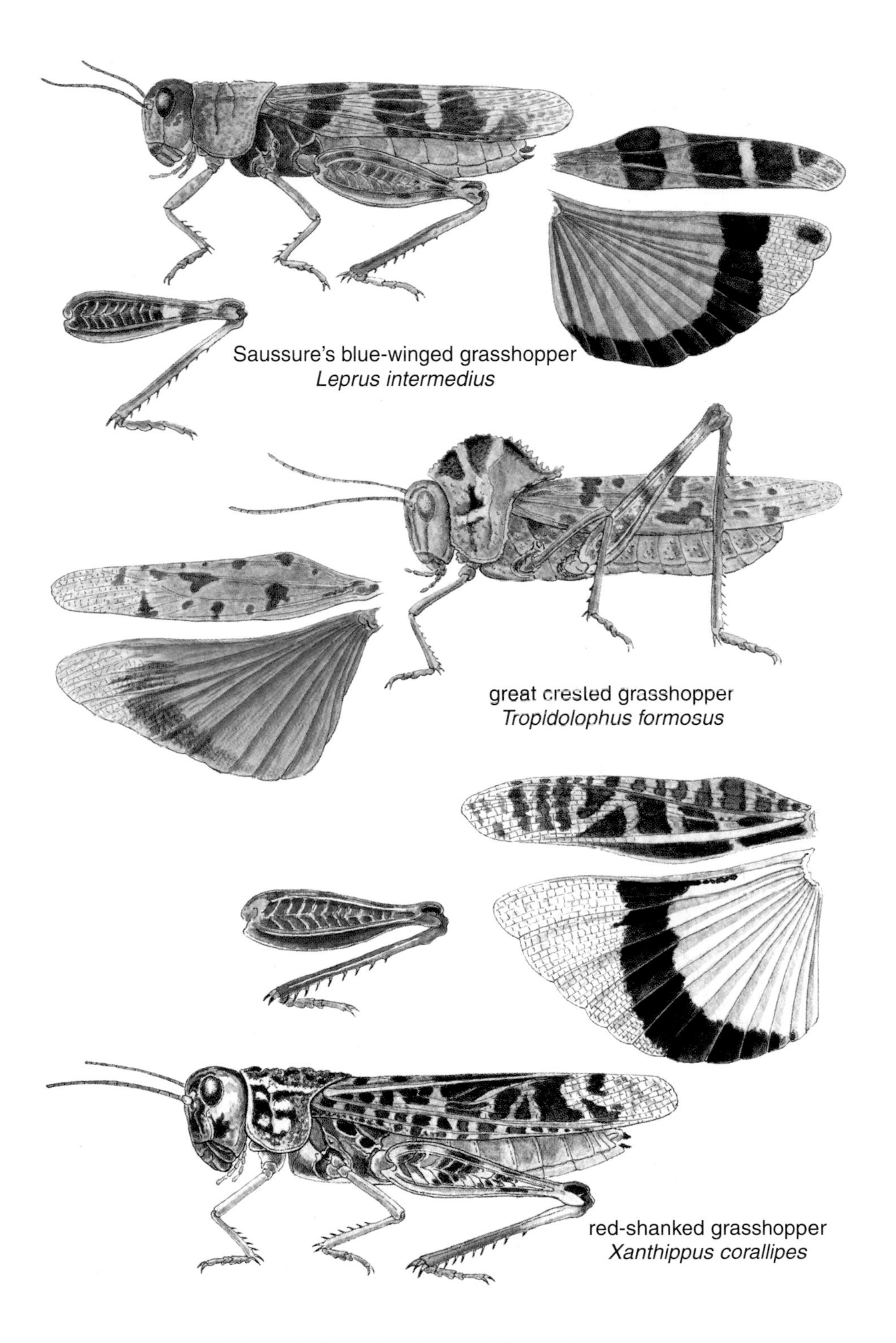

Plate 21: Band-winged Grasshoppers
subfamily Oedipodinae

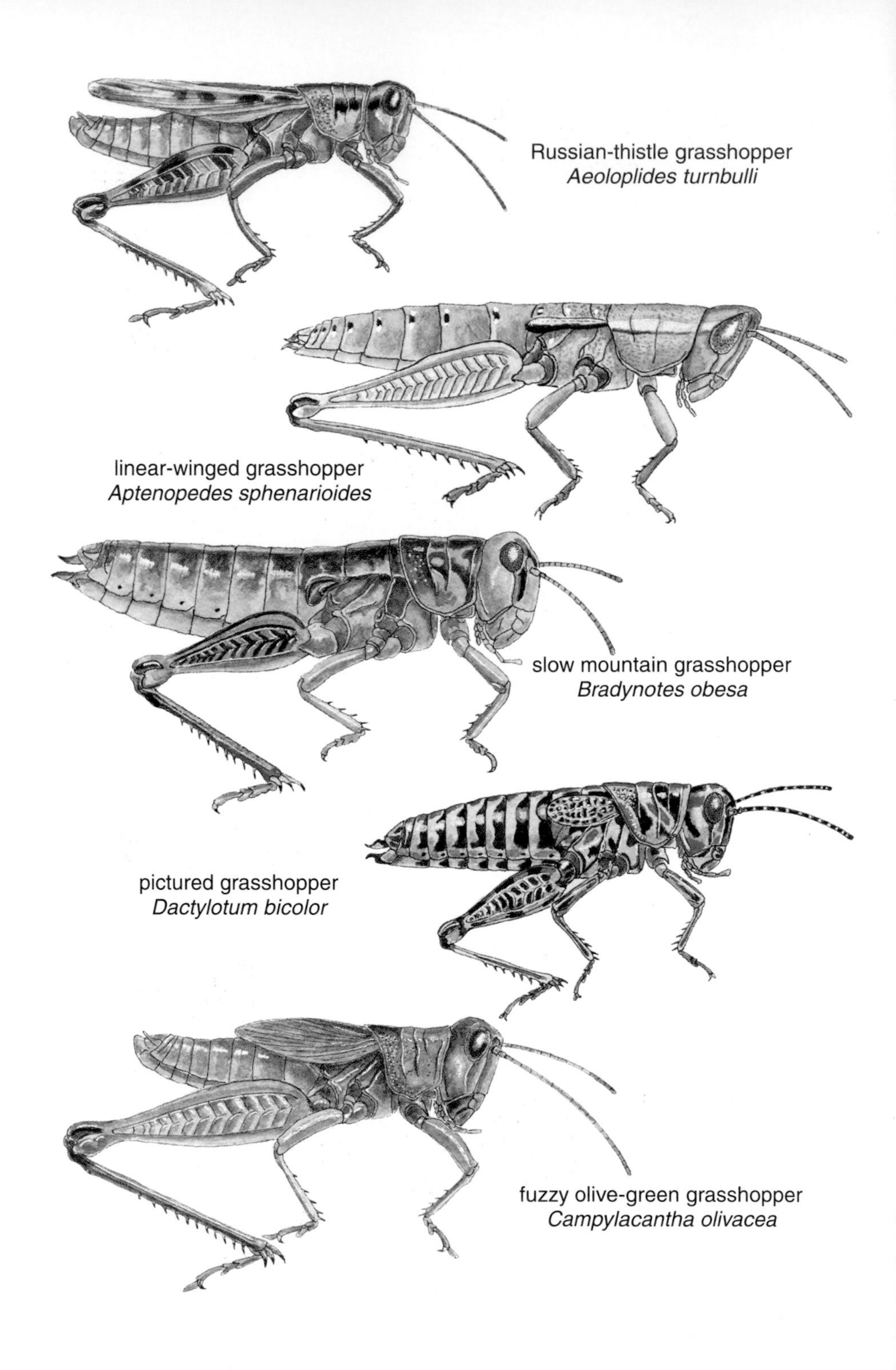

Plate 22: Spurthroated Grasshoppers
subfamily Cyrtacanthacridinae

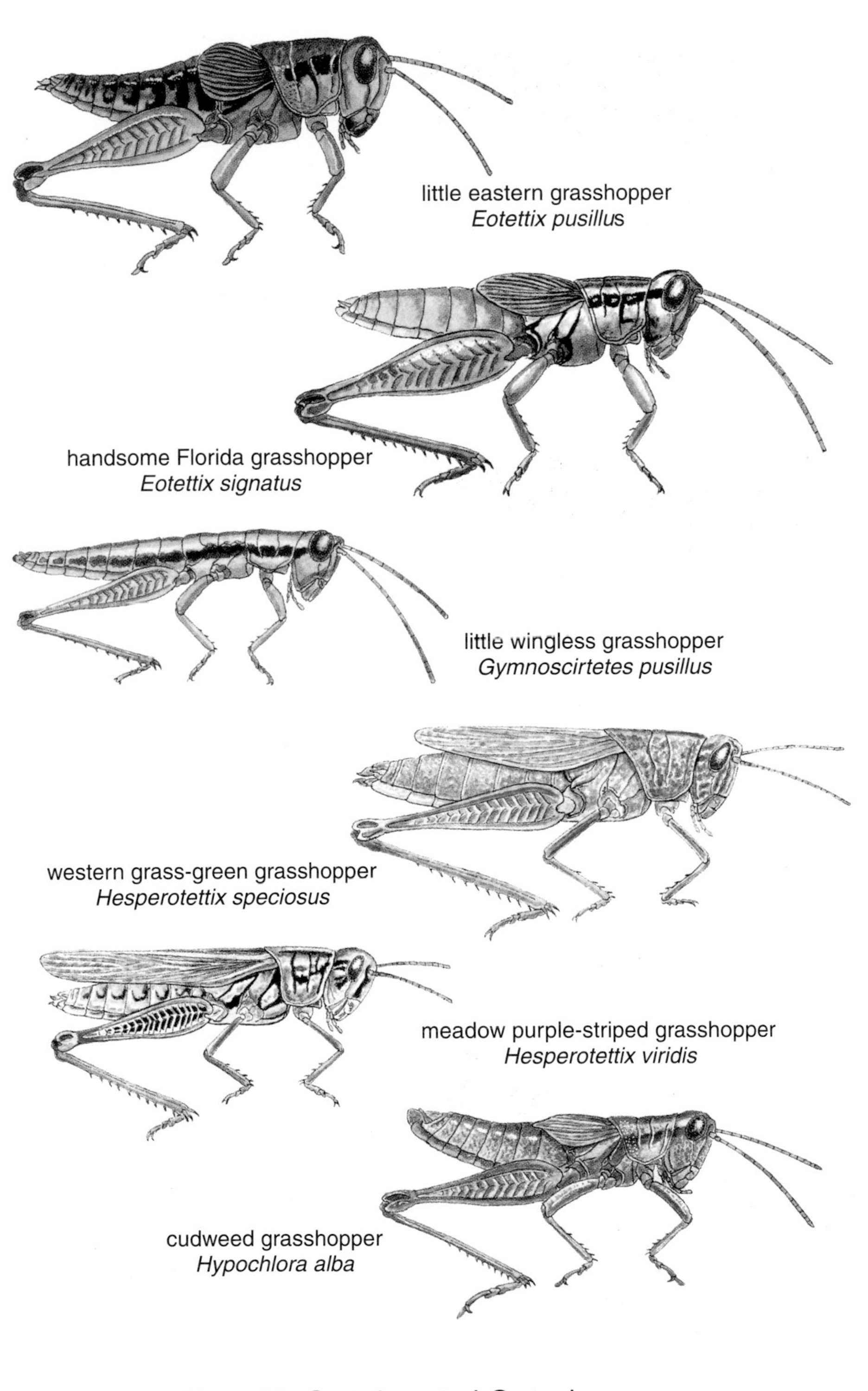

Plate 23: Spurthroated Grasshoppers
subfamily Cyrtacanthacridinae

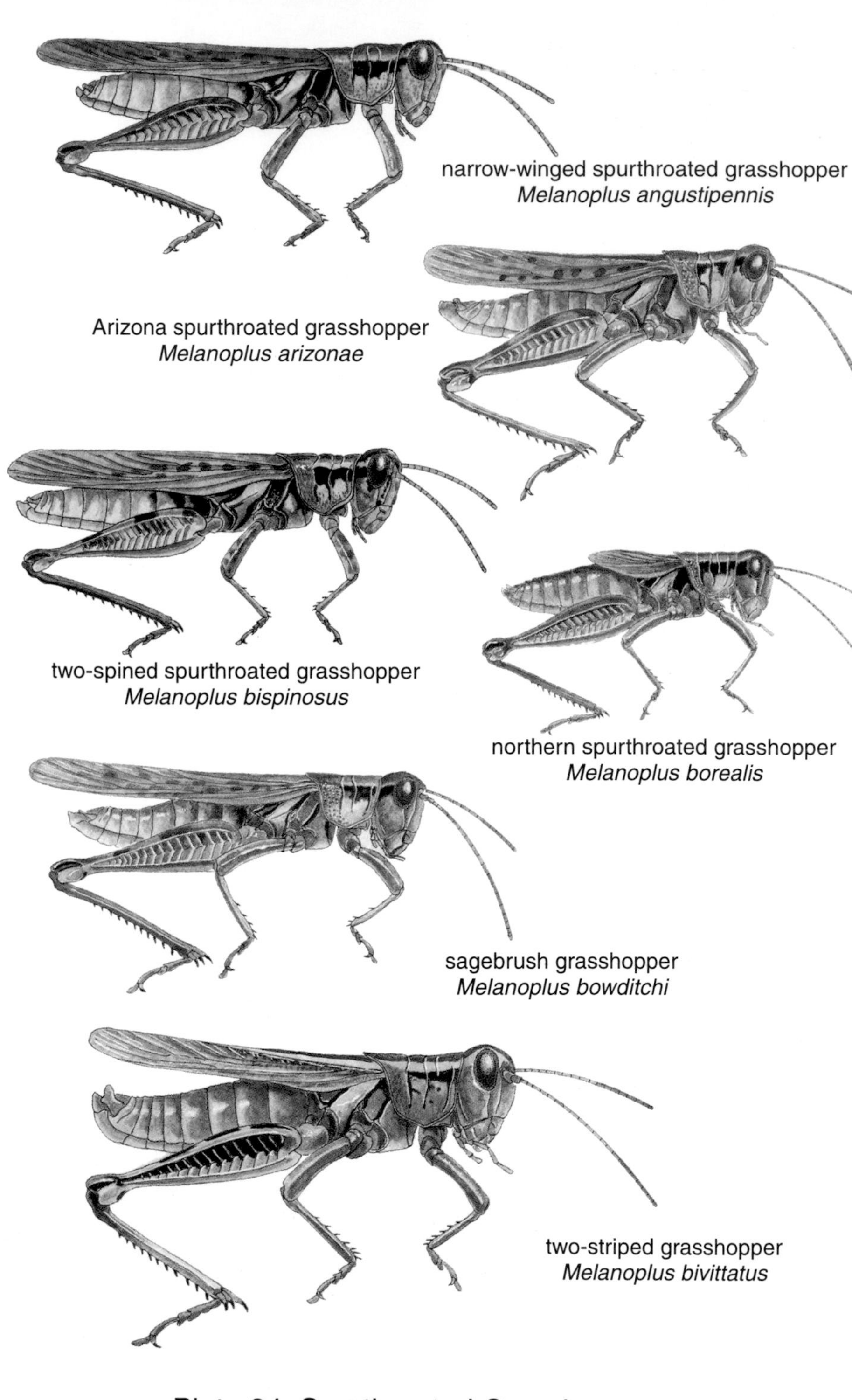

Plate 24: Spurthroated Grasshoppers
subfamily Cyrtacanthacridinae

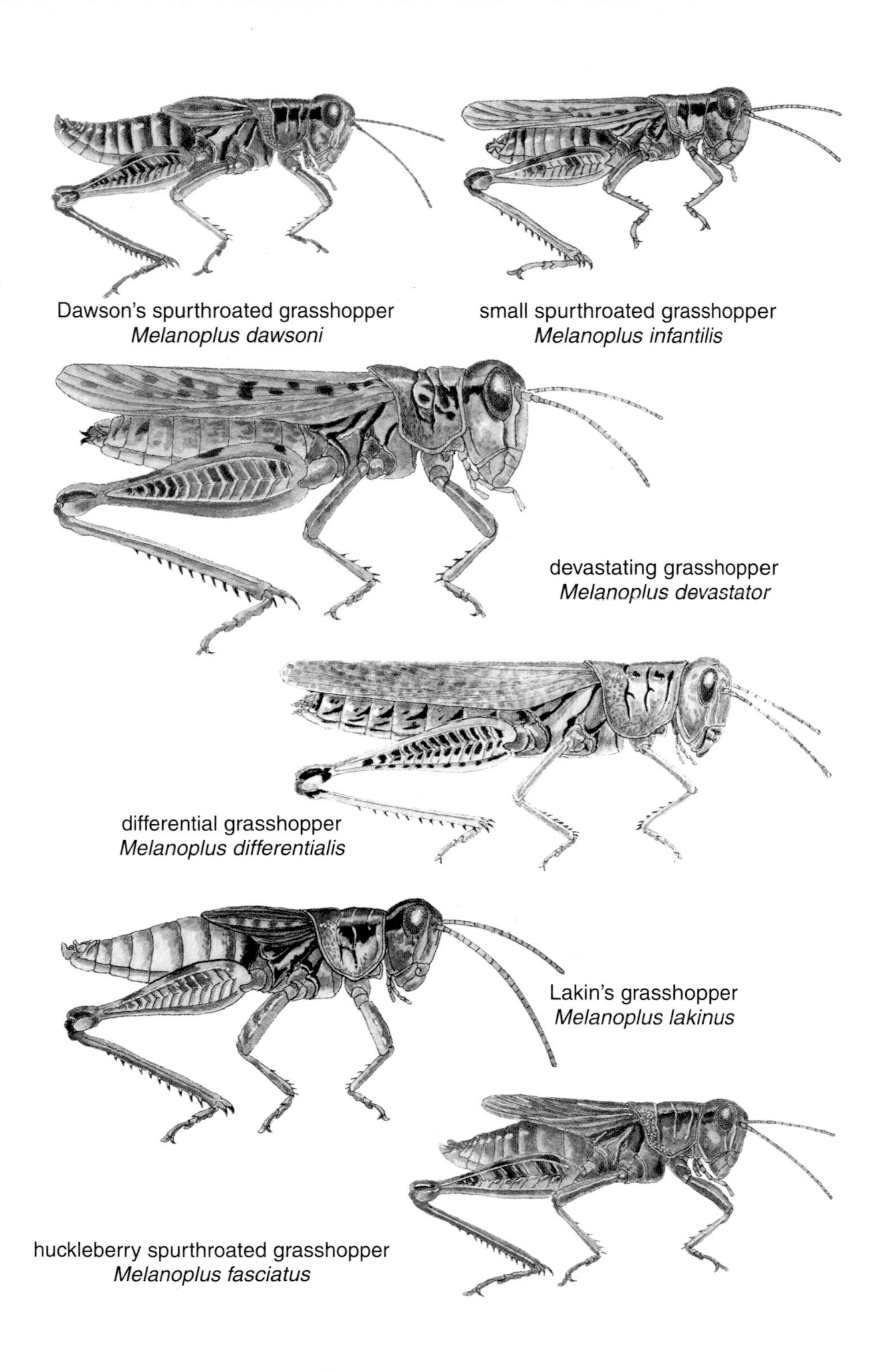

Plate 25: Spurthroated Grasshoppers
subfamily Cyrtacanthacridinae

red-legged grasshopper
Melanoplus femurrubrum

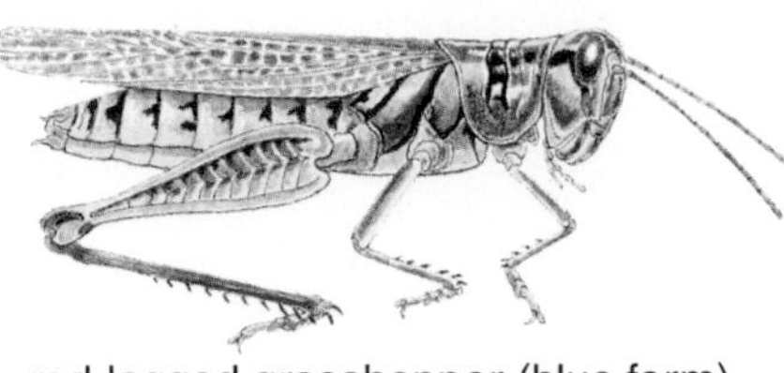

red-legged grasshopper (blue form)
Melanoplus femurrubrum

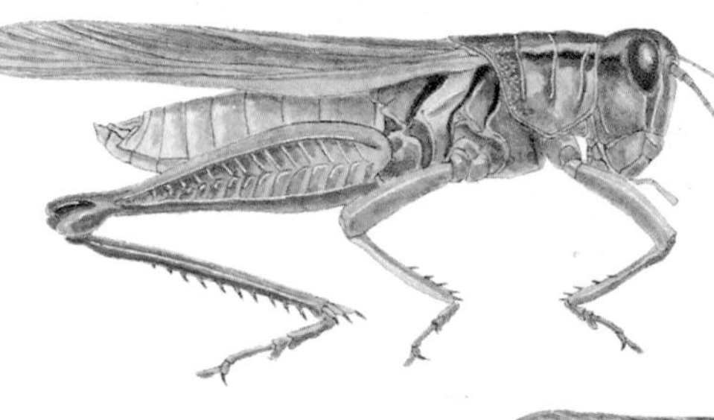

foedus grasshopper
Melanoplus foedus

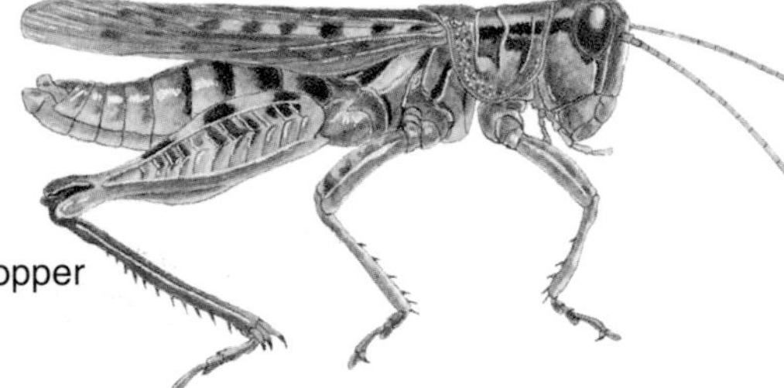

Gladston's spurthroated grasshopper
Melanoplus gladstoni

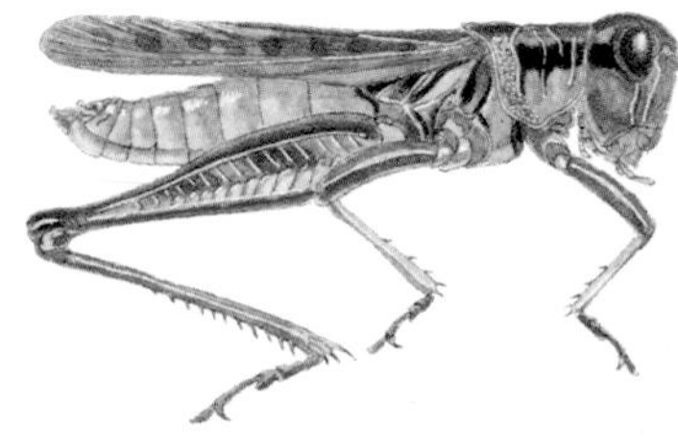

little pasture spurthroated grasshopper
Melanoplus confusus

Davis's oak grasshopper
Melanoplus davisi

Plate 26: Spurthroated Grasshoppers

subfamily Cyrtacanthacridinae

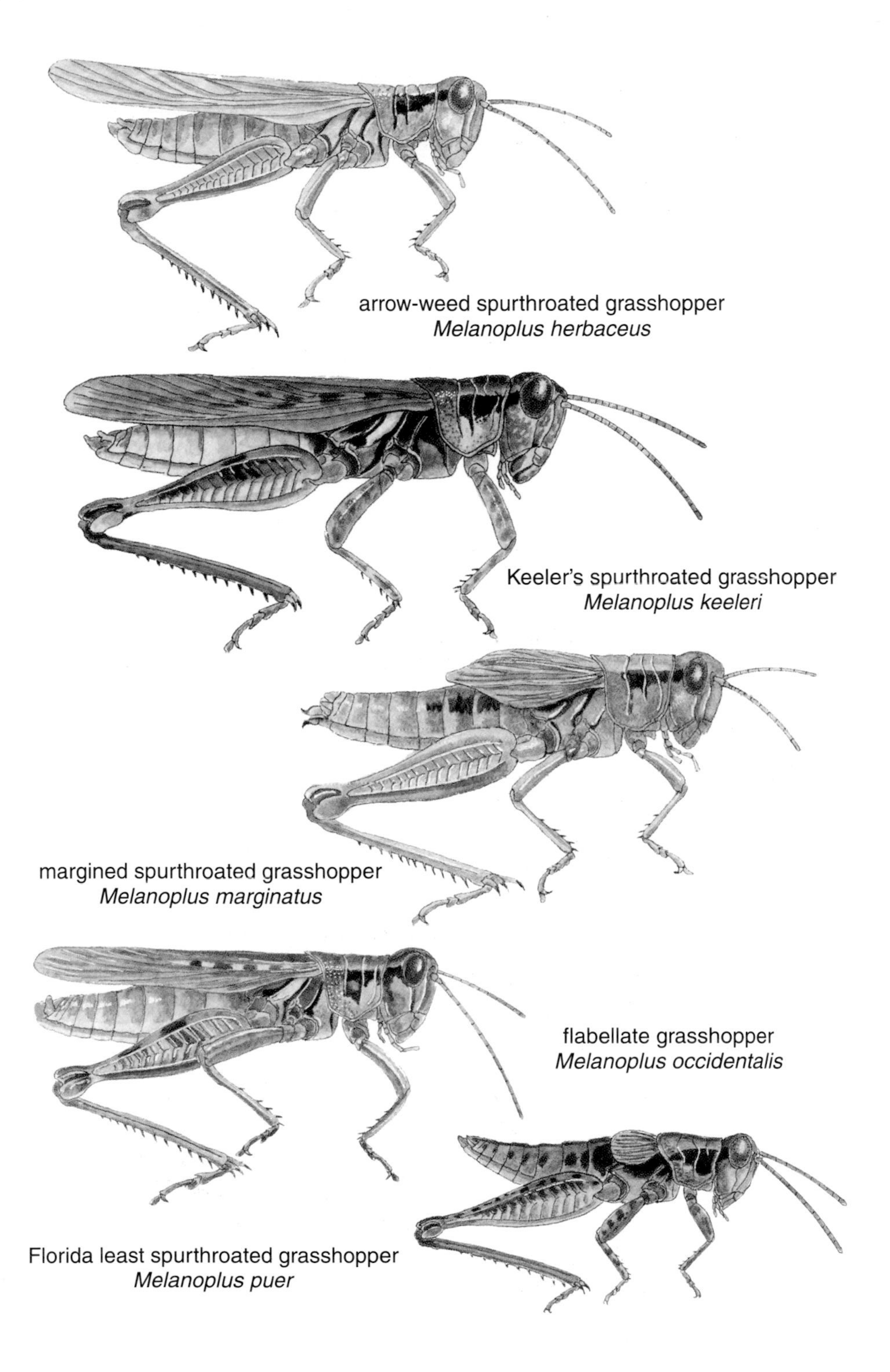

Plate 27: Spurthroated Grasshoppers
subfamily Cyrtacanthacridinae

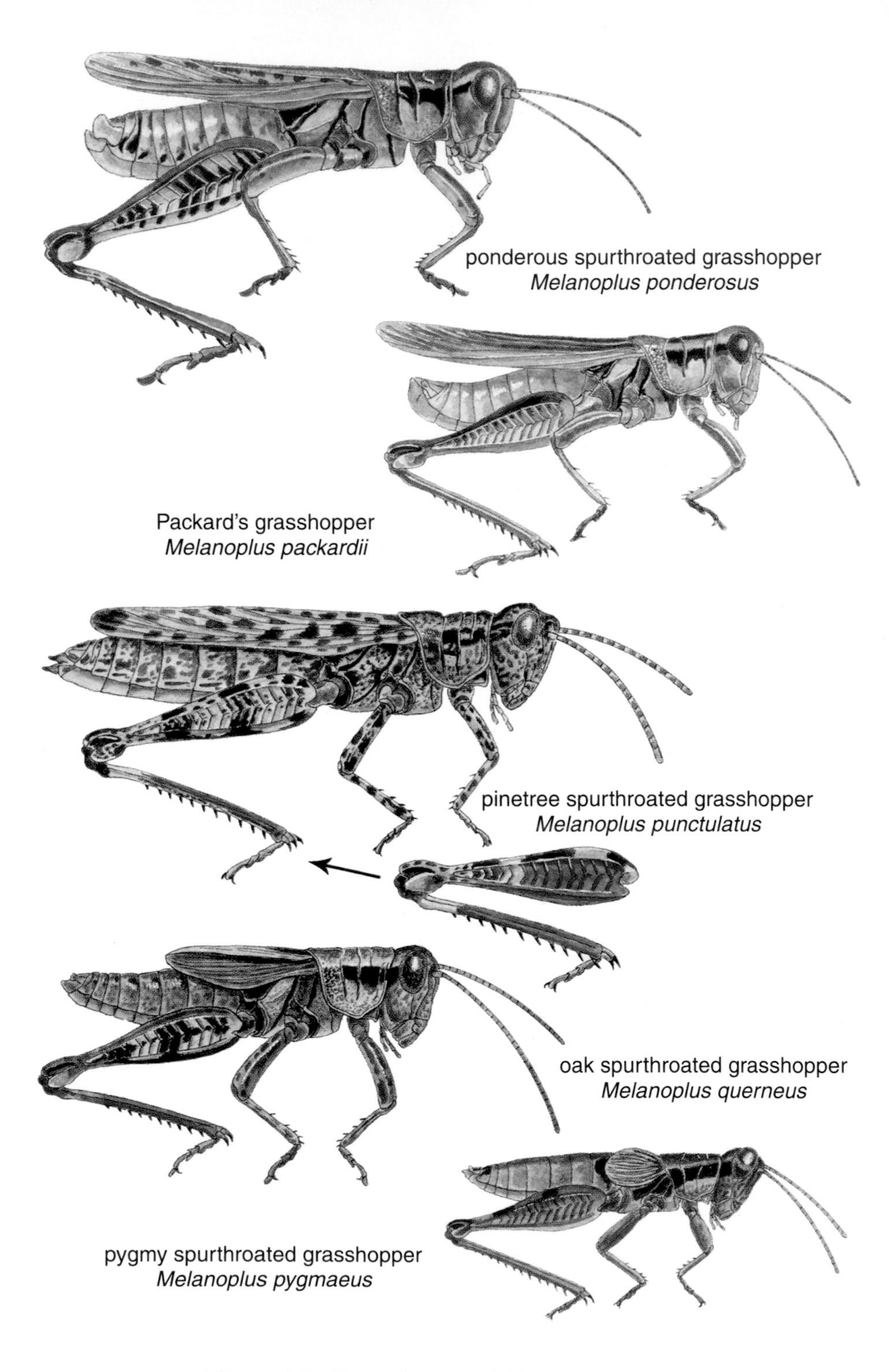

Plate 28: Spurthroated Grasshoppers
subfamily Cyrtacanthacridinae

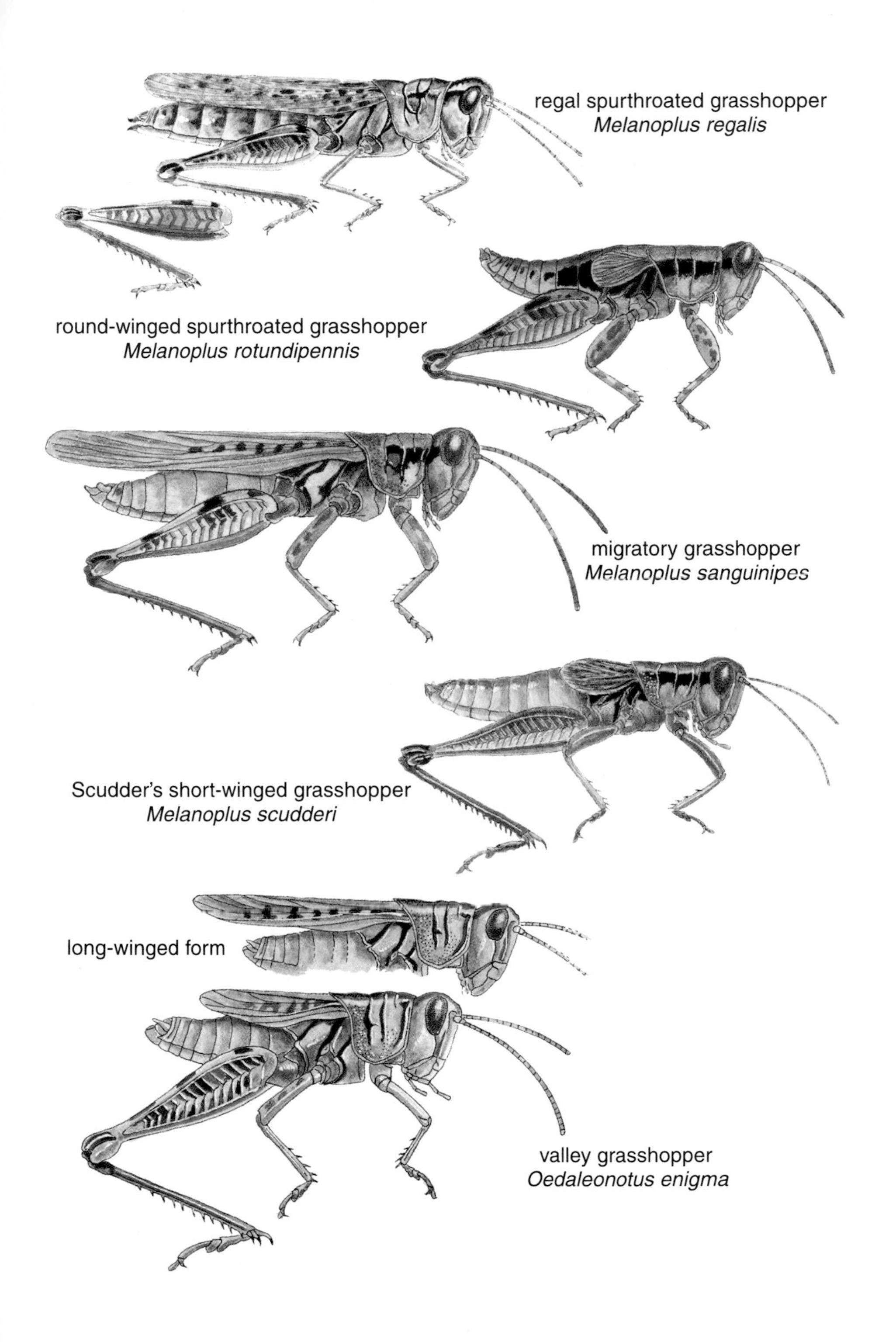

Plate 29: Spurthroated Grasshoppers
subfamily Cyrtacanthacridinae

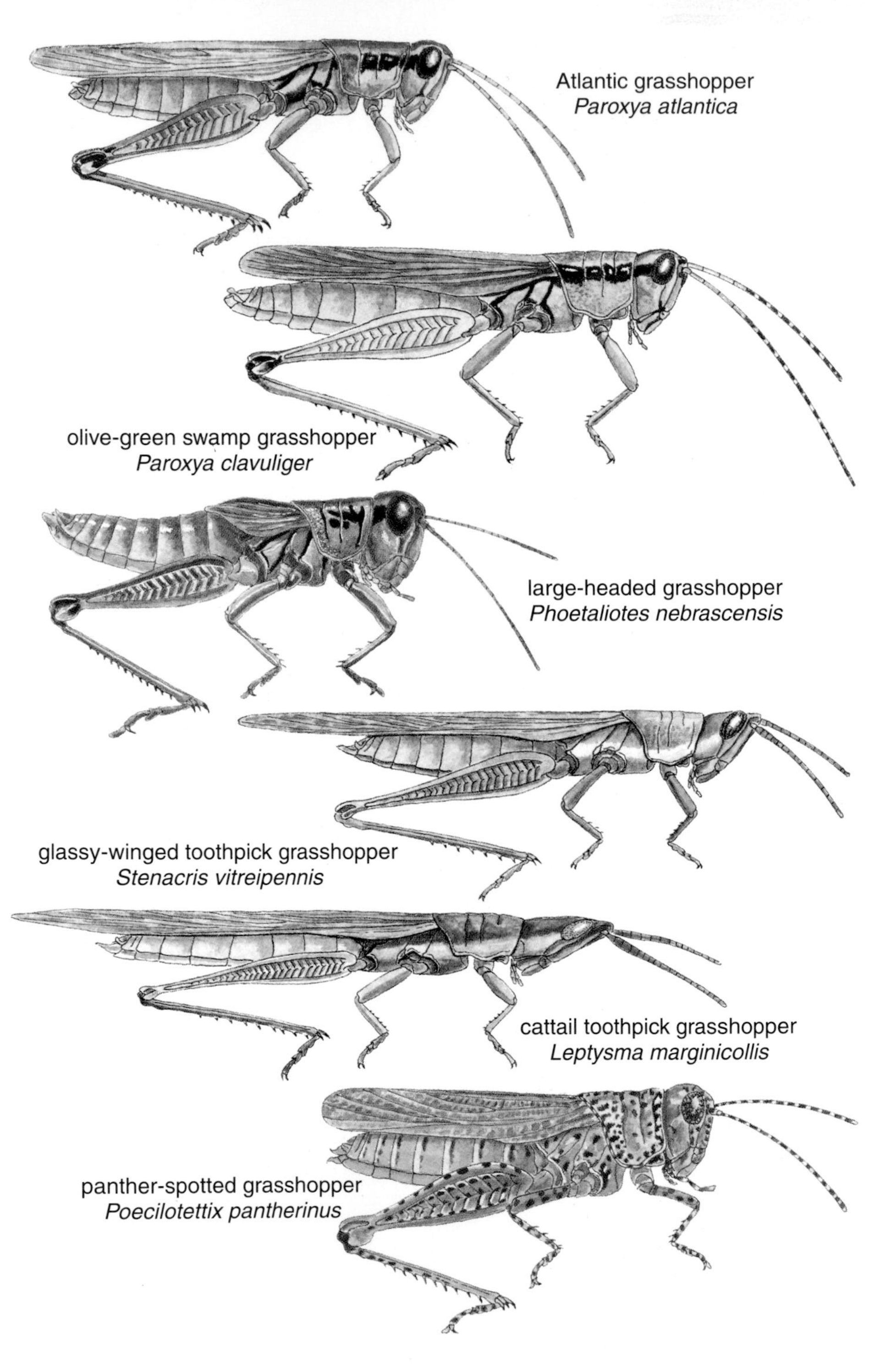

Plate 30: Spurthroated Grasshoppers
subfamily Cyrtacanthacridinae

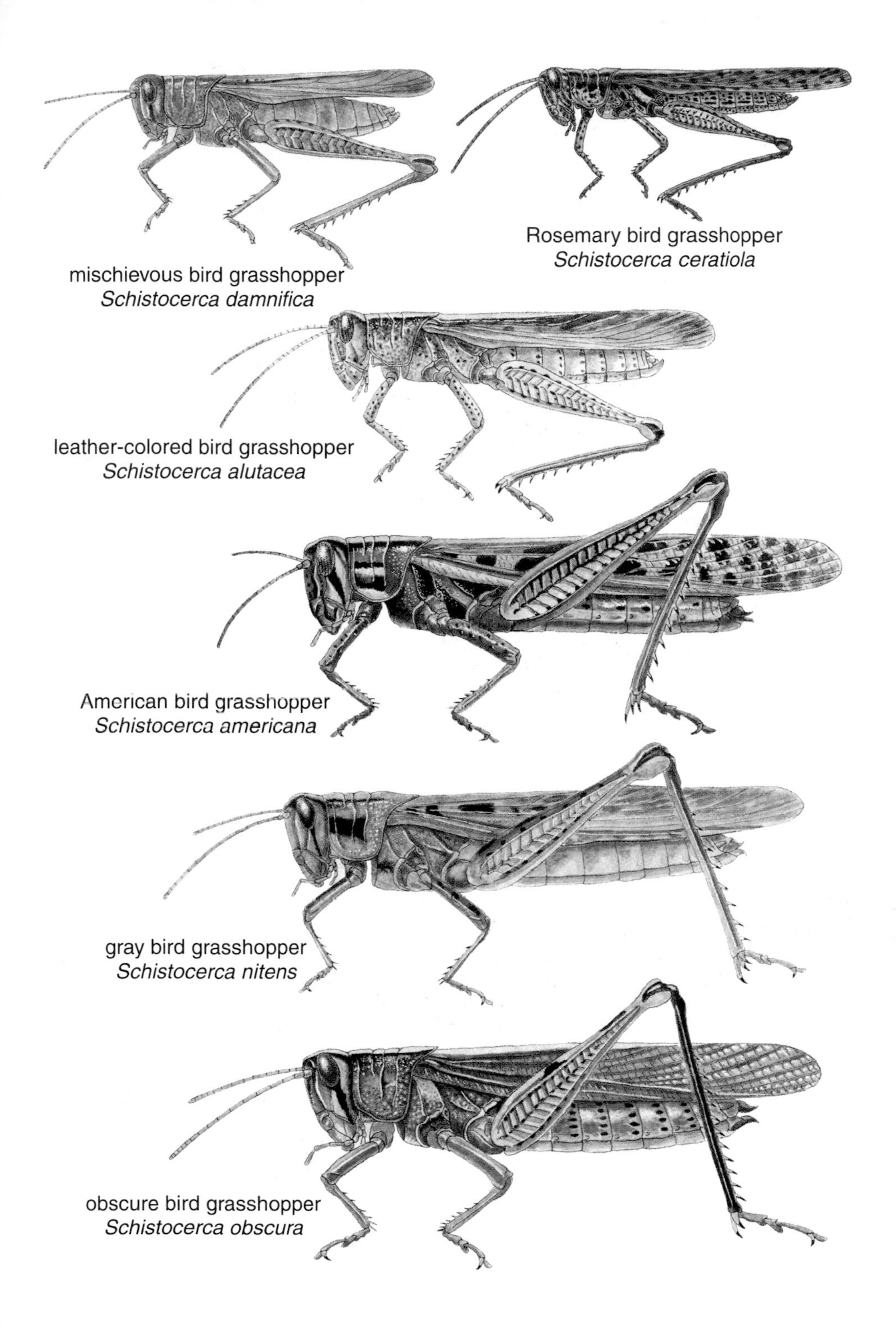

Plate 31: Spurthroated Grasshoppers
subfamily Cyrtacanthacridinae

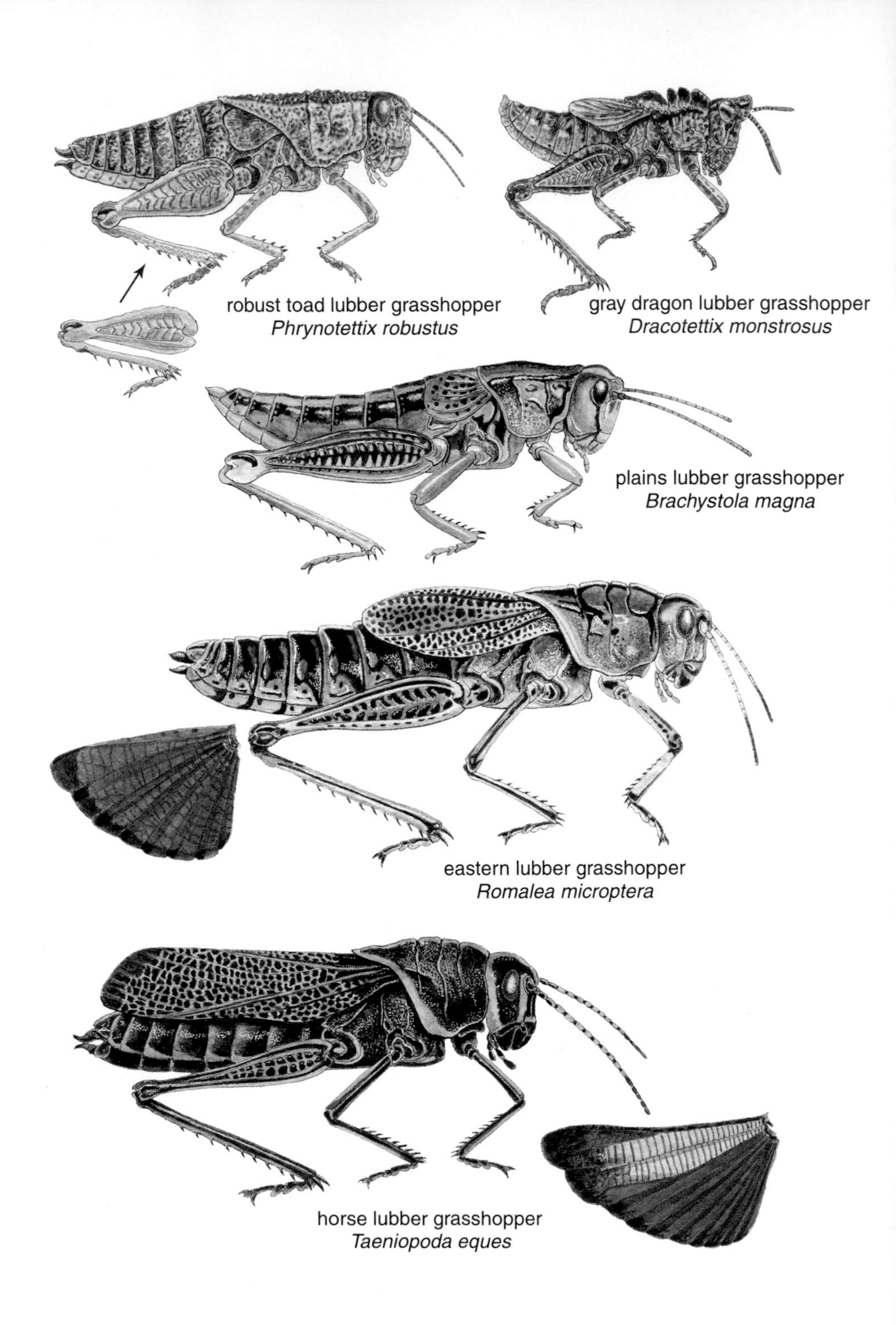

Plate 32: Lubber Grasshoppers
subfamily Romaleinae

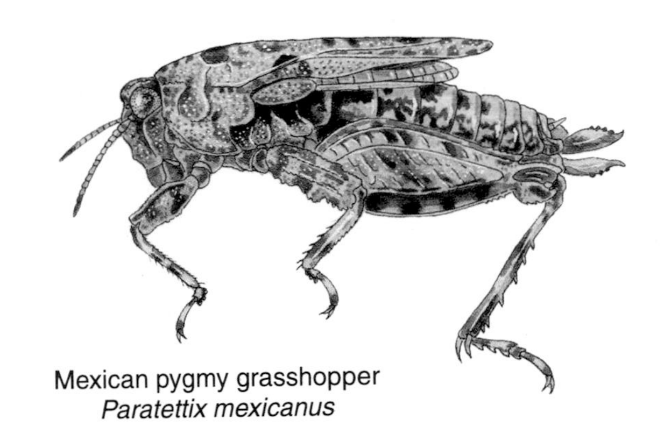

Mexican pygmy grasshopper
Paratettix mexicanus

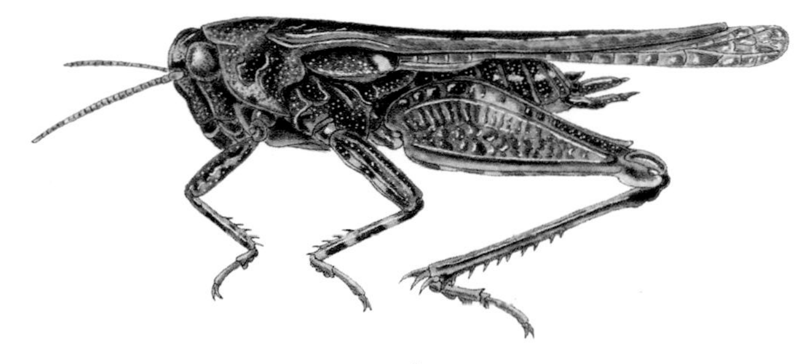

black-sided pygmy grasshopper
Tettigidea lateralis

subfamily Tetrigidae

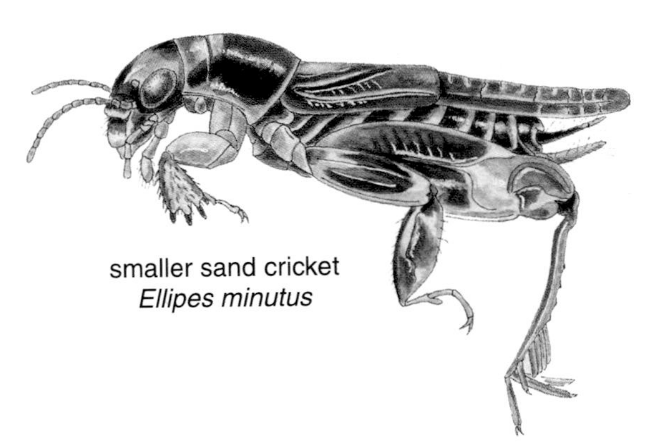

smaller sand cricket
Ellipes minutus

subfamily Tridactylidae

Plate 33: Pygmy Grasshoppers and Pygmy Mole Crickets

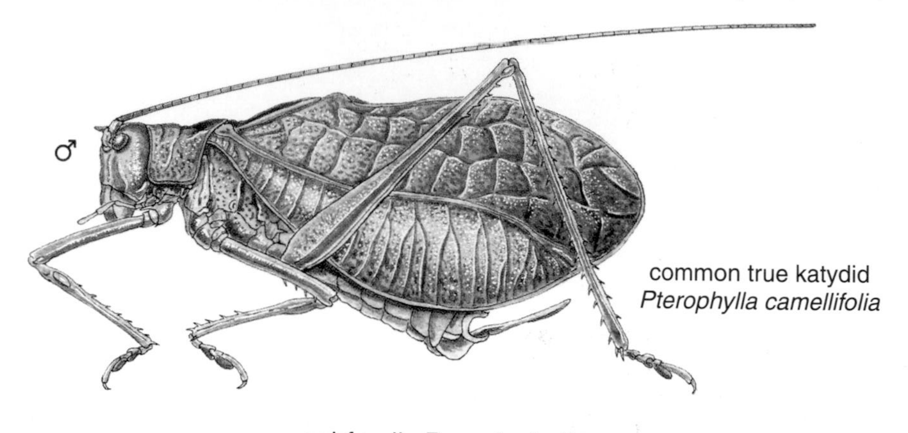

common true katydid
Pterophylla camellifolia

subfamily Pseudophyllinae

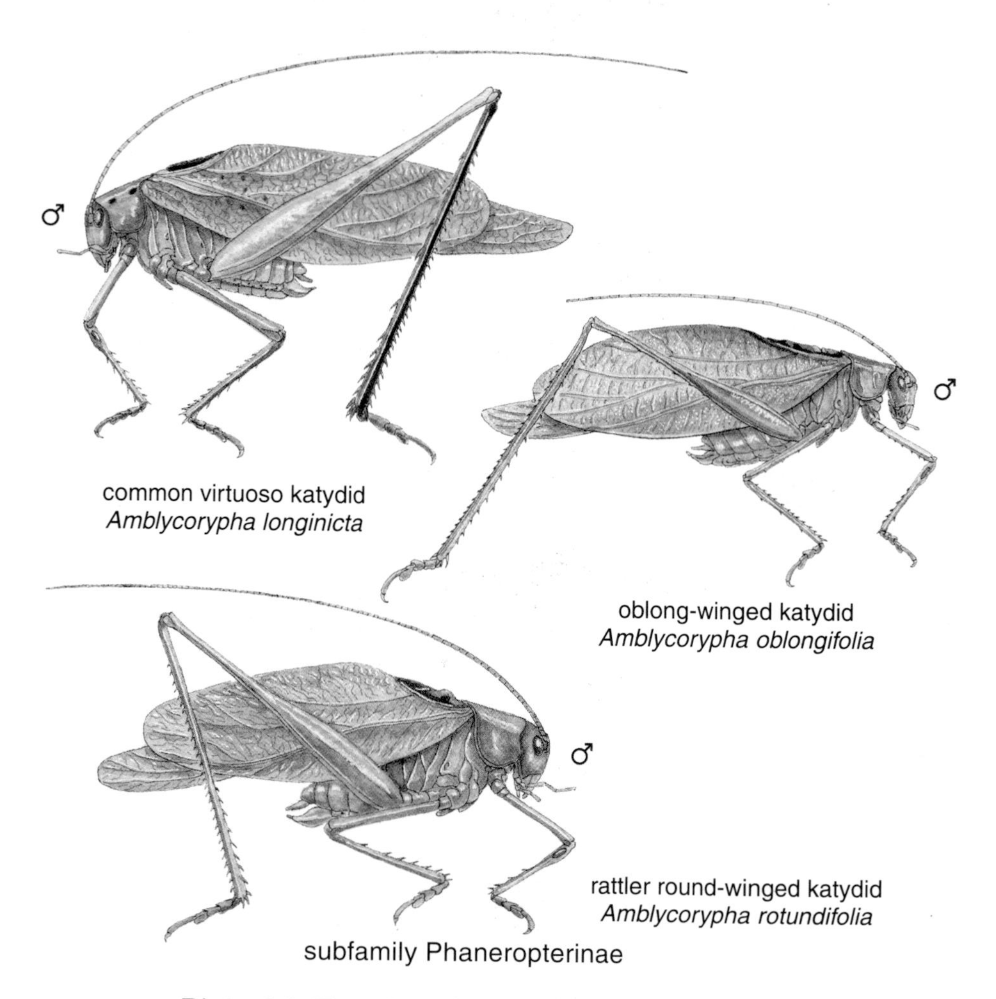

common virtuoso katydid
Amblycorypha longinicta

oblong-winged katydid
Amblycorypha oblongifolia

rattler round-winged katydid
Amblycorypha rotundifolia

subfamily Phaneropterinae

Plate 34: True Katydids and False Katydids

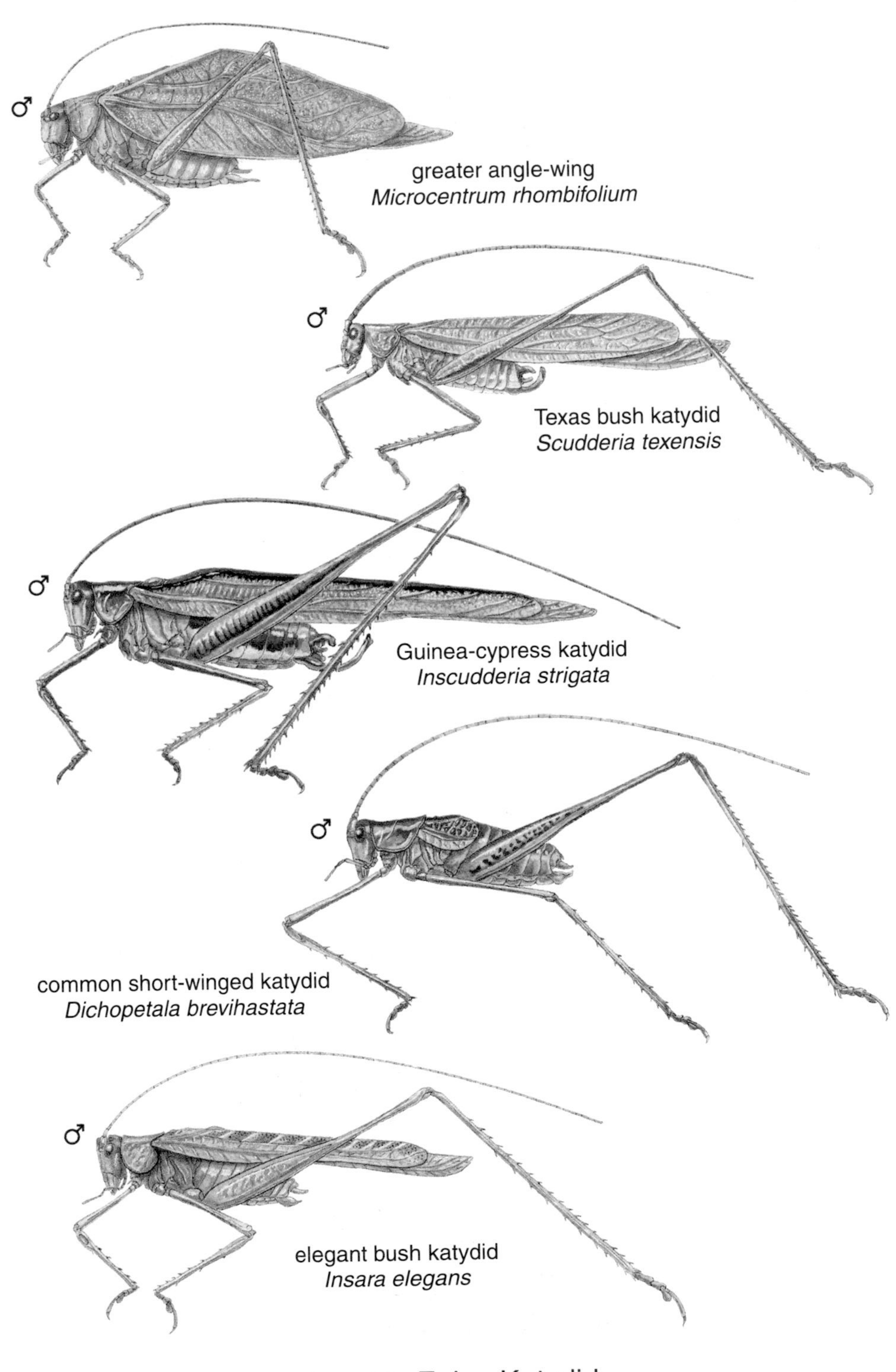

Plate 35: False Katydids
subfamily Phaneropterinae

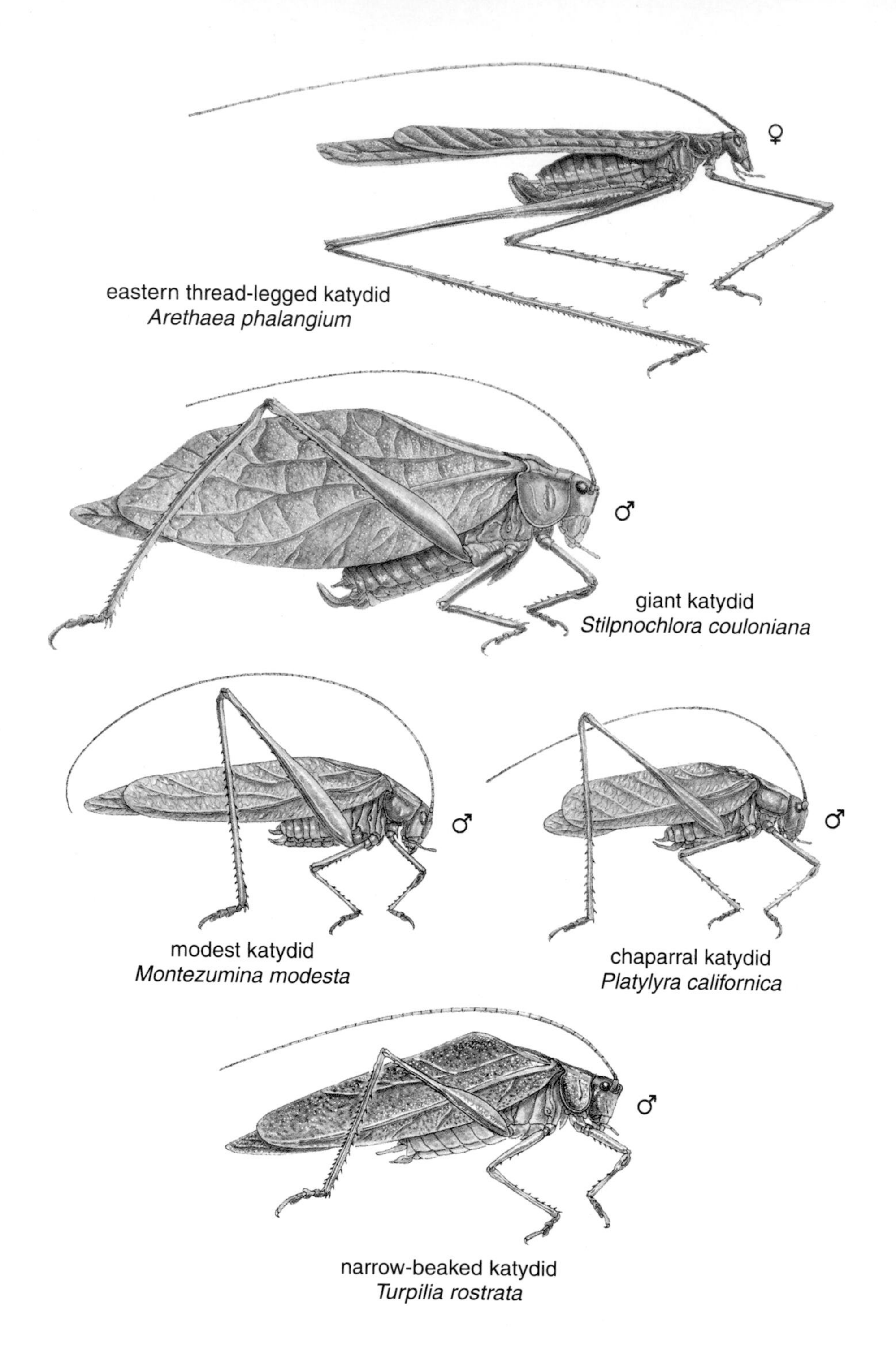

Plate 36: False Katydids
subfamily Phaneropterinae

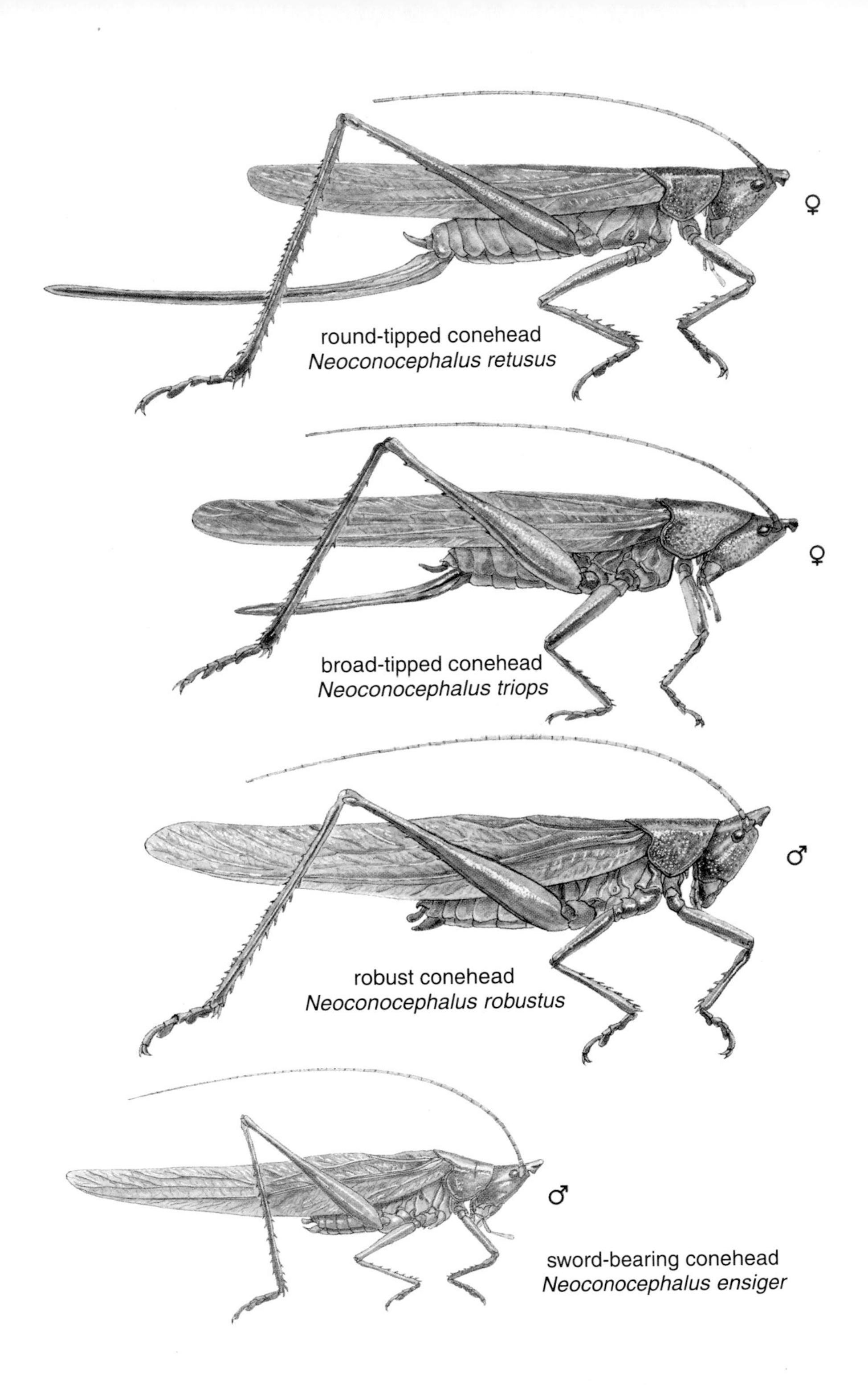

Plate 37: Coneheaded Katydids
subfamily Copiphorinae

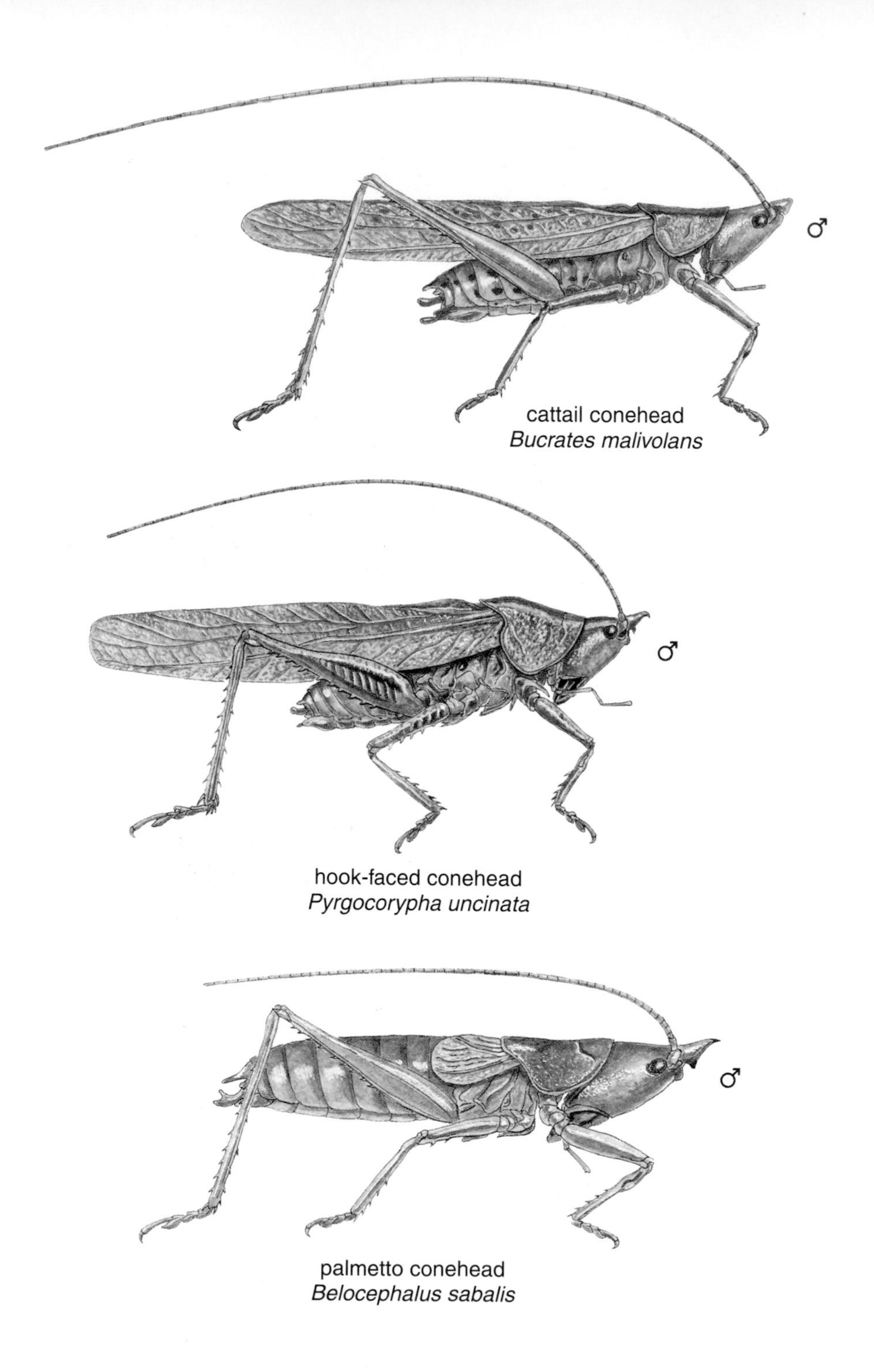

Plate 38: Coneheaded Katydids
subfamily Copiphorinae

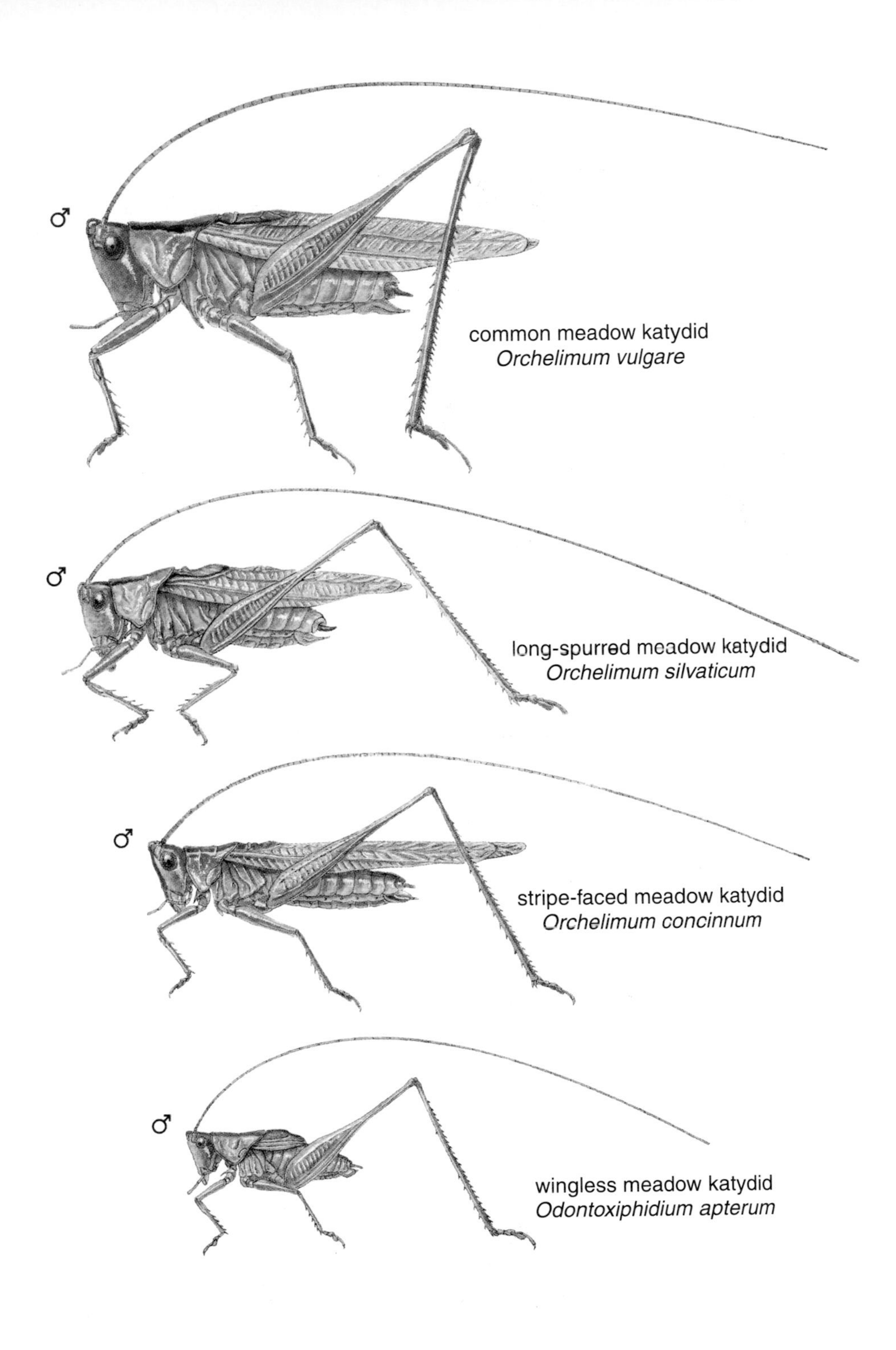

Plate 39: Meadow Katydids
subfamily Conocephalinae

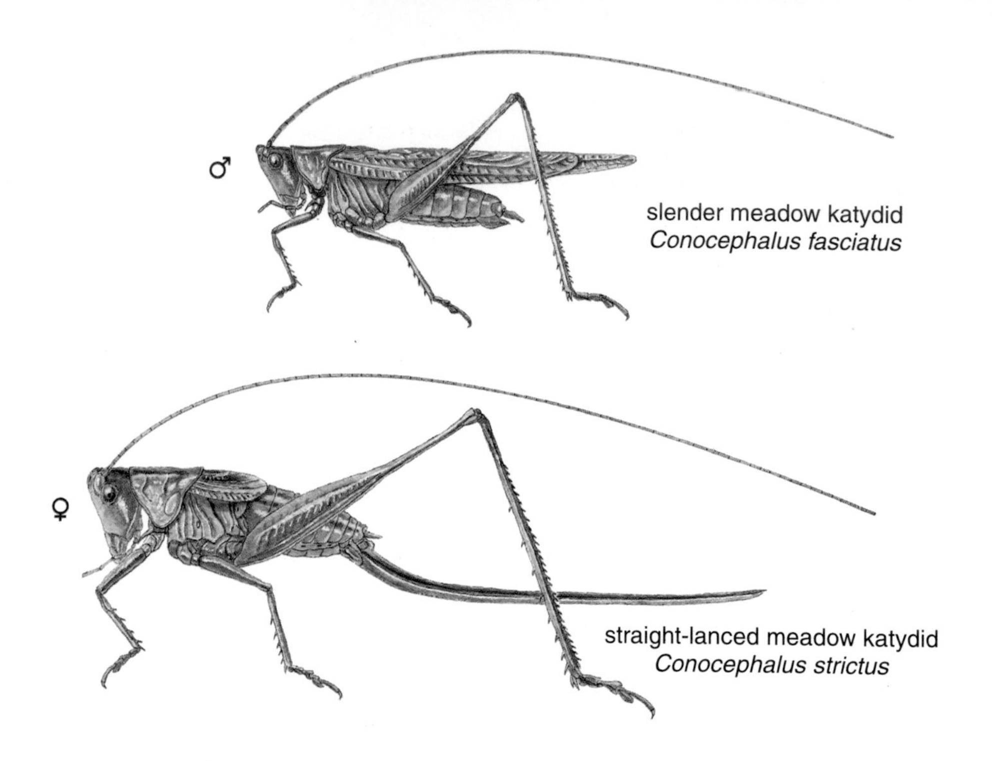

slender meadow katydid
Conocephalus fasciatus

straight-lanced meadow katydid
Conocephalus strictus

subfamily Conocephalinae

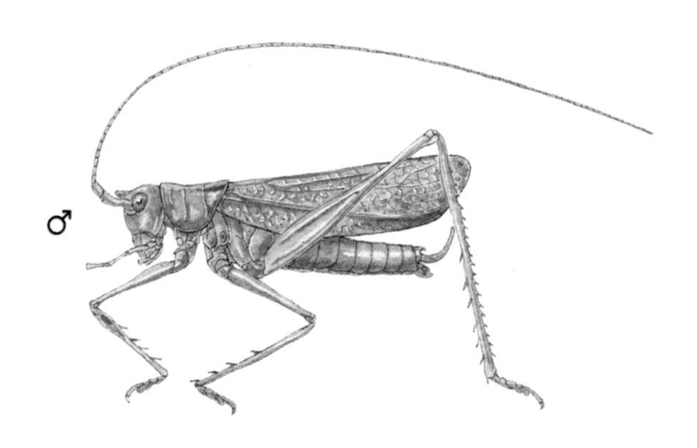

drumming katydid
Meconema thalassinum

subfamily Meconematinae

Plate 40: Meadow Katydids
and Quiet-calling Katydids

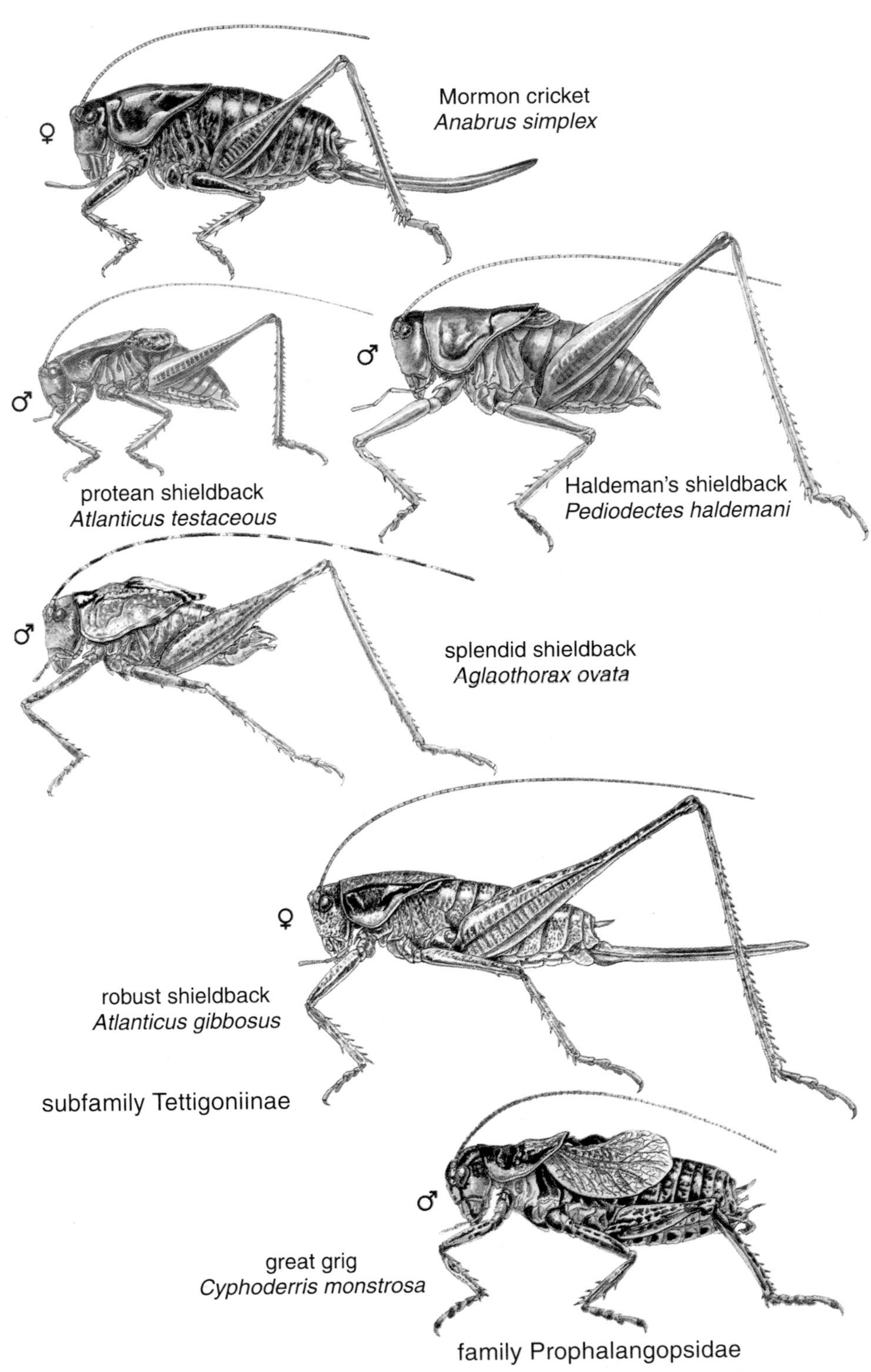

Plate 41: Predaceous Katydids and Hump-winged Grigs

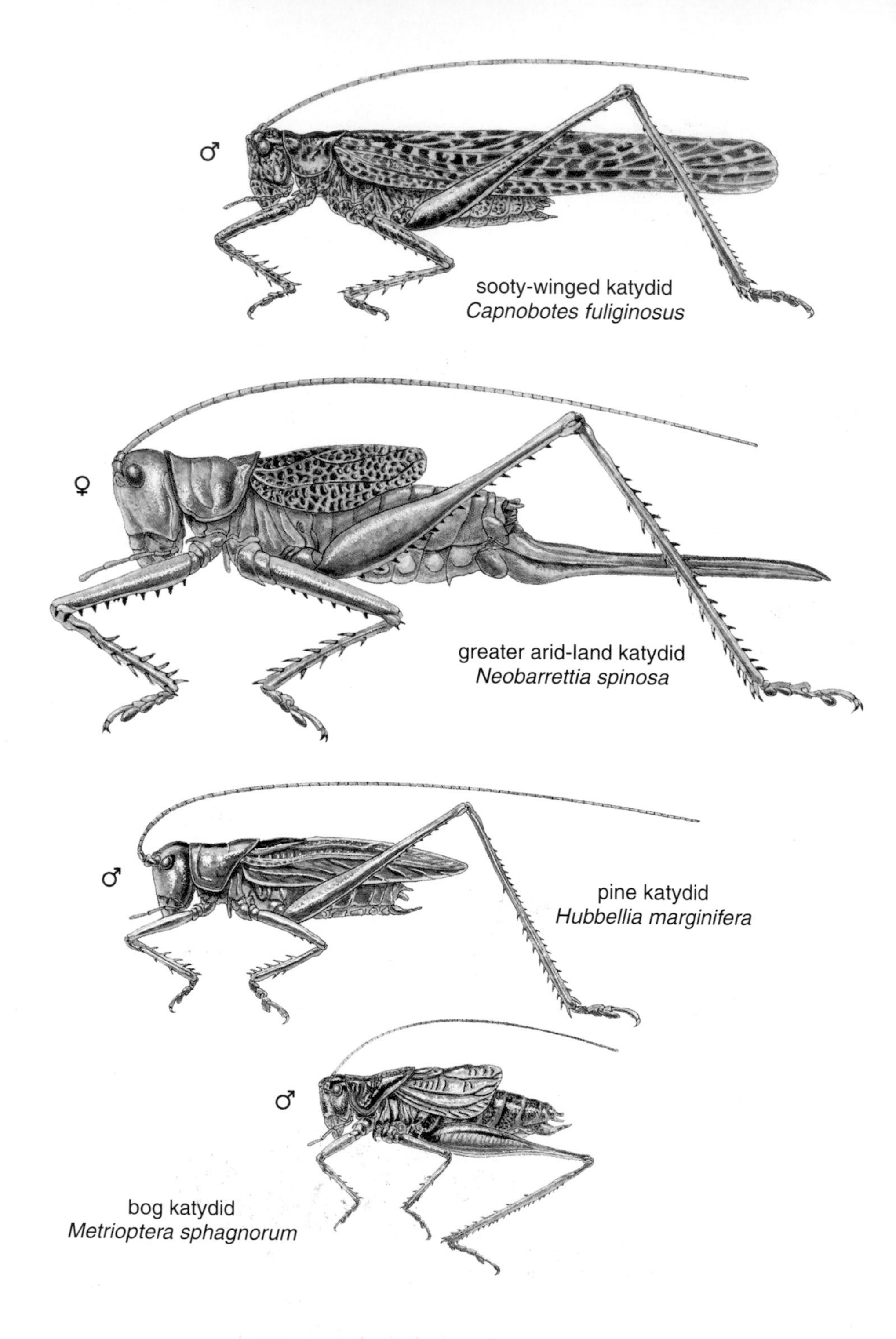

Plate 42: Predaceous Katydids
subfamily Tettigoniinae

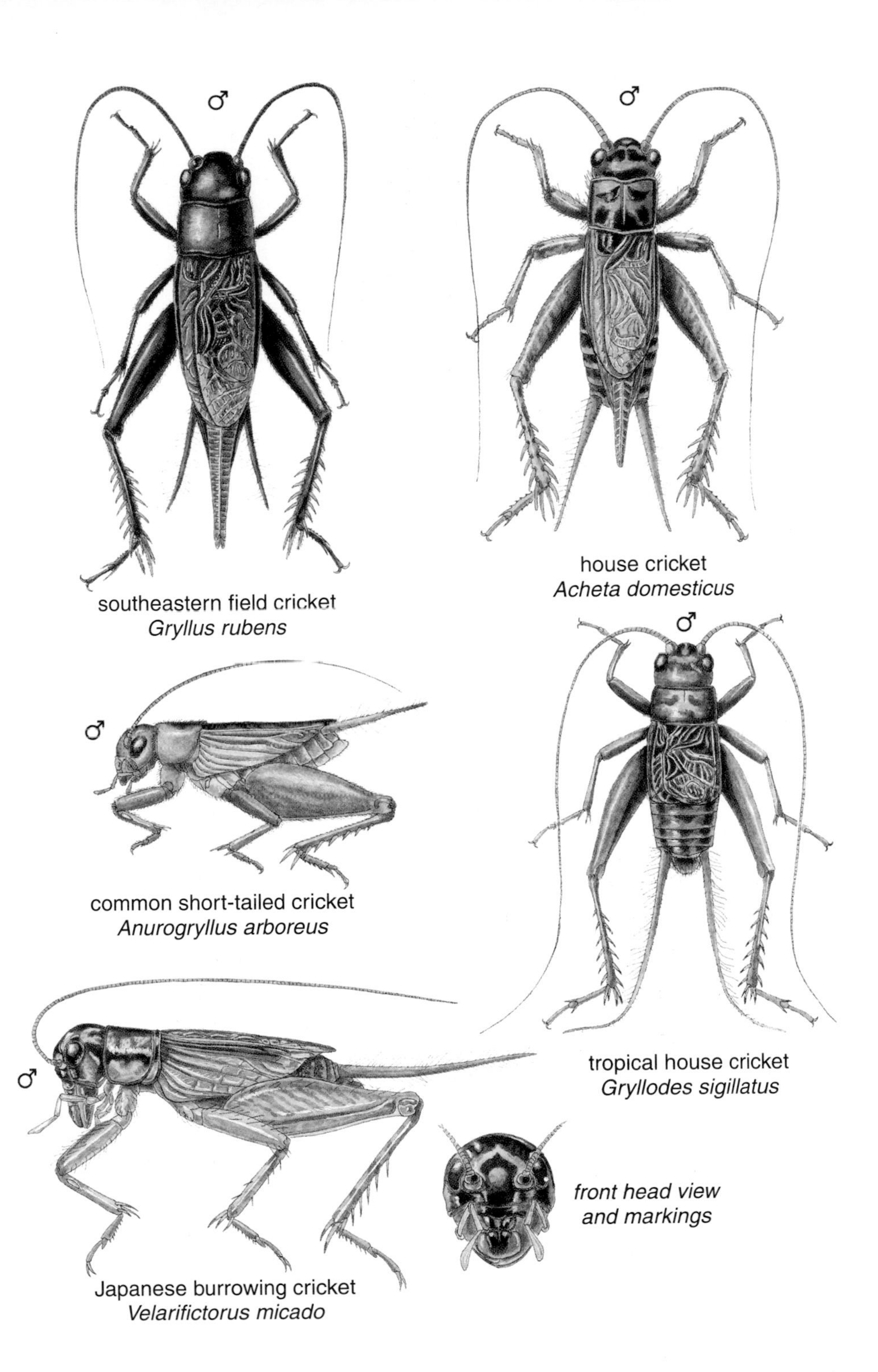

Plate 43: Field Crickets
subfamily Gryllinae

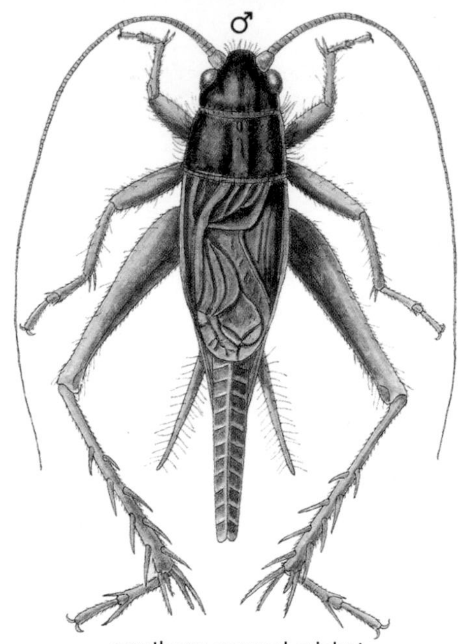

southern ground cricket
Allonemobius socius

subfamily Nemobiinae

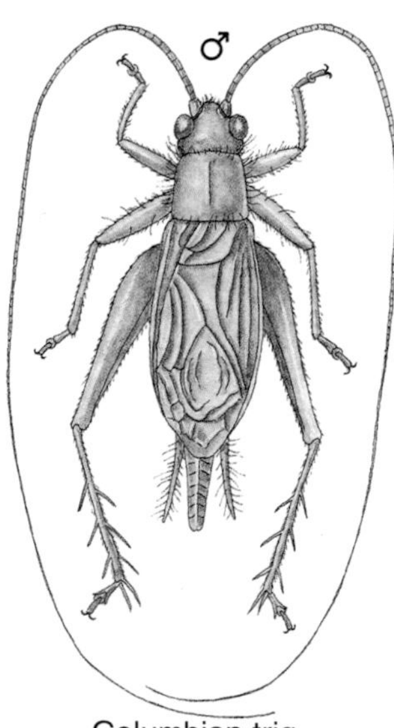

Columbian trig
Cyrtoxipha columbiana

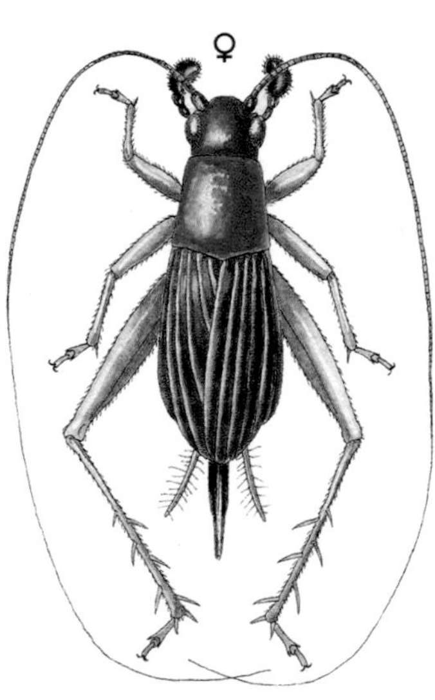

handsome trig
Phyllopalpus pulchellus

subfamily Trigonidiinae

Plate 44: Ground Crickets and Sword-tail Crickets

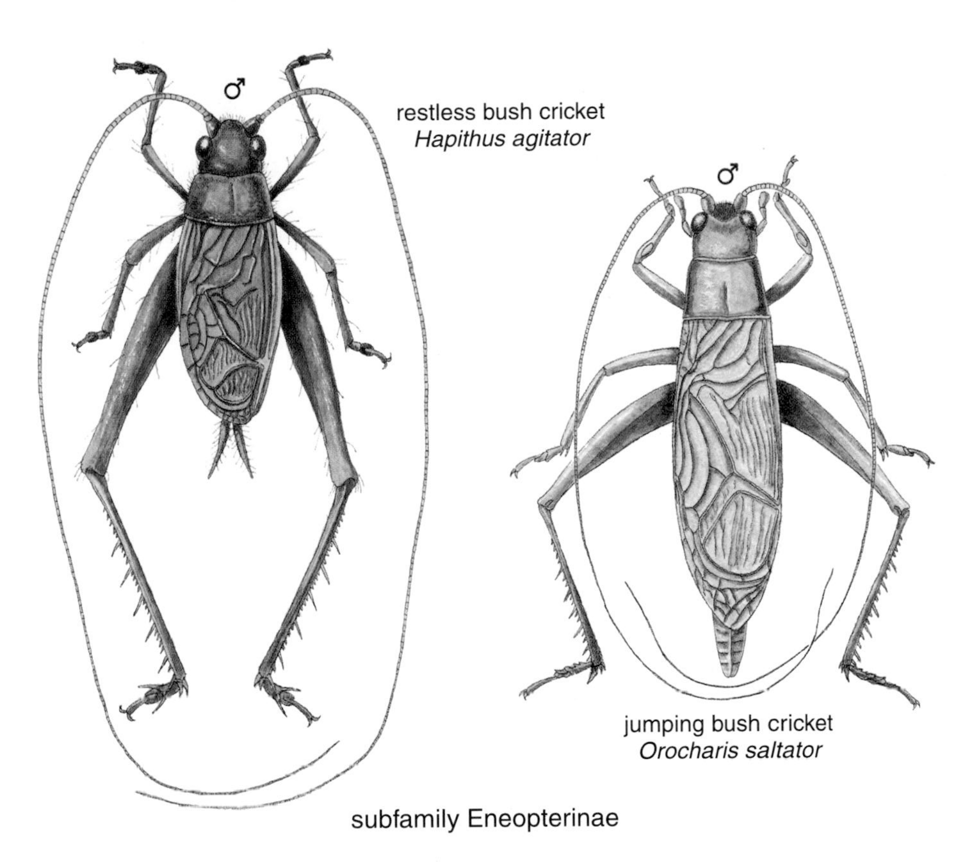

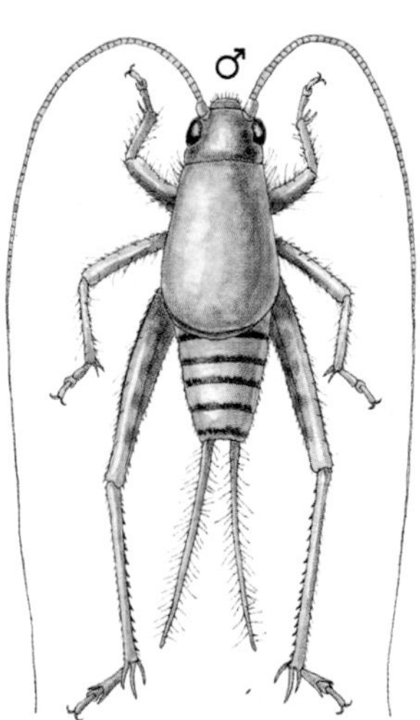

Slosson's scaly cricket
Cycloptilum slossoni

subfamily Mogoplistinae

Plate 45: Bush Crickets and Scaly Crickets

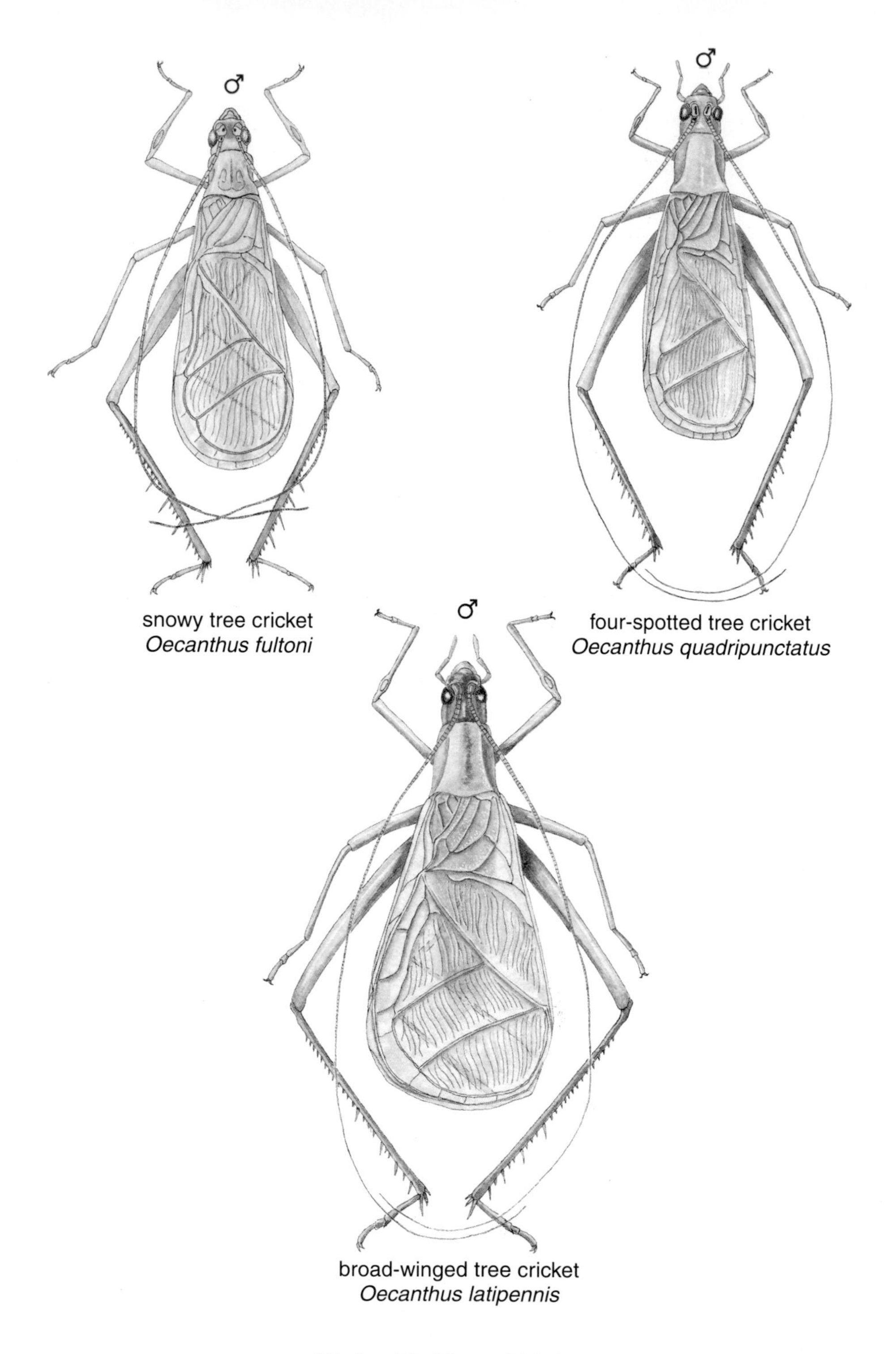

Plate 46: Tree Crickets
subfamily Oecanthinae

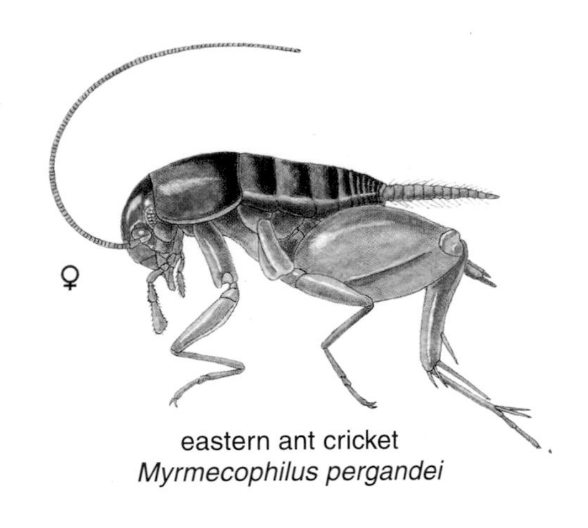

eastern ant cricket
Myrmecophilus pergandei

subfamily Myrmecophilinae

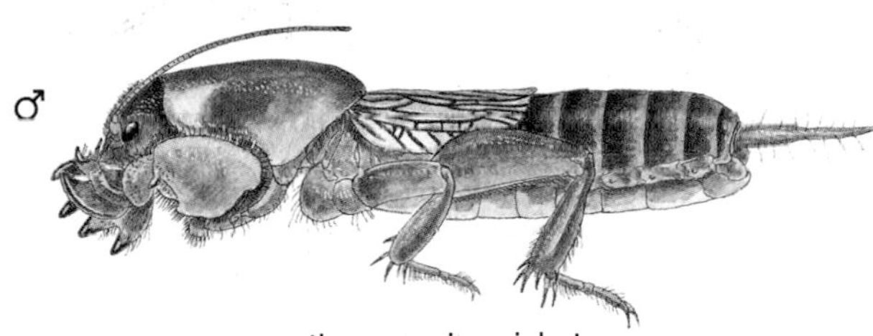

northern mole cricket
Neocurtilla hexadactyla

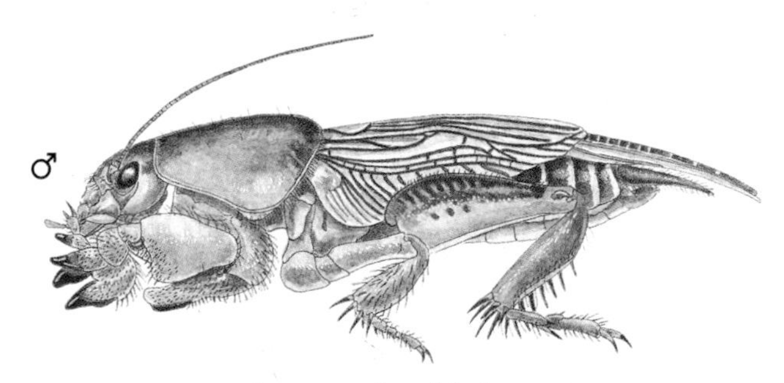

tawny mole cricket
Scapteriscus vicinus

family Gryllotalpidae

Plate 47: Ant Crickets and Mole Crickets

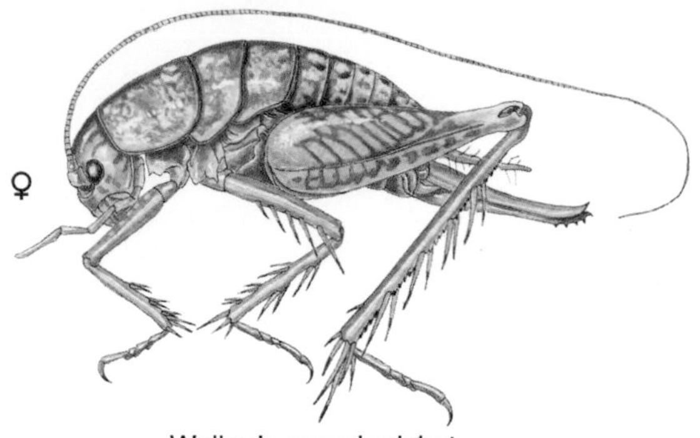

Walker's camel cricket
Ceuthophilus walkeri

family Rhaphidophoridae

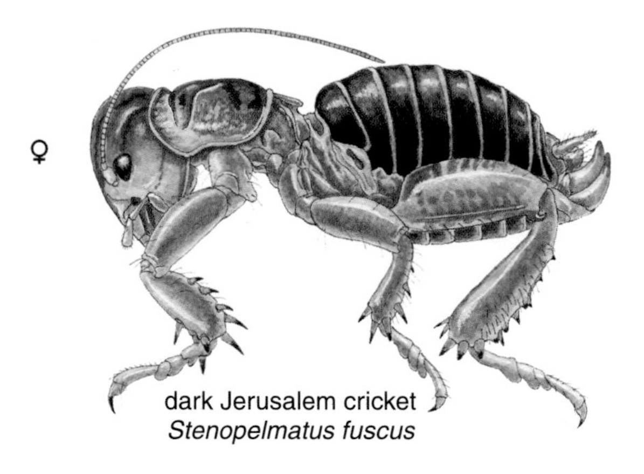

dark Jerusalem cricket
Stenopelmatus fuscus

family Stenopelmatidae

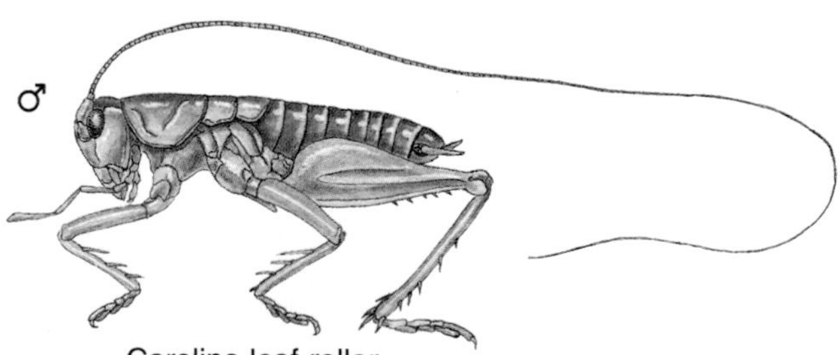

Carolina leaf-roller
Camptonotus carolinensis

family Gryllacrididae

Plate 48: Camel Crickets, Jerusalem Crickets, and Raspy Crickets

ends in a sharp point. The antennae are relatively short, as in the acridid grasshoppers (family Acrididae). They may be long or short winged, or wingless. Like other grasshoppers, they have enlarged hind femora. The front and middle tarsi are two-segmented, and the hind tarsi are three-segmented. Auditory and stridulatory organs are absent. They apparently feed on leaf debris and algae associated with soil. They often are found at the margins of water. Eggs are deposited in small clusters in the soil. They are difficult to collect unless special effort is made to sweep close to the soil. In some environments, they may be common. There are about 30 species in the United States and Canada, and they are difficult to distinguish.

MEXICAN PYGMY GRASSHOPPER

Paratettix mexicanus (Plate 33)

Distribution: Found widely in western states, and is known from Texas to Oregon and Idaho, and south to Central America. Its distribution also extends east along the Gulf Coast to Florida.

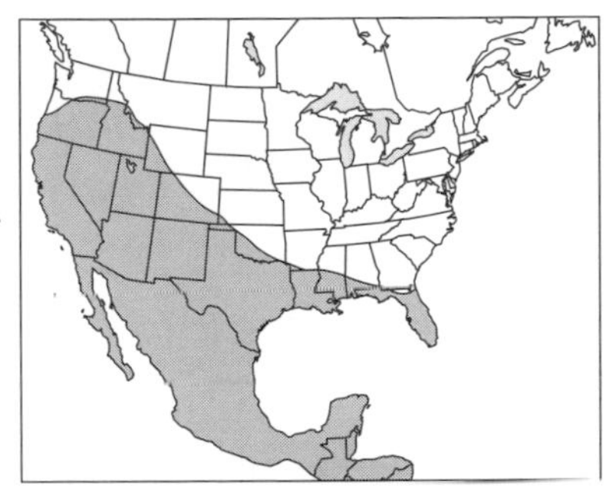

Mexican pygmy grasshopper

Identification: Variable in color pattern: may be grayish, brownish, or blackish. Pronotum is usually marked at the sides with black patches, and it may be elongate, extending to the tip of the abdomen. Males are about 8.5 mm long, females 8.8–12.4 mm.

Ecology: Typical for this family in that it is normally found along running or standing water, but otherwise, the habitat is quite varied. Two annual generations are reported over most of its range.

BLACK-SIDED PYGMY GRASSHOPPER

Tettigidea lateralis (Plate 33)

Distribution: Found widely east of the Rocky Mountains. It also occurs in the southern regions of eastern Canada, and in Mexico and Central America.

black-sided pygmy grasshopper

Identification: A common species with many different color forms. It may be uniform black, gray, or brown over its entire body, or lighter on the upper surface. In some specimens, the ridges on the pronotum are lighter, or a light spot is found on the hind femora. In the male, the lower part of the face usually is ivory white. The pronotum usually extends almost to the tip of the abdomen in southern populations, but is shorter in northern ones. Males are 7.9–10.5 mm long, females 11.5–13.4 mm.

Ecology: Occurs in extremely varied habitats, ranging from sand dunes to swamps, and from open areas to hardwood and pine forests. It has one generation annually in the north, and two in the south.

Pygmy Mole Crickets

Family Tridactylidae

These very small insects, measuring only 4 to 10 mm in length, are grasshoppers despite their name. The antennae are relatively short, as in acridid grasshoppers (family Acrididae); however, they possess some unusual features that differentiate them from other grasshoppers. They have front legs that are adapted for digging in soil and an arched pronotum, and so resemble mole crickets. The front wings are thickened and shorter than the hind wings. The tip of the abdomen bears a set of bristly appendages that resemble cerci, so they appear to have two sets of cerci. The hind tarsi possess plates that help them move on water, an important feature because they frequent the sandy edges of streams and ponds. They are quite good at walking on the water surface. They tend to be gregarious, and build tunnels 2 to 3 cm below the soil surface. These tunnels can end in brood chambers in which they deposit their eggs. Their diet apparently consists of organic material such as algae, often ingested along with sand particles.

SMALLER SAND CRICKET

Ellipes minutus (Plate 33)

Distribution: A wide-ranging species, known from New Jersey to California, and south through Central America.

Identification: A small insect, blackish or dark brown, but with whitish-yellow marks along the side and back margins of the pronotum, on the forewings and the hind femora, and sometimes elsewhere. Forewings are short, covering about half the length of the abdomen. Hind wings are long, reaching the tip of the abdomen or extending beyond. Front tarsi are broad and thin, bearing four fingers that aid in digging. Fringed organs or apical spurs are found at the tip of the hind tibiae. Hind tarsi are absent. Both males and females are only 4.0–5.0 mm long.

Ecology: Inhabits sandy areas along the margins of water, where it ingests algae.

Similar species: Apparently there is only one other species of this family in North America, the larger sand cricket, *Neotridactylus apicialis.* It occupies essentially the same geographic range but differs in size (measuring 5.5–10 mm) and bears hind tarsi.

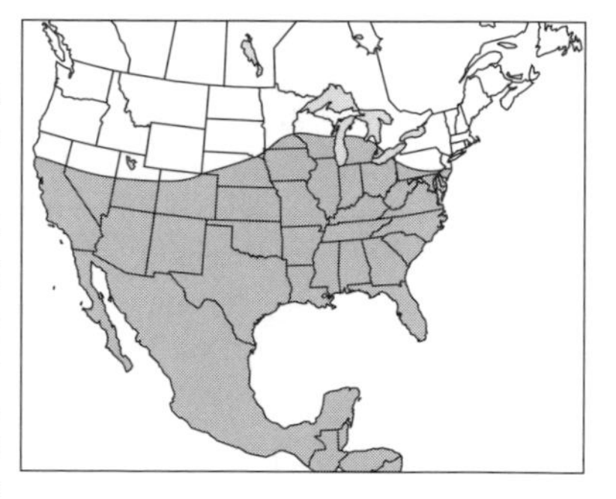

smaller sand cricket

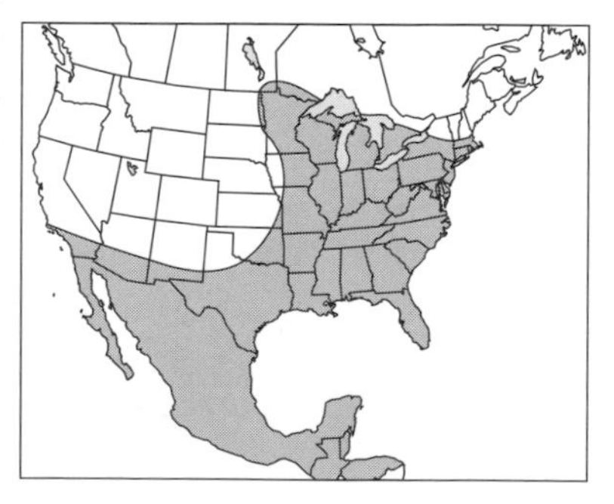

larger sand cricket

Katydids, Crickets, and Gryllacridoids

Suborder Ensifera

Ensiferans differ from members of the other suborder of Orthoptera, the Caelifera (primarily grasshoppers), in having long, thread-like antennae and a distinctive ovipositor in which the components are consolidated into a blade-like or needle-like structure. More so than with grasshoppers, sound production is a vital component of their biology (see the Introduction for a discussion of sound production in katydids and crickets).

All ensiferans belong to one of three superfamilies. Katydids belong to the superfamily Tettigonioidea, and "true" crickets belong to the superfamily Grylloidea. Insects belonging to the third superfamily, Gryllacridoidea, have no accepted common name that indicates their superfamily but can be referred to as gryllacridoids. They include camel crickets, Jerusalem crickets (neither are true crickets), and a variety of other groups, such as the wetas of New Zealand.

Katydids

Superfamily Tettigonioidea

Although many katydids (superfamily Tettigonioidea) look like grasshoppers (suborder Caelifera, superfamily Acridoidea), they are more closely related to crickets (superfamily Grylloidea). Like the antennae of crickets, and unlike those of grasshoppers, katydid antennae are at least as long as the body and usually considerably longer. Males of nearly all katydids produce species-specific calling songs. As in crickets (but not grasshoppers), the songs are made by rubbing together structures at the base of the forewings, and the ears are on the fore tibia. The song of the common true katydid, *Pterophylla camellifolia*, gives the group its name. Katydids are seldom confused with crickets. If you have doubts that are not resolved by the drawings in this book, count the number of segments in the tarsi (the last major component of each leg) (see Fig. 1). Katydids have four segments in each tarsus, whereas crickets have three. A group more likely to be confused with katydids is the gryllacridoids. These have four-segmented tarsi but no wings and no ears on the fore tibiae. Camel crickets, the most commonly encountered gryllacridoids, are humpbacked.

The distinctive calling songs of male katydids make many of them easy to detect and identify in the field. Indeed, species are often more readily identified by their songs than by their appearances. In cases where songs are important, in addition to our pictures and descriptions of songs, to resolve identification you can listen to samples of the songs posted on the Internet at http://buzz.ifas.ufl.edu/. This is a site where you also can learn about katydids not dealt with in this book and more about those that are.

True Katydids

Subfamily Pseudophyllinae

True katydids are the most often heard katydids, yet they are seldom seen. They are heard because their nocturnal calls are loud, raucous, and incessant. They are seldom seen because they inhabit the crowns of trees and do not fly to lights. When they are seen, they can be recognized by their forewings, which bulge out at the sides, and their stiff, fishing-pole-like antennae.

COMMON TRUE KATYDID

Pterophylla camellifolia (Plate 34)

Distribution: Found in the eastern United States west to the start of the Central Plains but not in southern Florida or in most of the northern tier of states.

Identification: The only species of true katydid (Pseudophyllinae) in most of its range. Its song is a repetition of loud raucous syllables, and neighboring individuals alternate their calls. In the Northeast and Midwest, the songs of true katydids can be rendered *ka-ty-did*, *she-did*, *she-didn't*. This rendering of the call is the basis of "katydid" being the usual common name for members of the family Tettigoniidae in the United States, Canada, and Australia. Elsewhere in the English-speaking world, members of the family are more often termed "bush-crickets." Common true katydids have regional dialects. The two- or three-pulse calls that can be rendered *she-did* or *ka-ty-did* are replaced to the southeast by three- to five-pulse calls at a much faster rate. (Yankee true katydids are the ones that drawl!) To the southwest, the pulse rate remains slow but the number of pulses drops to two or to one.

common true katydid

All North American true katydids have bulging forewings and stiff antennae. In the common true katydid, the sides of the pronotum are deeper than wide and the length of the pronotal disk is about equal to its rear width. Forewings are smoothly rounded behind. Length is 33–50 mm.

Ecology: Inhabits the crowns of deciduous trees and feeds on the foliage. Males seldom sing in branches low enough to be reached with a net. They do not fly, but if one is dislodged from its perch it uses its wings to flutter to the ground. It then runs or walks to the nearest tree trunk and ascends. Captured specimens differ strikingly from other katydids in their behaviors. They walk deliberately or run and hop rather than leaping or flying. They rear back when taunted and, by forewing stridulation, may squawk in protest. Only males call, but both males and females protest. Populations are often dense, and choruses may be loud enough to disturb human activities, for example, plays staged outdoors in wooded parks. Adults are first heard in June in Florida and in July farther north. Singing ends in October.

Similar species: Florida true katydids, *Lea floridensis*, have more elongate wings than common true katydids, and the lateral lobes of the pronotum are wider than deep. The song is a loud, hollow *chlonk* repeated at 2–3 s intervals. Truncated true katydids, *Paracyrtophyllus robustus*, have shorter, somewhat truncated forewings. Their song has 6 to 7 pulses produced in about 0.3 s (much too rapid to count).

Florida true katydid

truncated true katydid

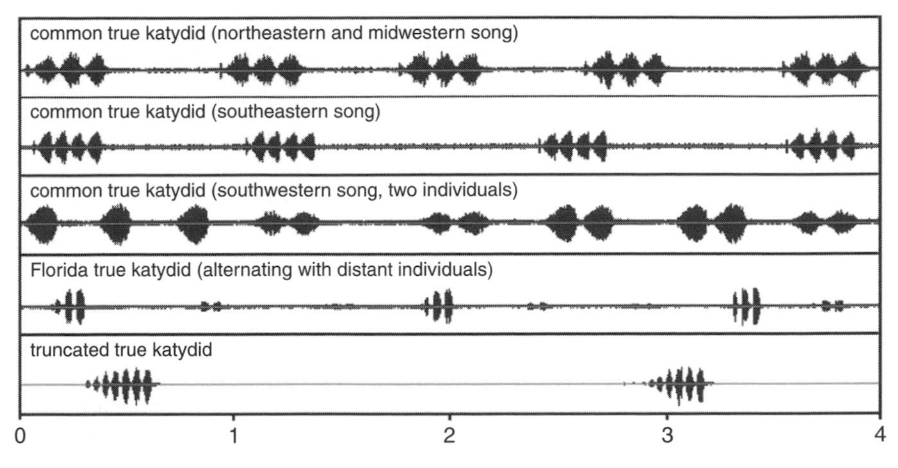

True katydid song comparisons

False Katydids

Subfamily Phaneropterinae

Most false katydids are like true katydids in being somewhat leaf-like, but the forewings do not bulge outward and the hind wings are longer than the forewings. All katydid subfamilies except Phaneropterinae have a pair of downward projecting spines between the bases of the forelegs. The ears of false katydids are "open," with the tympanum ("eardrum") fully exposed near the base of each fore tibia. The ears of other katydids, except for hump-winged grigs (family Prophalangopsidae) and drumming katydids (*Meconema thalassinum*), have the tympana concealed behind slit-like openings in the fore tibiae. False katydids also differ from other katydids in that females may answer the male song with an audible tick, in which case the male walks or flies to the female rather than the other way round. This is probably safer for the female. It also allows males who have not called but have heard the interchange to move to the answering female before the calling male does. Hence, it is possible for someone who can imitate the sound and timing of a sexually ready female to attract several males at once.

GREATER ANGLE-WING

Microcentrum rhombifolium (Plate 35)

Distribution: Found in the southwestern and eastern United States except for New England.

Identification: All angle-wing katydids (genus *Microcentrum*) have the forewings curved more sharply along the top than the bottom, and the distance between the antennal sockets is 1 to 2 times the width of the basal antennal segment. Hind femora do not reach the

greater angle-wing

rear one-fourth of the forewings. Ovipositors are short and upturned. The front margin of the pronotum has a slight, median, forward-projecting tooth. The area immediately in front of the male's stridulatory vein is green.

Males of this species produce two songs: a loud lisp repeated every 2–4 s and a series of ticks that sounds much like someone slowly running the thumbnail along the teeth of a pocket comb. Each lisp lasts less than 0.1 s and is made by a rapid stroke of the file. Each tick series, lasting 3–5 s, is also made by a stroke of the file, but this time the stroke is so slow that 20 to 35 individual tooth impacts are heard as ticks. Lisps attract distant females, and ticks elicit answering ticks from nearby females, allowing males to close the final distance. Length is 52–63 mm.

Ecology: Inhabits forests, shade trees, shrubbery, and fencerows. It is attracted to lights in the habitats where common true katydids chorus. Because greater angle-wings are the largest katydids collected in such places, they are sometimes falsely credited with being the source of the noise. However, neither of the songs they make is particularly intrusive. Ovipositing females glue their eggs single file along the margins of leaves. There is one generation annually, with adults occurring from July to October, except in Florida, where adults occur all year.

Similar species: Lesser angle-wings, *Microcentrum retinerve*, are usually smaller (44–53 mm), lack the pronotal tooth, and have a brown area in front of the stridulatory vein. They are abundant in the tops of broadleaved trees, as revealed by their loud, intermittent calls, and are often attracted to lights. California angle-wings, *Microcentrum californicum*, are smaller (41–51 mm) and lack the pronotal tooth. Their song is a brief, two-part lisp repeated at intervals of about 2.5 s.

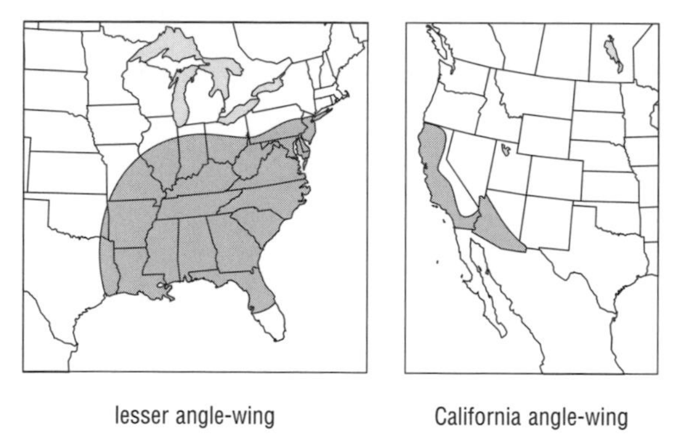

lesser angle-wing   California angle-wing

greater angle-wing (lisps)

greater angle-wing (tick series)

lesser angle-wing

California angle-wing

0 1 2 3 4

Angle-wing song comparisons

OBLONG-WINGED KATYDID

Amblycorypha oblongifolia (Plate 34)

Distribution: Found in the eastern United States west to the Great Plains but missing from southeastern Georgia and most of Florida.

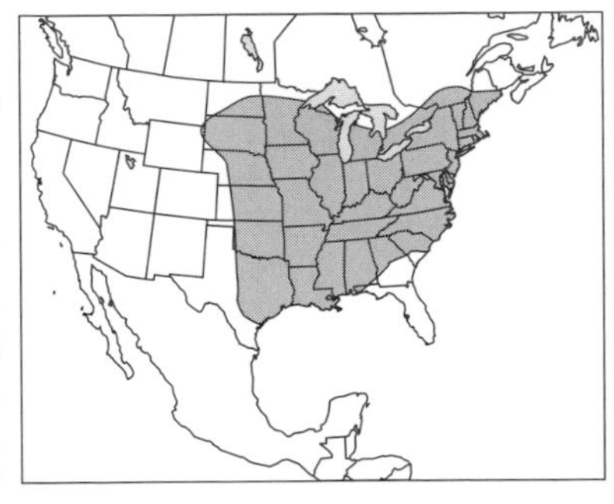

oblong-winged katydid

Identification: Like other round-headed katydids (genus *Amblycorypha*), they have broadly rounded heads with the distance between the antennal sockets 2 to 3 times the width of the basal antennal segment; forewings broad (length no greater than 4 times maximum width); and hind femora reaching near or beyond the ends of the forewings.

The subgenital plate of the male of this species ends with a V-shaped notch, and the stridulatory area of the male exceeds the area of the pronotal disk. The song can be rendered as *iz-zi-ZIK* repeated at intervals of several seconds. This katydid is nearly always green, but a bright pink form is occasionally encountered. It is the most widely distributed in a group of four *Amblycorypha* species that exceed 40 mm in length and have relatively broad forewings (length 3.1 to 4.0 times maximum width). Length is 42–52 mm.

Ecology: Inhabits the understory of deciduous forest, and shrubs and tall weeds in moist places. Eggs are laid in the soil, and, at least in the North, may require two or more years to hatch. There is a single peak of adult abundance each year, with the earliest dates ranging from late June to mid-August depending on latitude.

Similar species: Males of the three katydid species most similar to the oblong-winged katydid have stridulatory areas that are approximately equal in area to the pronotal disk. In carinate false katydids, *Amblycorypha carinata*, the lateral edges of the pronotal disk are angular, whereas in Florida false katydids, *A. floridana*, they are generally rounded anteriorly and somewhat angular posteriorly. The most reliable way to distinguish carinate and Florida katydids is by their songs. The former species produces a series of simple clicks, whereas the latter has a complex song that sounds somewhat like a slowly bouncing table-tennis ball that is made to buzz by bringing a paddle down on it: *tick, tick, tick, buz-zz-zz-zz*. This complicated sound is repeated several times, with each repetition more rapid than the one before. Males of the

carinate false katydid

Florida false katydid

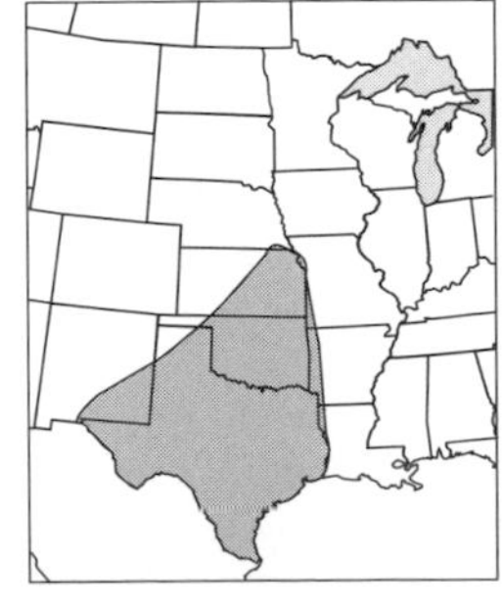

Texas false katydid

Texas false katydid, *A. huasteca*, lack the V-shaped notch in the male subgenital plate that males of the other species have. Their song is a series of 1- or 2-pulse ticks lasting about 0.3 s and repeated about every 4 s.

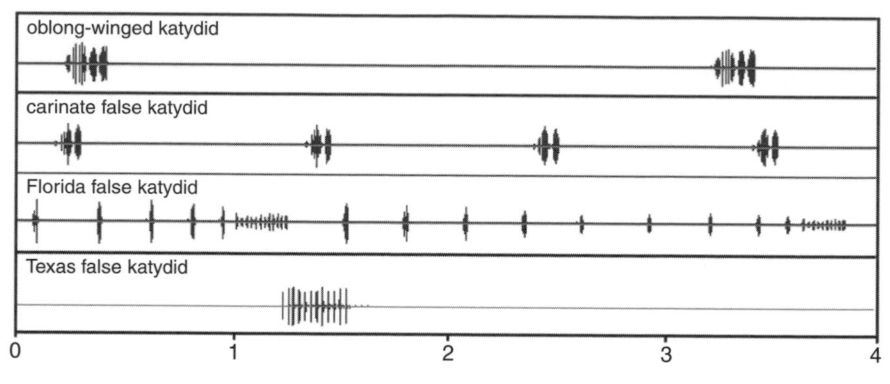

False katydid song comparisons

COMMON VIRTUOSO KATYDID

Amblycorypha longinicta (Plate 34)

Distribution: Found in the eastern United States south of the Great Lakes and east of the Great Plains.

common virtuoso katydid

Identification: Like other round-headed katydids (genus *Amblycorypha*), they have broadly rounded heads with the distance between the antennal sockets 2 to 3 times the width of the basal antennal segment; forewings broad (length no greater than 4 times maximum width); and hind femora reaching near or beyond the ends of the forewings.

This katydid is the most widely distributed species in a group of three *Amblycorypha* species that are small or intermediate in length (28–43 mm) and have relatively elongate forewings (length of forewing no less than 3.2 times maximum width). Males of this group can generally be distinguished from other small false katydids by the pronotal disk having a black dot or a longitudinal black line at each corner. The three species are known as virtuoso katydids because they have the most complex songs known for katydids. In each species, the song consists of four types of sounds (corresponding to four ways of stroking the stridulatory file), produced in stereotyped sequences. Depending on the species and temperature, a sequence requires 5–40 s. A sequence begins with a series of type 1 sounds that slowly increase in intensity. It is immediately followed by a series of type 2 sounds (a rattle) and a short series of type 3 sounds that rapidly diminish in intensity (*chuu*). Type 4 sounds are deliberately delivered ticks.

The song of the common virtuoso katydid lasts more than 20 s before it begins to repeat, which is longer than the songs of other virtuoso katydids. The short series of type 3 sounds (*chuu*) is repeated several times, sometimes with type 4 sounds in between, before the initial two types of sound are produced again (a soft buzz of slowly increasing intensity followed by a prolonged rattle). Length is 29–37 mm.

Ecology: Inhabits low herbaceous vegetation in old fields and along roadsides. The first adults appear as early as July or August, depending on the latitude.

Similar species: Sandhill virtuoso katydids, *Amblycorypha arenicola*, occur in habitats dominated by turkey oak and longleaf pines and are larger than other virtuoso katydids (37–43 mm). The males usually sing for a few minutes from perches several feet high and then fly and resume singing at another site. Cajun virtuoso katydids, *Amblycorypha cajuni*, occur in the rich bottomlands of the lower Mississippi River and are more sedentary singers. These two species have songs that begin to repeat after less than 10 s and are simple sequences of the four types of sounds (no repeated *chuu*).

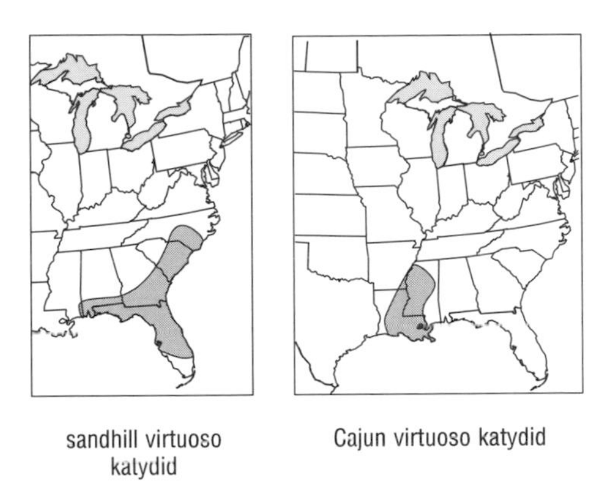

sandhill virtuoso katydid

Cajun virtuoso katydid

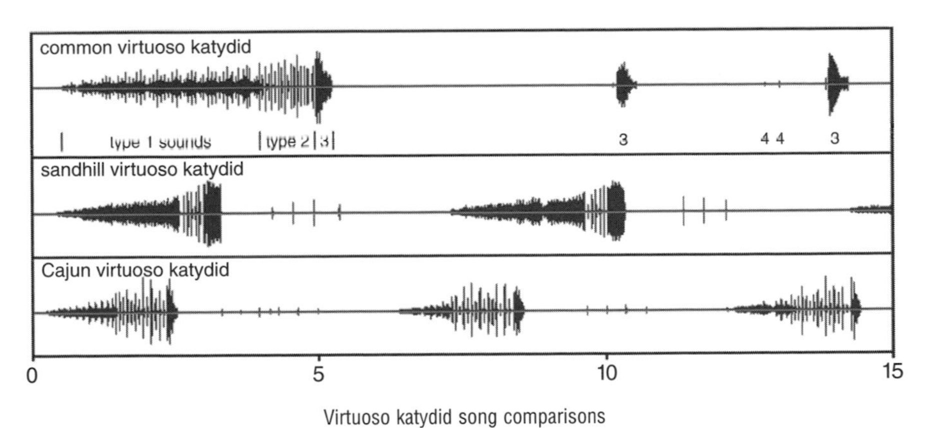

Virtuoso katydid song comparisons

RATTLER ROUND-WINGED KATYDID

Amblycorypha rotundifolia (Plate 34)

Distribution: Found from Illinois to New York and south to Tennessee and northern Georgia.

Identification: Like other round-headed katydids (genus *Amblycorypha*), they have broadly rounded heads with the distance between the antennal sockets 2 to 3 times the width of the basal antennal segment; forewings broad (length no greater than 4 times maximum width); and hind femora reaching near or beyond the ends of the forewings.

This katydid is the most widely collected species in a group of four *Amblycorypha* species that are small or intermediate (32–42 mm long) and have forewings that are usually broader than those of other members of the genus (length no more than 3.4 times maximum width). Three members of this group can be identified only by their song or habitat. The song of the rattler round-winged katydid is a series of prolonged, soft rattles. The initial rattles in a series last about a second, and the terminal or near-terminal rattle lasts several seconds. Length is 28–39 mm.

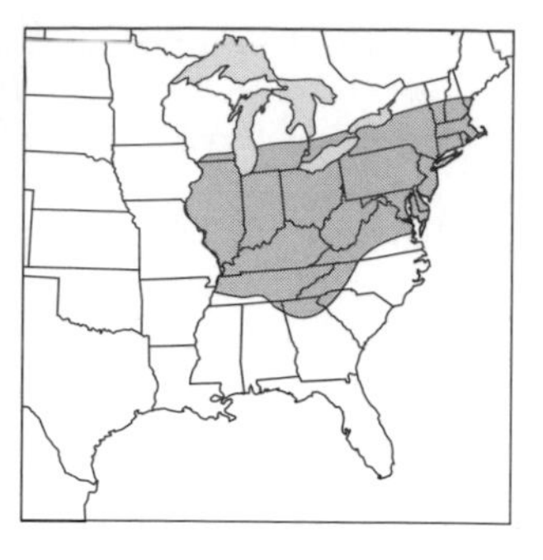

rattler round-winged katydid

Ecology: Inhabits weedy old fields, roadsides, and forest edges and openings. It sings, at night, on perches a few feet above the ground.

Similar species: Western round-winged katydids, *Amblycorypha parvipennis*, have hind wings that do not extend beyond the forewings at rest. Clicker round-winged katydids, *Amblycorypha alexanderi*, occur in the same habitats as rattler round-winged katydids, but their songs are a series of clicks produced at a rate of 3–5 per s. Bartram's round-winged katydids, *Amblycorypha bartrami*, occur in sandy habitats dominated by longleaf pine and turkey oak. Their songs are series of tick-like clicks produced at a rate of 7–10 per s.

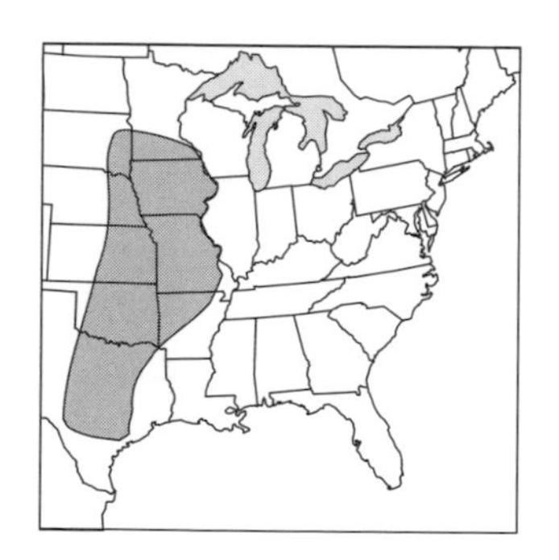

western round-winged katydid

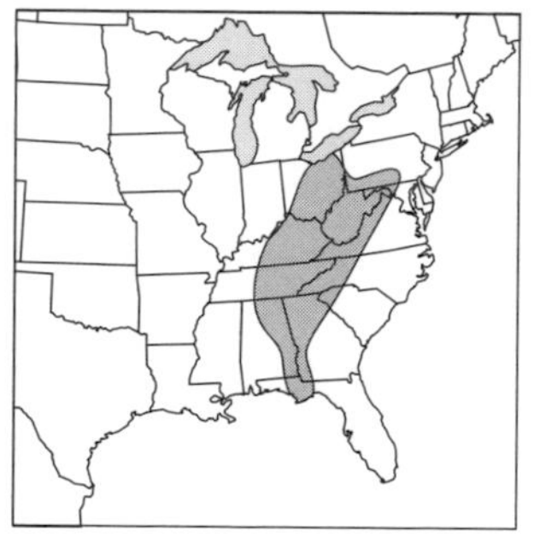

clicker round-winged katydid

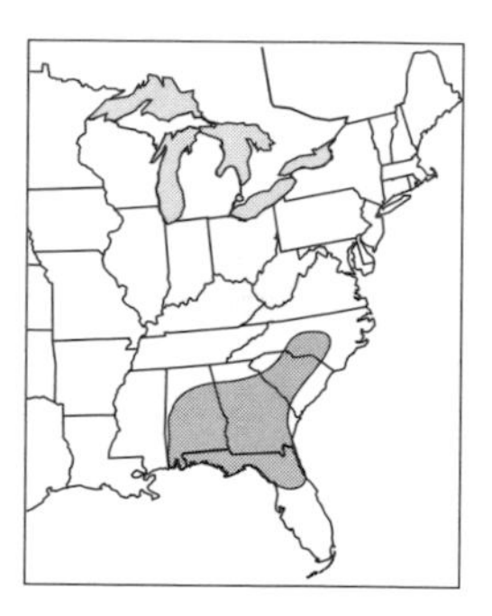

Bartram's round-winged katydid

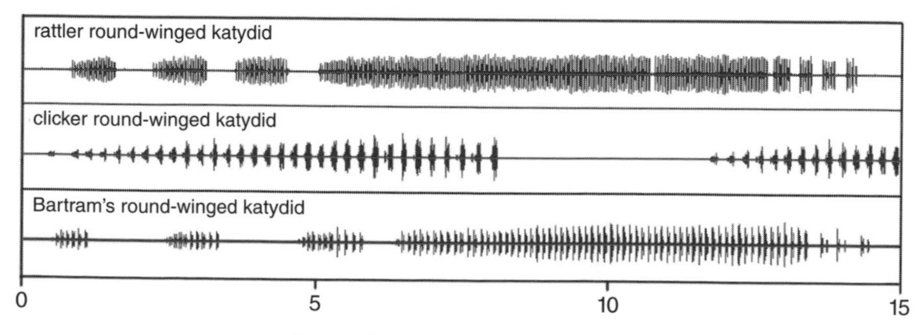

Round-winged katydid song comparisons

TEXAS BUSH KATYDID

Scudderia texensis (Plate 35, Fig. 58)

Distribution: Found throughout the eastern United States and adjacent Canada west to the western edge of the Great Plains.

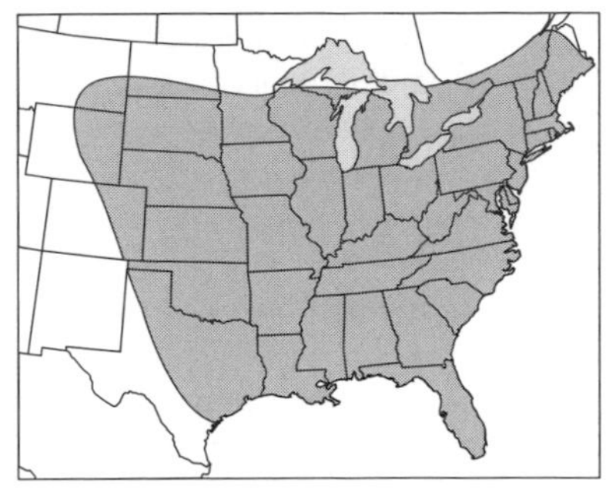

Texas bush katydid

Identification: Forewings not much broader at the middle than near the end; length of forewings about 5 times maximum width. In males, the subgenital plate has an up-curved *ventral process* that meets a dorsal extension of the supra-anal plate. This *dorsal process* is the key to identifying males of the genus *Scudderia* (see Fig. 58). In the Texas bush katydid, it ends with a pair of curved

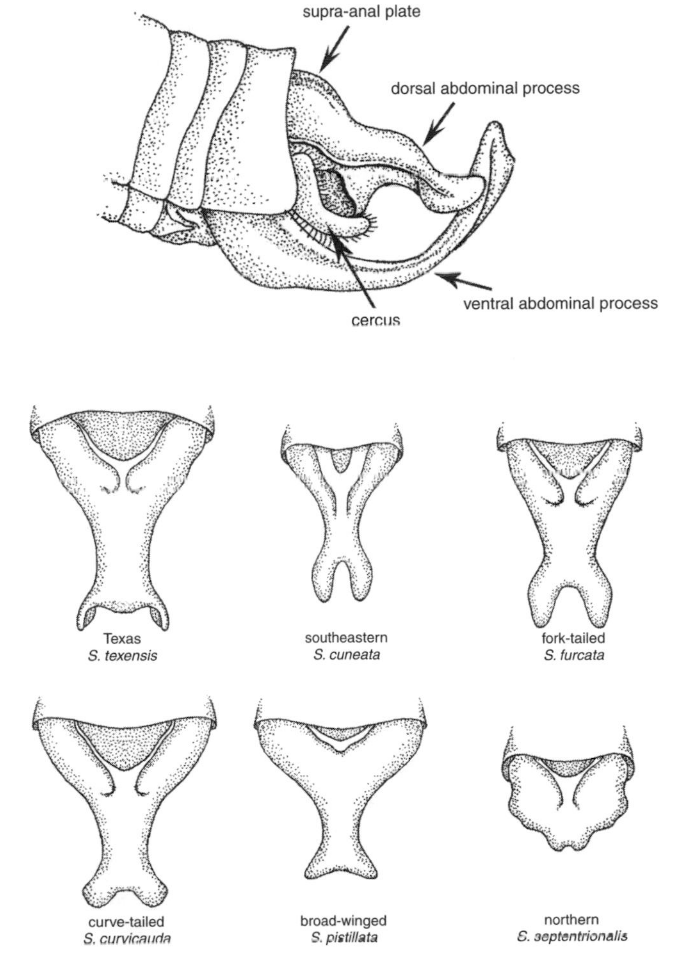

Figure 58. Dorsal abdominal processes of bush katydids, genus *Scudderia*. Above, lateral view of abdominal tip of a male, with parts labeled. Below, dorsal views of the supra-anal plates of six bush katydids.

indentations that define a small central tooth. In females, the upper margin of the basal portion of the ovipositor is at a right angle to the upper margin of the terminal portion. Solitary males of this species produce three types of songs: an irregular series of ticks (which may function in spacing males), a fast-pulsed song (which causes sexually responsive females to move toward the male), and a slow-pulsed song (to which females make answering ticks, which allow males to locate and move to the female). Length is 40–56 mm.

Ecology: Abundant in weedy old fields and roadsides. In the south, there are two generations annually with the first generation maturing in late spring and the second in early fall. In the north, there is a single generation that matures in late summer. Much of the size variation in this species is geographic and depends on the number of generations and the length of the growing season. In the two-generation portion of the range (Florida to North Carolina), individuals average about 44 mm in length. In Virginia, the southern extreme for one generation, average length is about 52 mm, whereas in Michigan, the northern extreme, the length again averages 44 mm. Thus, the size of adults is correlated with how fast they must mature in order to fully use the growing season while producing the maximum number of generations.

Similar species: Five other species of *Scudderia* are likely to be encountered in the eastern United States. *Scudderia* females are difficult to identify, but Texas bush katydids are the only species with a right angle between the basal and terminal portions of the ovipositor; in other species of the genus this angle is greater than 90 degrees. Males can be identified by their dorsal process, an extension of the supra-anal plate (see Fig. 58). The northern bush katydid, *S. septentrionalis*, nearly lacks the process. In the southeastern and fork-tailed bush katydids, *S. cuneata* and *S. furcata*, the processes end in simple, deeply cleft forks that differ in the widths of the side pieces. In the curve-tailed and broad-winged bush katydids, *S. curvicauda* and *S. pistillata*, the processes end with shallow bifurcations that differ in the shapes of the side pieces as viewed from above. These two species also differ in the widths of their forewings, with the broad-winged bush katydid having forewings that are less than 4 times as long as wide, whereas the curve-tailed and all other *Scudderia* have forewings more than 4 times as long as wide. *Scudderia* species also differ significantly in size, with the northern bush katydid being smallest (length 34–41 mm) and the curve-tailed bush katydid (length 38–54 mm) and the Texas bush katydid being the largest.

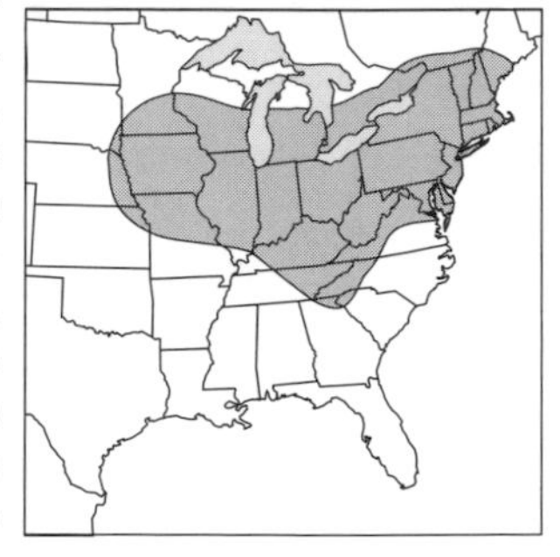

northern bush katydid

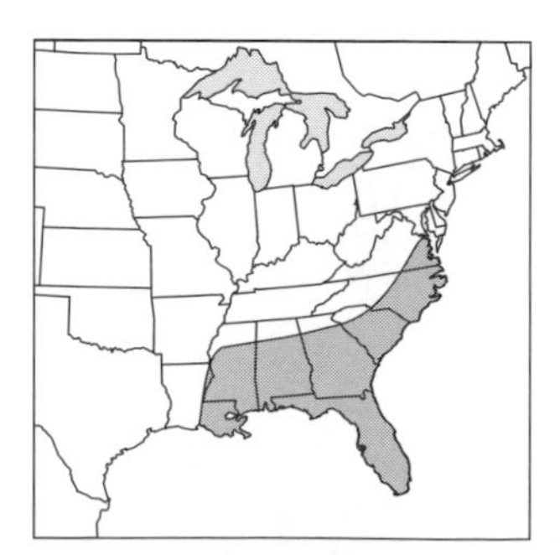

southeastern bush katydid

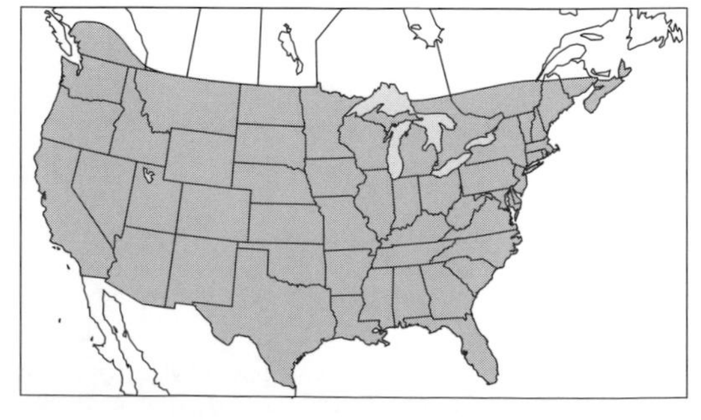

fork-tailed bush katydid

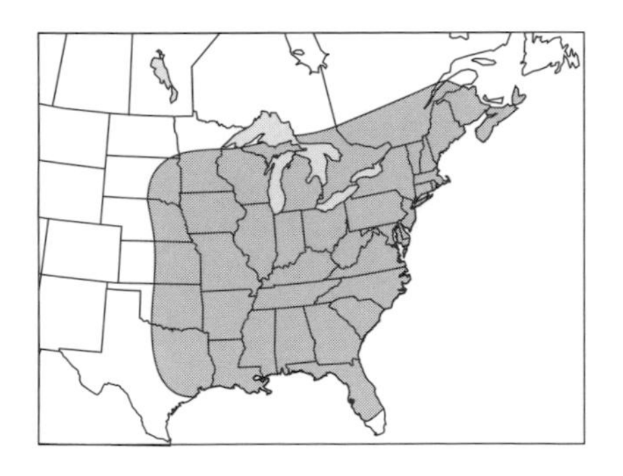

curve-tailed bush katydid

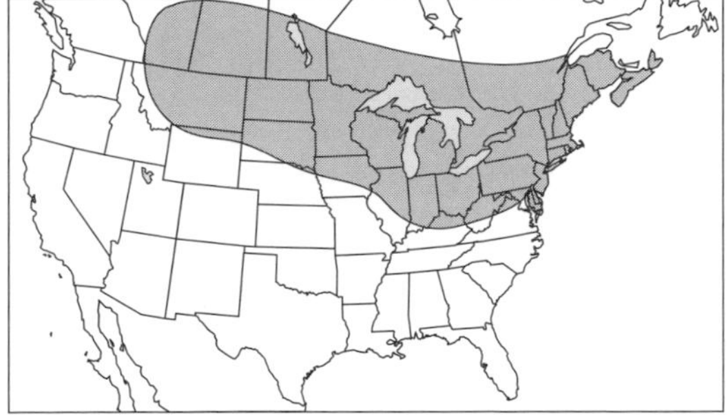

broad-winged bush katydid

GUINEA-CYPRESS KATYDID

Inscudderia strigata (Plate 35)

guinea-cypress katydid

Distribution: Found in Florida and in southern Georgia and South Carolina.

Identification: The most boldly patterned katydid in the Southeast. Viewed from above, the forewings have a tapered, orange-brown streak accented with a black spot at the anterior end and a black stripe on either side, with green beyond. Forewings are long and narrow (length 6.1 to 7.5 times maximum width). The male subgenital plate has an up-curved process like that of the Texas bush katydid and other species of *Scudderia*, but there is no dorsal extension of the supra-anal plate. Length is 41–47 mm.

Ecology: This species occurs on "guinea cypress" (*Hypericum fasciculatum*), a shrubby species of Saint-John's-wort that grows in thick stands around cypress (*Taxodium* sp.) ponds and in other poorly drained, open areas. Its nymphal and adult color patterns make individuals surprisingly difficult to detect on the host plants. Adults are found from July to October.

Similar species: Species of the genus *Scudderia* are never multicolored and boldly patterned, and their wings are less than 6 times as long as wide. Eastern cypress

eastern cypress katydid

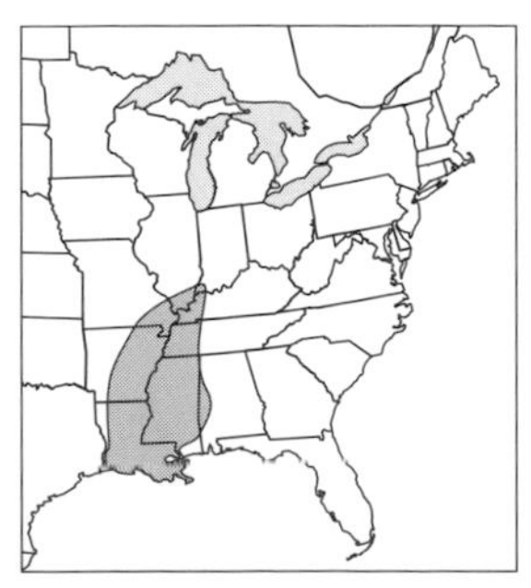

western cypress katydid

katydids, *Inscudderia walkeri*, and western cypress katydids, *Inscudderia taxodii*, are less robust (35–44 mm long), occur on cypress trees rather than on guinea cypress, and have oblique black marks along the upper edge of each forewing.

ELEGANT BUSH KATYDID

Insara elegans (Plate 35)

Distribution: Found from southeastern California to southern Texas north to southern Nevada and Colorado.

Identification: The length of the forewing is at least 8 times its maximum width, and the tip of the forewing is smoothly rounded. Pronotum is saddle-shaped except for ridges where the lateral lobes meet the pronotal disk. Hind femora reach almost to the tips of the forewings. Background color is a medium green. Viewed from above, a series of rear-pointing V's are formed by white veins and depigmented areas along the edges of the forewings where they meet at the midline. In some specimens, some of the wing cells between the depigmented areas are dark brown. The song is a soft whirring. Length is 29–37 mm.

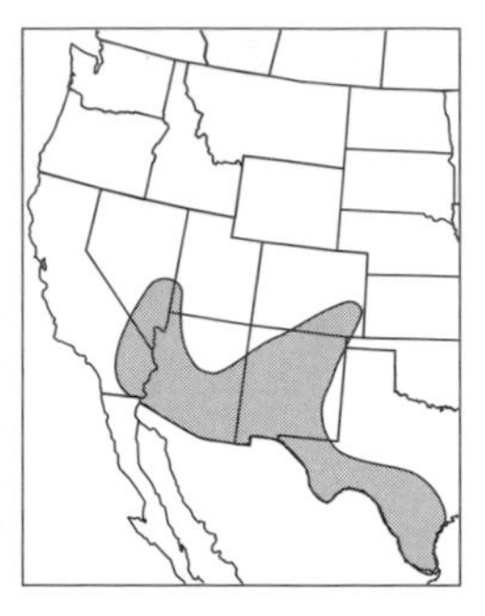

elegant bush katydid

Ecology: Primarily found on mesquite. It is sometimes attracted to lights.

Similar species: There are five other species of western bush katydids (genus *Insara*), but only the creosote bush katydid, *I. covilleae*, is widely distributed and collected. It occurs only on creosote bushes and has large white or pale green spots on the forewings.

creosote bush katydid

EASTERN THREAD-LEGGED KATYDID

Arethaea phalangium (Plate 36)

Distribution: Found in Florida and north to central Alabama and southern South Carolina.

Identification: No other eastern katydid has such long, slender legs. Length of the front femur is about 2.5 times the length of the pronotum. Length of the forewing is more than 8 times its maximum width. Top of the head protrudes forward into a point that does not end flush with the face. Pronotum is saddle-like with no well-defined disk. Eyes are elongate. The song is high pitched and difficult to detect in the field. Length is 42–53 mm.

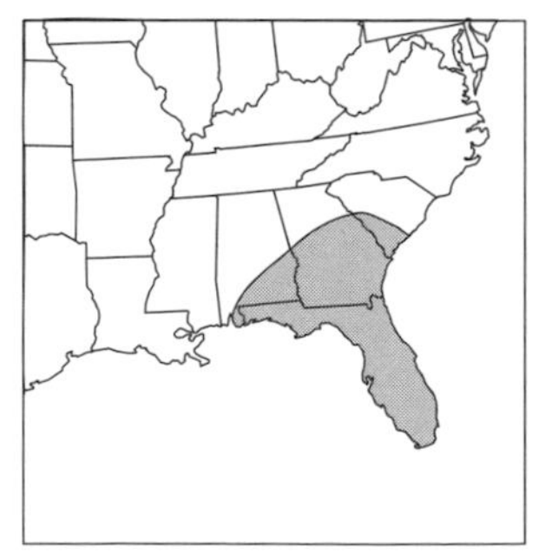

eastern thread-legged katydid

Ecology: Inhabits weeds and grasses in well-drained old fields and undergrowth in open, well-drained pinewoods. Adults are found from May through September. Their slender legs and wings make them difficult to spot among grasses and weeds.

Similar species: Fourteen additional species of thread-legged katydids (genus *Arethaea*) occur in the Southwest. All *Arethaea* are characterized by long legs and narrow forewings. Females of some of the southwestern species have abbreviated forewings.

GIANT KATYDID

Stilpnochlora couloniana (Plate 36)

Distribution: Found in peninsular Florida as far north as Gainesville.

giant katydid

Identification: No other U.S. katydid is as large, and no other has two "angles" along the upper margins of its forewings. Pronotal disk is concave, with its rear lateral margins wrinkled and elevated. Ovipositor is disproportionately small. Unlike other North American false katydids (Phaneropterinae), the tibial tympana are visible through a wide slit rather than completely exposed. The song is a prolonged, high-pitched *lisssp*. Length is 66–87 mm.

Ecology: A common inhabitant of the crowns of broadleaved trees in peninsular Florida that is sometimes attracted to lights. In most of its range (south Florida, Cuba, and the Bahamas), adults occur year-round. In northern peninsular Florida, it seems to have but one generation annually with principal adult activity in fall and winter.

Similar species: Greater angle-wings, *Microcentrum rhombifolium* (Plate 35), have only a single angle to their forewings and are less than 64 mm long.

MODEST KATYDID

Montezumina modesta (Plate 36)

Distribution: Found throughout the southeastern United States.

modest katydid

Identification: This species has moderately narrow forewings (length between 4.0 and 4.4 times maximum width) and oblong-oval eyes (length 1.5 times width). Top of the head extends into a narrow, grooved beak that meets a narrow extension of the face between the antennal sockets. Femora are armed beneath with four or more spines. Two pale yellow lines start at the front of the pronotum, then extend rearward along the edges of the pronotal disk and on either side of the stridulatory area to the top edges of the forewings at rest. Males sing principally at twilight and make brief soft lisps, each lasting about 20 or 30 msec with few intermediates. Each lisp may be followed by a tick. Length is 31–37 mm.

Ecology: Modest katydids are found in small numbers in a wide variety of habitats. In north peninsular Florida they are in dry habitats such as turkey oak woods and sand pine scrub, as well as wet ones such as shrubby flatwoods and cypress heads. In states to the north, many specimens come from bottomland forests. Adults are captured mostly from June through August.

Similar species: Katydids of the genus *Scudderia* have globular rather than oblong-oval eyes, and most exceed 37 mm in length.

NARROW-BEAKED KATYDID

Turpilia rostrata (Plate 36)

Distribution: Found in the coastal areas of central and south peninsular Florida.

Identification: The distance between the antennal sockets is about one-third the width of the basal segment of the antenna. The forewing is more strongly curved along the top than the bottom, and the hind femur does not reach the rear one-fourth of the forewing. The song consists of various series and mixtures of ticks and lisps. Length is 36–44 mm.

Ecology: This arboreal species occurs in mangroves and other tropical hardwoods in coastal south Florida. Adults occur at all times of year.

narrow-beaked katydid

Similar species: Greater angle-wings, *Microcentrum rhombifolium* (Plate 35), share the sharp curve in the upper margin of the forewings and the short hind femora, but the antennal sockets are much more widely separated than in narrow-beaked katydids.

CHAPARRAL KATYDID

Platylyra californica (Plate 36)

Distribution: Found only in California, west of the Sierras.

Identification: The wings are short but fully developed, with the length of the forewing less than 4 times its maximum width. The sides of the pronotum are longer than deep. At the base of the male's forewings, the stridulatory area is substantially wider than the rear of the pronotum. Males answer one another as they make a loud, frequently repeated *zwick*. Length is 19–26 mm.

chaparral katydid

Ecology: Nymphs appear in early spring and are found on a variety of herbaceous and woody shrubs. Adults inhabit the tops of broadleaved trees, where their presence is revealed by the calls of males. This species is often attracted to lights at night.

Similar species: Scudder's bush katydids (genus *Scudderia*) have more elongate forewings.

COMMON SHORT-WINGED KATYDID

Dichopetala brevihastata (Plate 35)

Distribution: Found in southern Arizona, New Mexico, and Texas.

Identification: Forewings of males cover only about one-third of the abdomen; forewings of females are scale-like. The outside lobe at the end of the male cercus is short and simple and does not cover more than half of the inner lobe. The subgenital plate of the male has a V-shaped notch. Length is 13–18 mm.

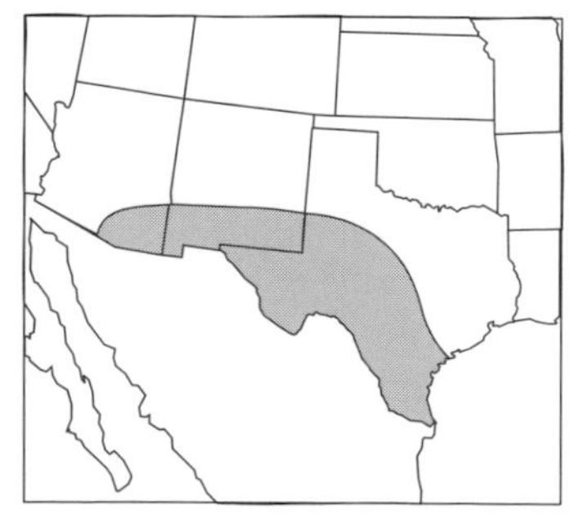

common short-winged katydid

Ecology: Inhabits various low bushes and grasses and green weedy plants. In some localities, adults first appear in late July, and in others not until early September.

Similar species: There are seven other species of short-winged katydid (genus *Dichopetala*). Males of the species that overlap the range of the common short-winged katydid can be distinguished as follows. The subgenital plate of the chestnut short-winged katydid, *D. castanea*, has a U-shaped notch. The outside cercal lobes of the emarginate and mountain-dwelling short-winged katydids, *D. emarginata* and *D. oreoeca*, are flattened or spoon-like.

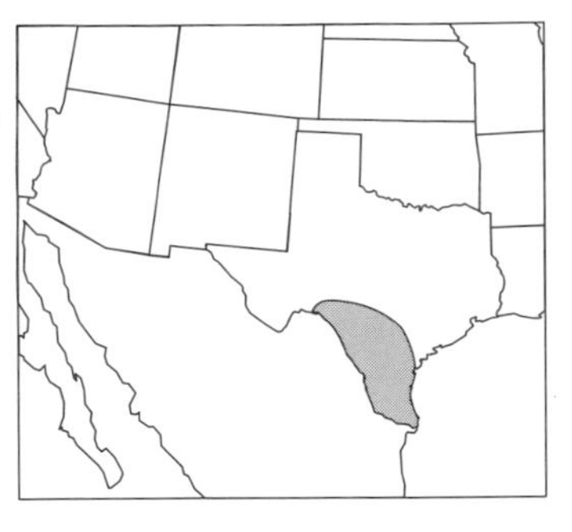
chestnut short-winged katydid

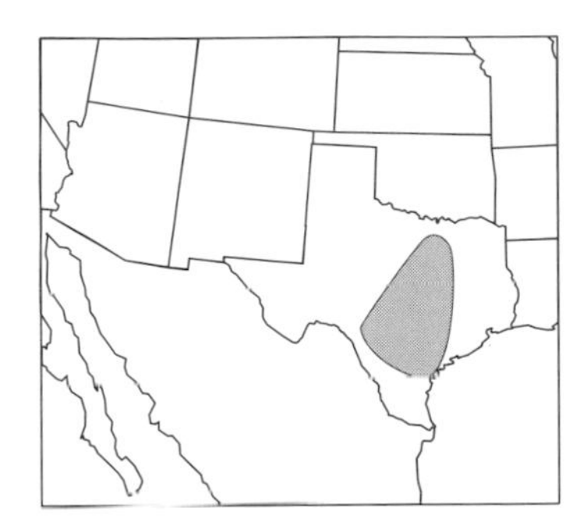
emarginate short-winged katydid

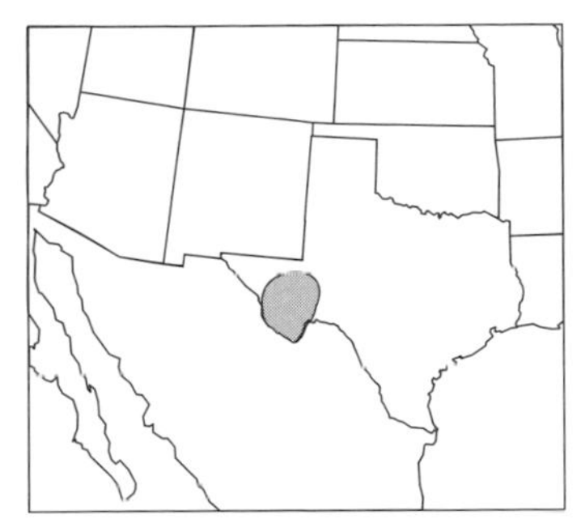
mountain-dwelling short-winged katydid

Coneheaded Katydids
Subfamily Copiphorinae

Coneheads are medium to large grasshopper-like katydids with oversized jaws. They are the only katydids with the head developed into a pointed or rounded cone that projects beyond the basal antennal segments. Most species have long, narrow forewings and, with the aid of concealed hind wings, are strong fliers; a few species have abbreviated forewings and are flightless. The songs of males are generally loud and penetrating, making them among the most often heard katydids. Most species call from low enough in the vegetation to be easily accessible. Calling males are generally wary, however, and will either fly when disturbed or dive head first into the ground cover, where they may resemble just another blade of grass.

All North American species occur in both brown and green color phases. The proportions of the two phases vary with the species and sometimes with the sex and season. One effect is that color-sensitive insect-eating or insect-collecting vertebrates find it difficult to spot the coneheads they are seeking. It is easier to find items of one color than items of either of two colors. Female coneheads are generally much larger than males, so in some cases their lengths are given separately.

The simplest ways to identify coneheads are by the shape and markings of their cones and by the calling songs of the males. Figure 59 shows the cones of 12 species. Coneheads have strong jaws and can inflict painful bites when handled carelessly.

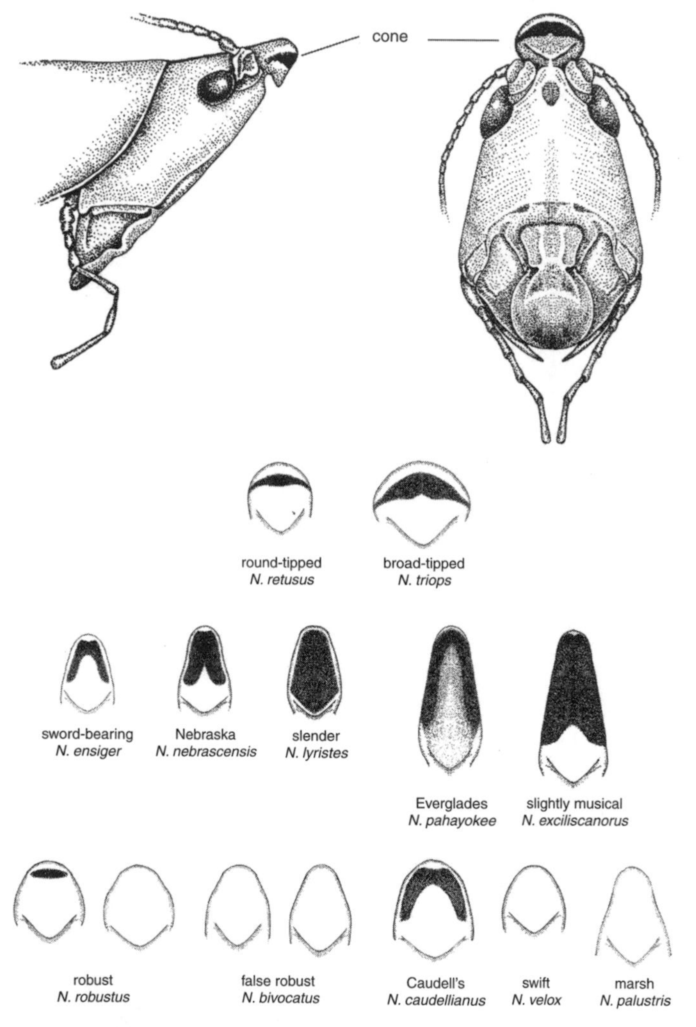

Figure 59. Cones of coneheaded katydids, genus *Neoconocephalus*. Above, lateral and ventral views of head showing the position of cone. Below, ventral views of cone of twelve species, showing distinctive shapes and markings.

ROUND-TIPPED CONEHEAD

Neoconocephalus retusus (Plate 37, Fig. 59)

Distribution: Found in the eastern United States except for the northern tier of states and southern Florida.

Identification: A small conehead with a cone scarcely longer than wide. Forewings extend less than 11 mm beyond the hind femora. Ovipositor is slender and 30–

40 mm long, much longer than the hind femora. The calling song is a rather weak, beady, continuous buzz heard from within tangles of vegetation, during the afternoon as well as at night. Length is 37–52 mm.

Ecology: Inhabits dry to fairly wet grassy or weedy open areas, such as roadsides, old fields, and the edges of marshes. Adults are found from August to October.

Similar species: The broad-tipped conehead, *Neoconocephalus triops* (Plate 37), is larger and has a broader cone and a shorter ovipositor. When its hind femora are extended directly rearward, the forewings extend more than 11 mm beyond the femoral tips.

round-tipped conehead

BROAD-TIPPED CONEHEAD

Neoconocephalus triops (Plate 37, Fig. 59)

Distribution: Found in the southeastern United States west to southern California.

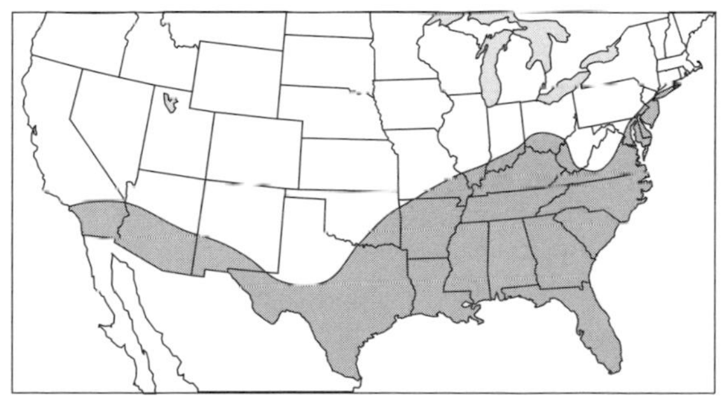

broad-tipped conehead

Identification: No other conehead has a cone that is wider than long. Forewings extend more than 11 mm beyond the hind femora. Ovipositor is about as long as the hind femur. This species calls in early spring, more than a month before any other common conehead (genus *Neoconcephalus*). Its springtime song begins with a buzz interrupted at regular intervals of about 1 s and then shifts to a continuous buzz. The song of the summer generation of adults (where one occurs) differs from that of the spring generation in being entirely like the beginning phase of the spring song—that is, a buzz that is continuous except for regular, momentary interruptions about once per second. Males are 43–60 mm long, females 51–67 mm.

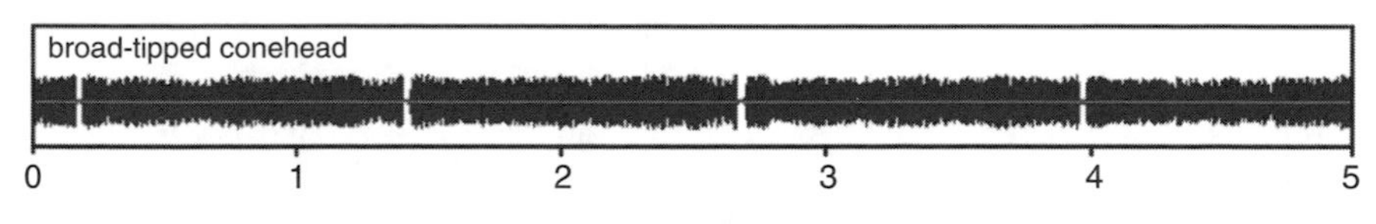

Broad-tipped conehead waveform

Ecology: Juveniles and feeding adults inhabit grassy areas; overwintering adults occur in thickets and woods. Males call from perches near ground level to high in tree tops in a variety of habitats. Both sexes are strong fliers and come to lights.

Unlike other *Neoconocephalus* spp. treated here, which overwinter as eggs, this species overwinters only in the adult stage. Males that have overwintered become reproductively active in early spring—for example, in February in north Florida and in April in North Carolina. In the northern and western portions of its range, this species has a single generation annually. Elsewhere, a partial second generation matures and calls in midsummer. The progeny of this generation mature in late fall and overwinter as adults. Progeny of the spring generation mature in either midsummer or early fall. Those that mature in early fall do not call but remain in a reproductively dormant state until the following spring. Each spring generation is a complete generation that consists of both progeny and grand-progeny of the previous spring generation.

The brown/green color dimorphism of this conehead varies with sex and season. In the overwintering generation, more than 95% of males are brown, whereas about 70% of females are green. In the summer generation, more than 80% of males and females are green.

Similar species: The round-tipped conehead, *Neoconocephalus retusus* (Plate 37), has a more rounded cone, the forewings extend less than 11 mm beyond the hind femora, and its song never has regular, momentary interruptions. It is reproductively active in late summer and early fall.

SWORD-BEARING CONEHEAD

Neoconocephalus ensiger (Plate 37, Fig. 59)

Distribution: Found in the northeastern United States, adjacent Canada, and the northern Great Plains.

Identification: A slender species with a cone of medium length edged in black beneath. The stridulatory vein is long and weakly swollen. The song is distinctive, being a continuous series of lisps at a rate of 10 per s, sounding like a distant, fast-moving steam locomotive (to those who have had the fortune to hear one). Males are 45–55 mm long, females, 52–64 mm.

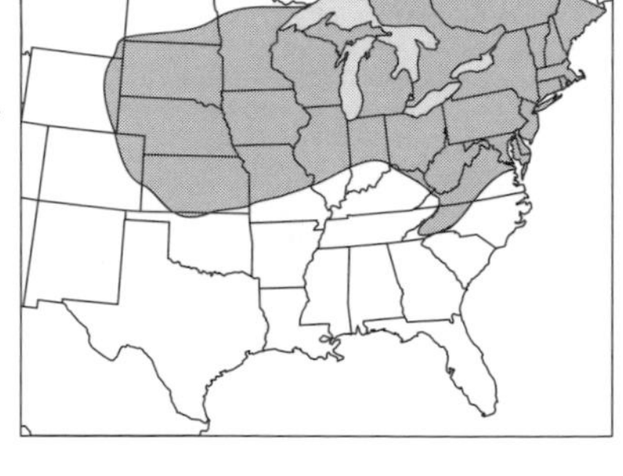

sword-bearing conehead

Ecology: Inhabits wet grassy areas, moist fields, and roadsides. It calls from July through September.

Similar species: Four other *Neoconocephalus* species have slender cones with black beneath. The Nebraska conehead, *N. nebrascensis*, and the slender conehead, *N. lyristes*, are most similar, but the lower surface of the cone is wholly black or nearly so, and the stridulatory vein is thicker and has more pronounced subsidiary veins. The former species calls with a loud *bzzzzz* lasting more than 1 s repeated every 2 s (at 77 °F). The latter produces a high-pitched, smooth, continuous buzz. The slightly musical conehead, *N. exiliscanorus*, and the Everglades conehead, *N. pahayokee*, have much longer cones that are mostly black beneath. The

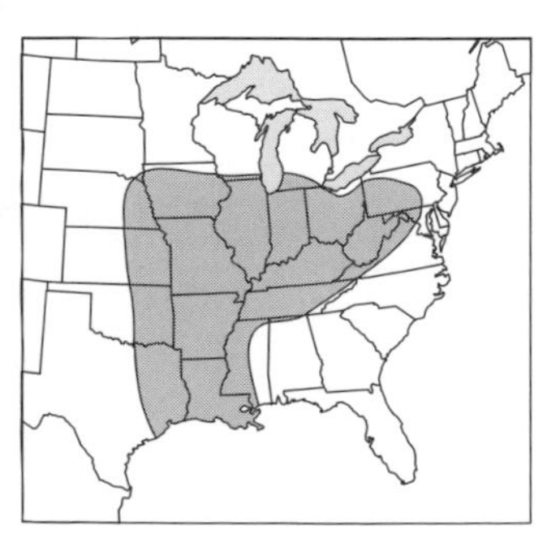

Nebraska conehead

slightly musical conehead calls with sequences of loud, raspy, brief (0.1 s) buzzes made at a rate of about 3 per s (at 77 °F), whereas the Everglades conehead produces a high-pitched, continuous buzz.

slender conehead

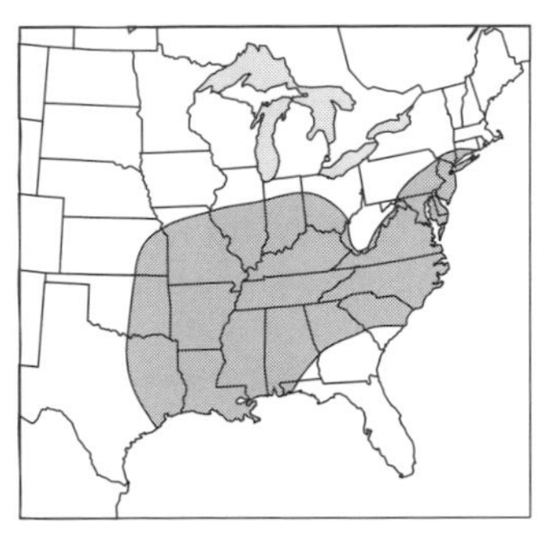

slightly musical conehead

Everglades conehead

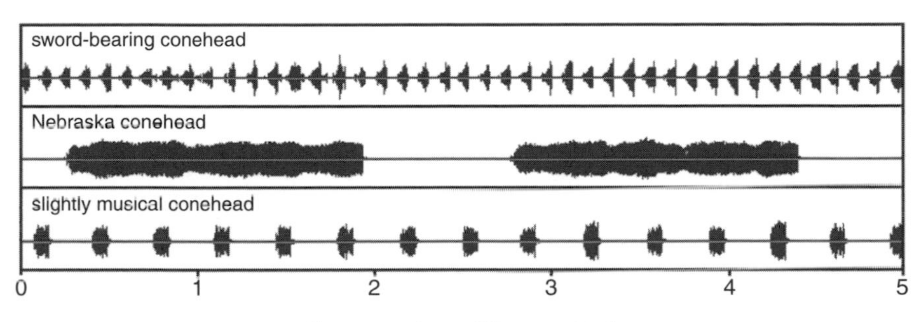

Song comparisons of three coneheads

ROBUST CONEHEAD

Neoconocephalus robustus (Plate 37, Fig. 59)

Distribution: Found east of the Rocky Mountains and in the Central Valley of California.

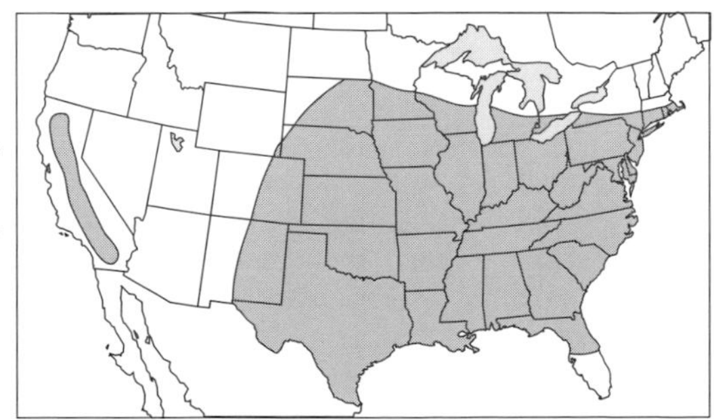

robust conehead

Identification: This is the largest, loudest conehead north of Mexico. The width of the stridulatory area at the base of the forewings is 4.9 mm or greater. The cone is immaculate beneath or has a transverse black mark near the tip. Ovipositor is 1.0 to 1.1 times the length of the hind femur. Males make an intense continuous buzz that has a whining quality at a distance and up close is dominated by a low-pitched hum. Under favorable conditions the song can be heard as far as 500 m away. Length is 53–74 mm.

Ecology: Inhabits tall, rank vegetation, such as moist upland prairies, cornfields, wet areas behind coastal dunes, and the edges of salt marshes. Adults are found from July through September. They are strong fliers, which perhaps accounts for the species range extending in the Southwest to areas that are mostly desert. The disjunct population in California may be a result of an inadvertent introduction from the East.

Similar species: The false robust conehead, *Neoconocephalus bivocatus*, is most easily distinguished from the robust conehead by its song, which is a loud buzz rather than a piercing whine. The basis of the difference is that the sound-producing wingstrokes are produced in pairs rather than at a uniform rate. Caudell's conehead, *N. caudellianus*, has more black on its cone, and its song is not continuous but a regular series of powerful half-second buzzes produced about once per second. Neighboring males synchronize their buzzes. The swift conehead, *N. velox*, is a slender species compared with the previous three; its song is a weak version of that of the robust conehead. The marsh conehead, *N. palustris*, has a cone with a slightly pinched-in tip and is smaller. Males are 37–46 mm long, females, 40–62 mm.

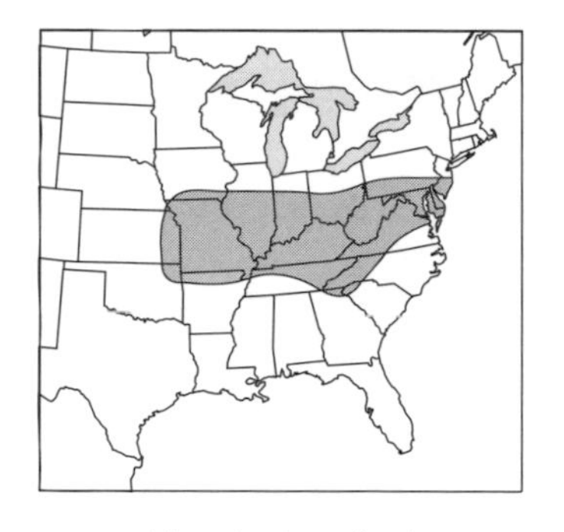

false robust conehead

Caudell's conehead

swift conehead

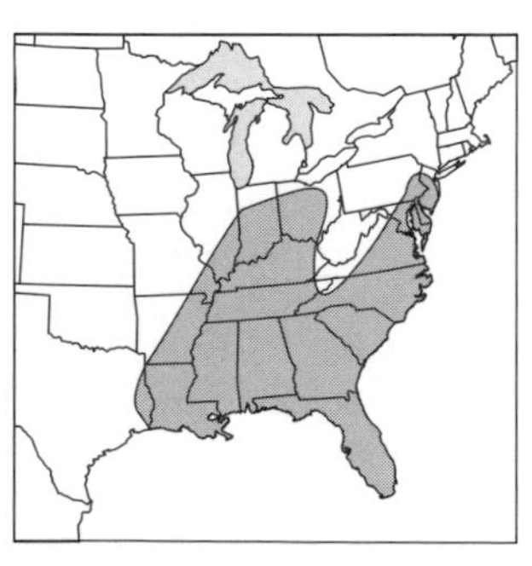

marsh conehead

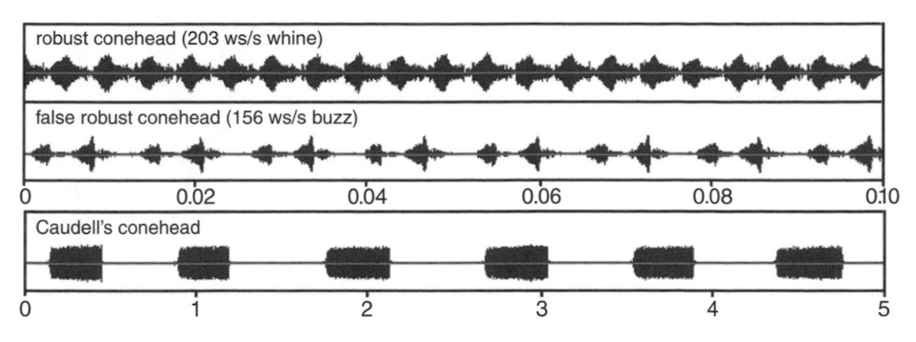

Song comparisons of three coneheads

CATTAIL CONEHEAD

Bucrates malivolans (Plate 38)

Distribution: Found near the coast from New Jersey to Texas and throughout Florida and southern Louisiana.

cattail conehead

Identification: This species has a rounded tip to its immaculate cone, and the cone joins the face ventrally with no gap. Forewings either extend well beyond the end of the abdomen or end near it. Both long-winged and short-winged forms can be either brown or green. Ovipositor is broad and longer than 30 mm. The song is a brief, coarse *bzzzz* repeated individually or at a nearly regular rate of about 2 per s for short to medium sequences. Neighbors do not synchronize. Length of short-winged forms is 29–46 mm; of long-winged forms, 46–70 mm.

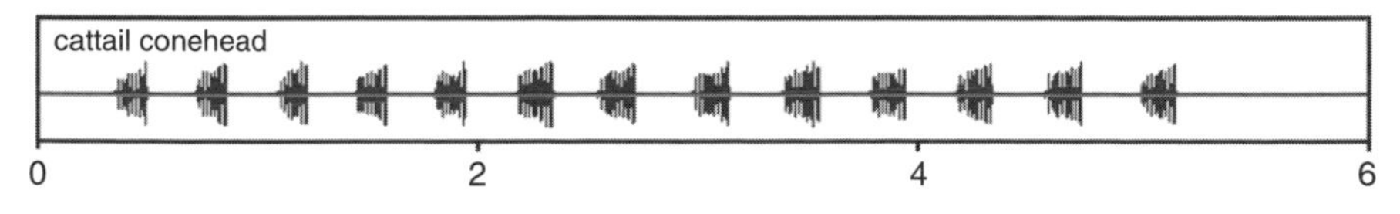

Cattail conehead waveform

Ecology: Inhabits freshwater marshes, where it is found on cattails, sawgrass, and other tall grasses; moist thickets; and tangled vegetation along wet or flooded ditches. Adults appear no earlier than June, except in south Florida, where they are found as early as April and as late as January, and breeding may be continuous throughout the year.

With two sexes, two colors, and two wing lengths, cattail coneheads could exhibit eight different forms; however, no long-winged, green males are known. Few males are green (<5%), but nearly 50% of females are. The short-winged form is predominant in both sexes (about 70% vs. 30% long-winged forms).

Similar species: Common coneheads (genus *Neoconocephalus*) have a gap between the face and the ventral surface of the cone, and the cone has a small ventral tooth at its attachment.

HOOK-FACED CONEHEAD

Pyrgocorypha uncinata (Plate 38)

Distribution: Found in the southeastern United States.

Identification: Forewings extend beyond the abdomen, and the cone ends in a sharp, downturned point. The song, heard in the spring, is a high-pitched, ringing hiss modulated momentarily 4 to 5 times per s. Length is 47–62 mm for Florida specimens; 44–54 mm northward.

hook-faced conehead

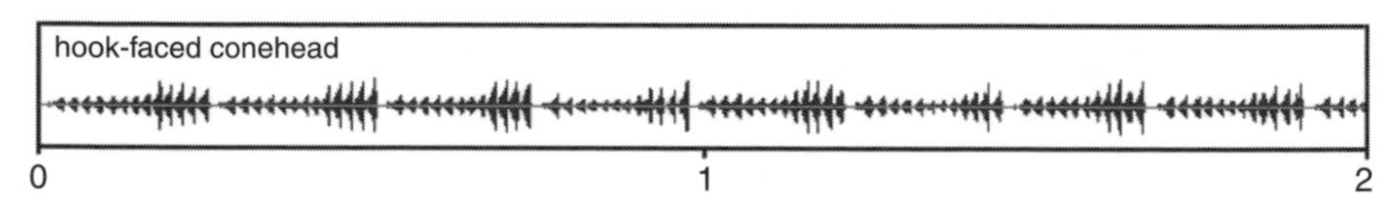

Hook-faced conehead waveform

Ecology: Juveniles may feed and develop on grasses. Males sing mostly from hardwood trees and from woodland undergrowth. Adults mature in late summer or fall and do not become reproductively active until the following spring (March to May in peninsular Florida, and April to May in North Carolina). Singing occurs later on the Florida Keys (April to July) than elsewhere, possibly as an adaptation that allows juveniles to avoid the spring dry season.

Males are always brown and females occur in approximately equal numbers of brown and green—a sex-influenced color dimorphism similar to that of overwintering adults of the broad-tipped conehead, *N. triops.* This and the broad-tipped conehead are the only coneheads that occur both in the West Indies and in temperate North America. They are also the only ones that do not produce eggs that overwinter.

Similar species: Common coneheads (genus *Neoconocephalus*) do not have cones that end in sharp or down-turned points. Short-winged coneheads (genus *Belocephalus*) have forewings that extend less than half the length of the abdomen.

PALMETTO CONEHEAD

Belocephalus sabalis (Plate 38)

Distribution: Found in south and central peninsular Florida and up the east coast to extreme south Georgia.

Identification: The cone is sharp-pointed and does not turn down abruptly at the tip; its length as measured from the ventral tooth to the tip exceeds the distance between the eyes. Forewings of the male cover about one-third of the abdomen; forewings of the female

palmetto conehead

are inconspicuous small flaps on either side that just reach the rear of the second abdominal segment. The song is a continuous rattle with the units of the rattle produced at a rate of about 12 per s at 77 °F. Length is 37–50 mm.

Ecology: Most frequently found on saw palmetto and cabbage palm; however, it sometimes calls from other trees, from shrubby undergrowth, and from grasses. It has a single generation annually with adults occurring from June through January.

Similar species: Both Davis's conehead, *Belocephalus davisi*, and the half-winged conehead, *B. subapterus*, are smaller (24–39 mm) than the palmetto conehead. Davis's occurs to the north and west of the range of the palmetto conehead. The half-winged conehead has a shorter cone with no sharp point or a sharp point that turns down abruptly. Its song, unlike that of Davis's and palmetto coneheads, consists of phrases that increase in length and intensity.

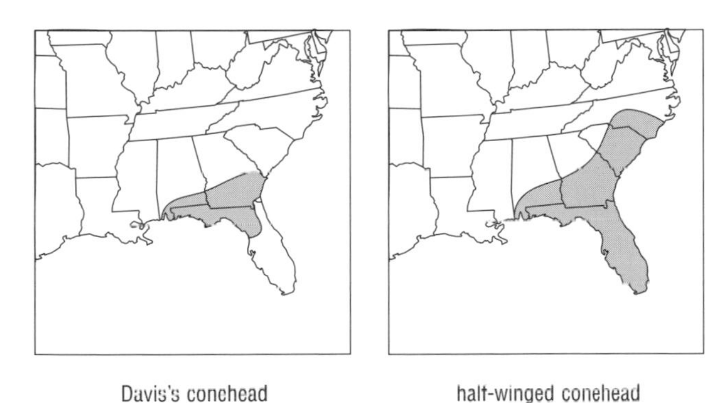

Davis's conehead half-winged conehead

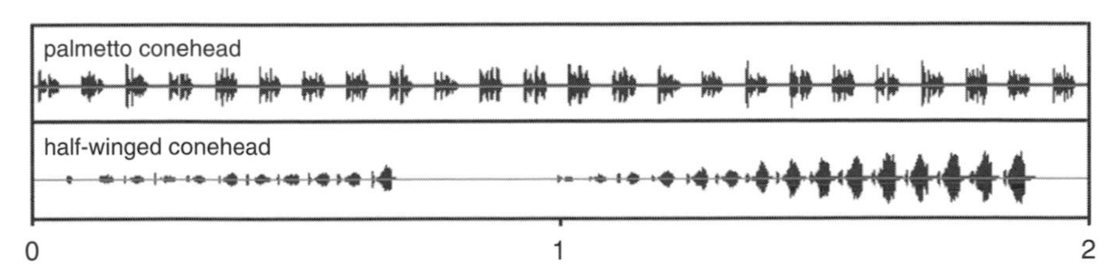

Palmetto conehead and half-winged conehead song comparison

Meadow Katydids

Subfamily Conocephalinae

This subfamily is one of several in which the tibial tympana are concealed behind slit-like openings. Meadow katydids differ from coneheaded katydids (subfamily Copiphorinae) in lacking cones and from predaceous katydids (subfamily Tettigoniinae) in having no spines on the dorsal surface of the fore tibia. As the name implies, meadow katydids are usually found on herbaceous vegetation. Most species call both day and night, and their songs generally consist of series of brief ticks and prolonged buzzes.

There are 39 species of meadow katydids in the United States and Canada. Of these, 19 species that are relatively large and robust belong to the genus

Orchelimum, the larger meadow katydids (body length usually more than 18 mm). Nineteen others that are smaller and slimmer belong to the genus *Conocephalus*, the smaller meadow katydids (body length usually less than 17 mm). The remaining species, the wingless meadow katydid, *Odontoxiphidium apterum*, is small and has wingless females and males with forewings just large enough to maintain their calling function.

Identifying meadow katydids to species is difficult. The male cerci are the most distinctive morphological features (Fig. 60), but besides being of no help in identifying females, they are tiny and best viewed from the top—which is often made difficult by overlying wings. The following accounts and comparisons will help identify 16 commonly encountered species.

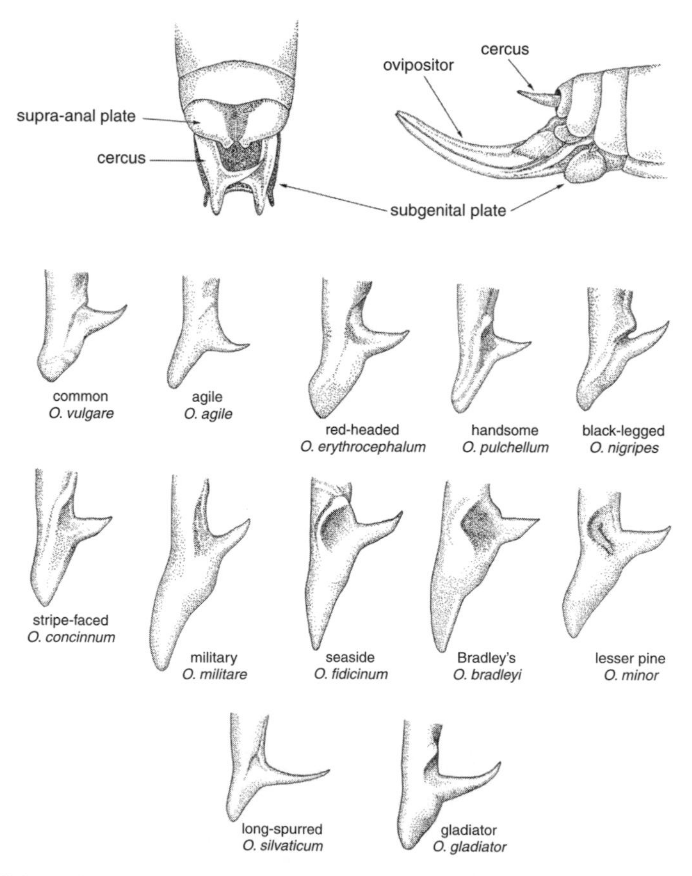

Figure 60. Abdominal structures of meadow katydids. Above, dorsal view of rear of male abdomen (left) and lateral view of rear of female abdomen (right). Below, dorsal views of the left male cercus of twelve meadow katydids, genus *Orchelimum*.

COMMON MEADOW KATYDID

Orchelimum vulgare (Plate 39, Fig. 60)

Distribution: Found in the eastern Great Plains and in the eastern United States and adjacent Canada except for Florida and the Coastal Plain.

common meadow katydid

Identification: The tooth of the male cercus is shorter than the distal cercal shaft (measured from the tooth's base to the cercal tip) but is longer than half the length of the distal cercal shaft. The distal cercal shaft is thick and ends with a slight bump followed by a somewhat flattened tip. The ovipositor is distinctly curved upward and less than half the length of the hind femur. The song is an alternation of long buzzes and groups of ticks. The buzzes are loud, last for 3–7 s, and increase in intensity before they end abruptly. The ticks, made between buzzes, are delivered irregularly and too rapidly to count. Length is 22–40 mm.

Ecology: Common inhabitant of pastures, marshes, fencerows, roadsides, and old fields—by far the most commonly encountered meadow katydid in its range. Adults first appear in July and are present until frost.

red-headed meadow katydid

Similar species: In four other *Orchelimum* species the tooth of the male cercus is shorter than the distal cercal shaft but longer than half its length. The red-headed meadow katydid, *O. erythrocephalum*, has a reddish face; the agile meadow katydid, *O. agile*, has a white face and a plain cercal tip; the handsome and black-legged meadow katydids, *O. pulchellum* and *O. nigripes*, have a prominent fold that extends from the base of the cercal tooth to near the cercal tip. Both of these species have black tibiae, but the latter has a longer cercal tooth.

agile meadow katydid

handsome meadow katydid

black-legged meadow katydid

STRIPE-FACED MEADOW KATYDID

Orchelimum concinnum (Plate 39, Fig. 60)

Distribution: Found in coastal areas from Massachusetts to Texas, in peninsular Florida, and in the central Midwest. A likely hypothesis to account for the disjunct midwestern populations is that they spread from the Atlantic coast by way of the Hudson-Mohawk glacial outlet at the end of the last North American glaciation.

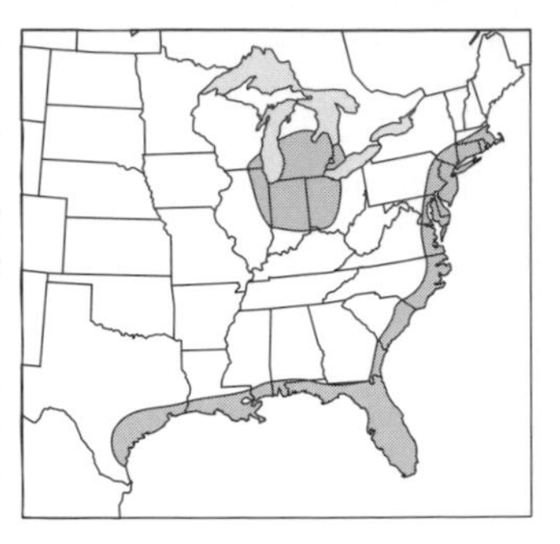

stripe-faced meadow katydid

Identification: Slender in form with a reddish-brown, vertical facial stripe. The distal cercal shaft, from the base of the tooth to the cercal tip, is about twice as long as the tooth. The ovipositor is curved upward and is about half the length of the hind femur. The song is an alternation of 1 to 5 ticks delivered at 2–4 per s and buzzes lasting 1–2 s. Length is 22–30 mm.

Ecology: Inhabits freshwater marshes and saltwater marshes including black-rush marshes. In its midwestern distribution, it is restricted to relict marl bogs and other alkaline situations. In most of its range, there is a single annual generation with adults occurring from July to September; however, in peninsular Florida, adults occur as early as April and as late as November, suggesting at least two generations.

Similar species: Four other larger meadow katydids (genus *Orchelimum*) have long distal portions to their cercal shafts. One of these, the military meadow katydid, *O. militare*, sometimes has a reddish-brown facial stripe, but its cercal shaft is much longer than that of the stripe-faced meadow katydid and its ovipositor is straight and nearly as long as the hind femur. The others lack the facial stripe—the seaside meadow katydid, *O. fidicinum*, which occurs only in saltmarshes; Bradley's meadow katydid, *O. bradleyi*, which occurs only in freshwater marshes and has unusually massive cerci; and the lesser pine katydid, *O. minor*, which occurs only in pine trees and usually well beyond net's reach. The lesser pine katydid can nonetheless be identified by the brief, beady phrases of the male's calling song, produced at night at uniform intervals of about 1.8 s.

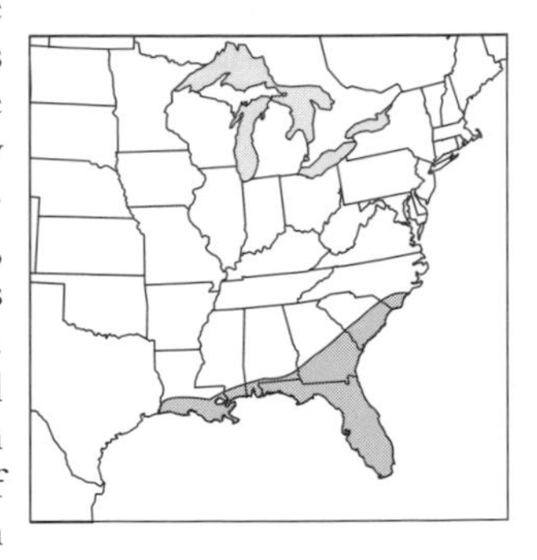

military meadow katydid

seaside meadow katydid

Bradley's meadow katydid

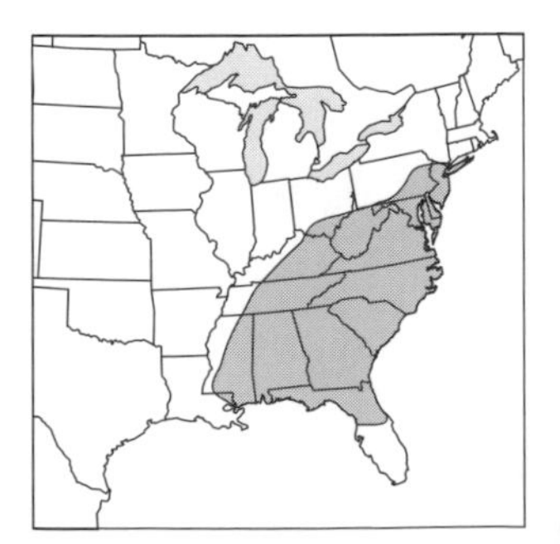

lesser pine katydid

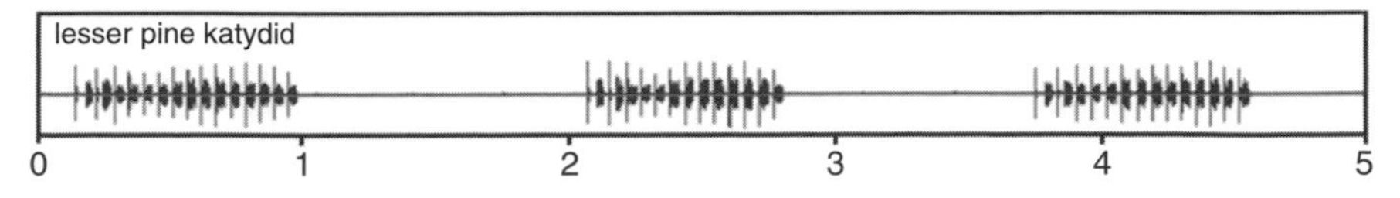

Lesser pine katydid waveform

LONG-SPURRED MEADOW KATYDID

Orchelimum silvaticum (Plate 39, Fig. 60)

Distribution: Found from Kansas to Texas east to Ohio and Alabama.

Identification: The cercal tooth in this species is much longer than the distal portion of the cercal shaft. The song lacks ticks and is a series of buzzes, each of which lasts several seconds and begins with a slow wingstroke rate that soon shifts to a wingstroke rate more than twice as fast. Length is 20–30 mm.

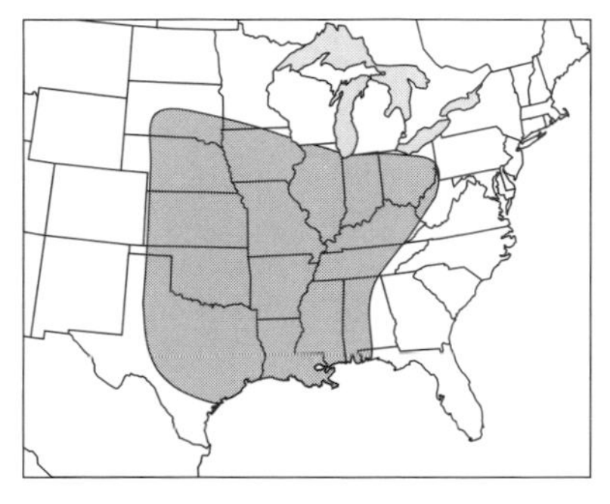

long-spurred meadow katydid

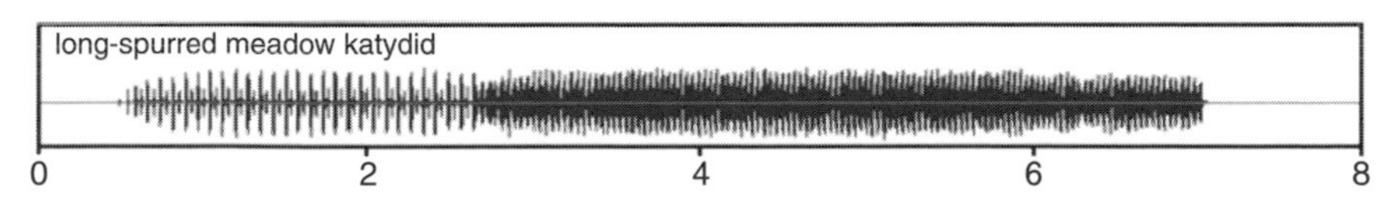

Long-spurred meadow katydid waveform

Ecology: Most commonly inhabits deciduous trees. It is also found in shrubby and herbaceous vegetation, often at the edges of woods. Adults appear as early as July.

Similar species: The gladiator meadow katydid, *Orchelimum gladiator*, is the only other species that has an unusually long cercal tooth. It differs from the long-spurred meadow katydid in having a swelling on the cercal shaft just distal of the tooth.

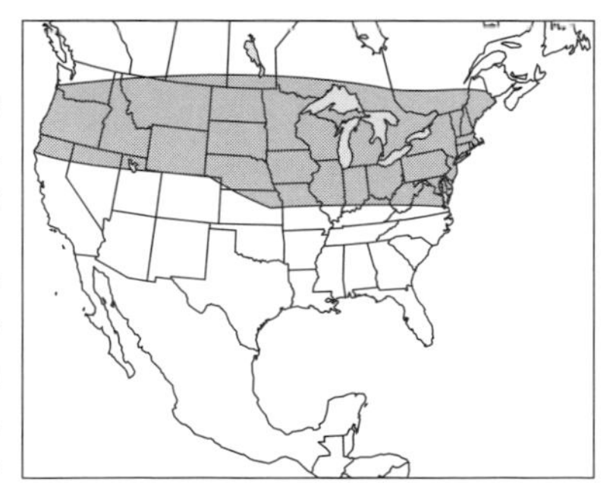

gladiator meadow katydid

SLENDER MEADOW KATYDID

Conocephalus fasciatus (Plate 40)

Distribution: Found east of the Continental Divide in the United States and southern Canada.

Identification: A slender, little meadow katydid with forewings that are longer than the abdomen and hind wings that exceed the forewings by 2–3 mm. The cerci of males are green and have a stout tooth on the inner border and a tip that is weakly flattened. The ovipositor is straight and about two-thirds the length of the hind femur. The song is very soft and consists of sequences of ticks alternating with buzzes. The buzzes vary in duration from 1–20 s. Length is 18–26 mm.

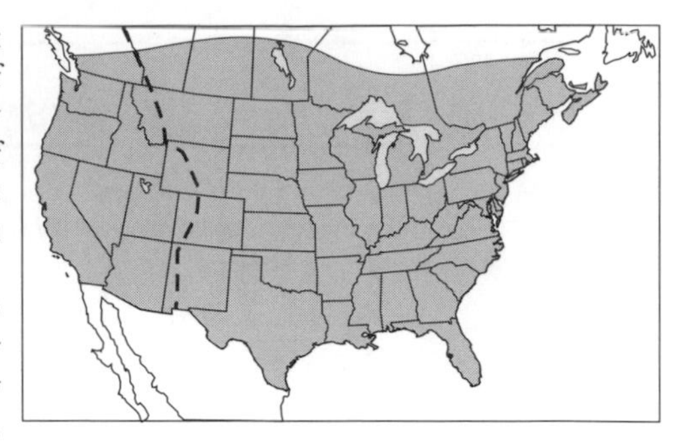
slender/western meadow katydid

Ecology: This species is by far the most widespread and commonly encountered of the lesser meadow katydids (genus *Conocephalus*). Occurs in grassy and weedy vegetation in a wide variety of habitats, such as pastures, open pine woods, and roadsides. It is the earliest meadow katydid to mature, and adults are found until frost, or, in Florida, until midwinter. In Michigan, adults first occur in late July; in Virginia, late June; and in Florida, late April. There are two generations annually as far north as North Carolina.

Similar species: The western meadow katydid, *Conocephalus vicinus*, occurs in the Pacific drainage and is so similar that many consider it a geographic race of the slender meadow katydid rather than a separate species. The Caribbean meadow katydid, *C. cinereus* is known only from peninsular Florida and has male cerci that are golden, with decidedly flattened tips.

Caribbean meadow katydid

STRAIGHT-LANCED MEADOW KATYDID

Conocephalus strictus (Plate 40)

Distribution: Found from the Rocky Mountains eastward except for North Dakota, New England, and Florida.

Identification: Robust for a lesser meadow katydid (*Conocephalus*). Females are easily recognized by their nearly straight ovipositor that much exceeds the body in length. Male cerci are more than 2 mm long, tapered beyond the tooth, and flattened in their final third. In most individuals, the forewings cover only one- to two-thirds of the abdomen and the hind wings are completely concealed by the forewings. However, a few have forewings that extend rearward to the tips of the femora, with the hind wings extending even farther. The song is a very faint, continuous fluttery buzz, sometimes alternating long periods of a fast, regular pulsation rate with shorter periods of a slower, somewhat irregular, pulsation rate. Length is 13–22 mm in the short-winged form, 21–30 mm in the long-winged form.

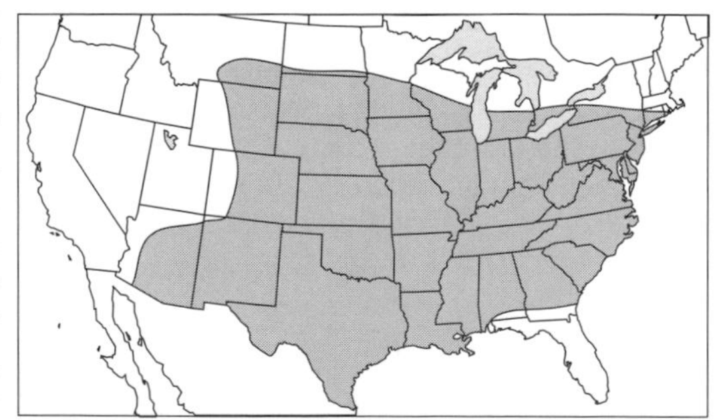
straight-lanced meadow katydid

Ecology: Most commonly inhabits dry open grassland and dry grassy areas along roads and fences. Adults first appear in late summer.

Similar species: The smaller species of larger meadow katydids (genus *Orchelimum*) are of similar size, but their ovipositors and the distal portions of their cerci are not nearly as long.

WINGLESS MEADOW KATYDID

Odontoxiphidium apterum (Plate 39)

Distribution: Found in the southeastern United States.

Identification: Both males and females lack hind wings. Forewings of the male are shorter than the pronotum and are convex. Those of the female are covered by the pronotum and so tiny as to be easily overlooked. The male cercus is unique among meadow katydids in having a small tooth proximal to the main tooth. Ovipositor is straight and about the length of the hind femur. The song consists of highly irregular sequences of brief and prolonged beady buzzes. Length is 9–19 mm.

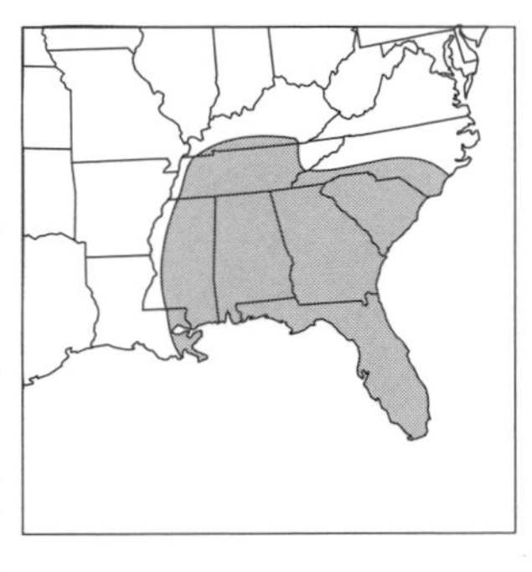

wingless meadow katydid

Ecology: Found on herbaceous vegetation and low bushes, often in pine woods and old fields. In most of its range there is a single generation annually with adults first appearing in July. In Florida, there are apparently two generations with adults occurring from May until midwinter.

Similar species: Lesser meadow katydids (genus *Conocephalus*) have longer forewings, often with well-developed hind wings.

Quiet-calling Katydids

Subfamily Meconematinae

This subfamily is not native to North America, but one species has been introduced from Europe. About 200 species occur worldwide, including three that make songs that are wholly ultrasonic in range. The U.S. species has no male stridulatory apparatus of the usual kind. There are, however, minute teeth on the forewings that may substitute in a quiet way. In addition, it has another method of calling, as indicated by its common name, drumming katydid.

DRUMMING KATYDID

Meconema thalassinum (Plate 40)

Distribution: Native to Europe, this species was first recognized as established in the New World when it was discovered on Long Island, New York, in 1957. It has since extended its range to Rhode Island and central New York.

drumming katydid

Identification: A tiny katydid with a tympanum fully exposed on each fore tibia. Forewings are longer than the hind wings, and no stridulatory area is apparent at the base of the male forewings. Males call at night not by tegminal stridulation but by rapidly tapping either of the hind tarsi on the substrate, such as the surface of a leaf. The sound varies with the substrate but under favorable conditions can be heard at 3.5 m. A bout of drumming consists of several bursts, the initial ones being brief and the later ones lasting about 1 s. Length is 14–19 mm.

Ecology: Found in deciduous trees and on the vegetation beneath. It lays its eggs in crevices in bark and may have been imported to the United States as eggs on woody ornamental plants.

Similar species: Meadow katydids (subfamily Conocephalinae) have the tympanum visible only through slits in the fore tibia. False katydids (subfamily Phaneropterinae) are larger, and the hind wings are often longer than the forewings.

Predaceous Katydids

Subfamily Tettigoniinae

Most North American species of this subfamily have a shield-like pronotum that extends rearward over all or part of much reduced forewings. Among these "shieldbacks," the forewings of females are often tiny, concealed pads; those of males are developed for stridulation but are still very short. Whereas shieldbacks lack hind wings and are flightless, other species in this subfamily have long wings and are fully flight capable. Common features of these two very different-looking groups of tettigoniines are one or more spines on the upper surface of the fore tibia, tympana that lie behind slits in the fore tibia, and space between the antennae that is broad and flat.

Of the 255 species of katydids found in the United States and Canada, nearly half—122 species—belong to this subfamily. Of the 122, only 10 occur east of the Great Plains. The others are western species, most of which are seldom encountered because of their limited distributions and secretive habits.

Unlike the previous katydid subfamilies, which feed principally on plant material and do not generally subdue and eat insects, many tettigoniines are active predators. Most also scavenge dead insects and feed on plant material, including leaves, flowers, and fruits. Most species seem to prefer animal foods but of necessity include vegetable matter in their diets. Mormon crickets, for example, feed on a great variety of plants and damage gardens and grain crops, but they feed greedily on the carcasses of their comrades crushed by traffic as they crossed roads, and they capture and consume other insects.

In keeping with their predaceous tendencies, many tettigoniines inflict painful bites when collected by hand. A few species occasionally draw blood.

ROBUST SHIELDBACK

Atlanticus gibbosus (Plate 41)

Distribution: Found in Florida, Alabama, Georgia, and South Carolina.

Identification: This is the largest of the seven species of the genus *Atlanticus* (eastern shieldbacks). The unusually long hind femur measures 29–35 mm. Forewings of the male are completely concealed by the pronotum or very nearly so. Pronotum is less than twice as long as its greatest width and has no median ridge, and its lateral ridges are indis-

robust shieldback

tinct. The male cercus has a long, slender, curved inner tooth near its base. The female has a straight ovipositor that is about four-fifths the length of the hind femur. The song consists of groups of 3 to 5 brief buzzes. Length is 28–37 mm.

Ecology: Found on the ground and understory in a variety of environments, including pine and hardwood forests and oak and palmetto scrub. In captivity, adults will readily pounce on and consume live ground crickets. Adults are found from July to October in the north and from May to November in Florida. There is a single peak of adult abundance each year.

Similar species: The American shieldback, *Atlanticus americanus*, is smaller, and the forewings of males project from beneath the pronotum by a distance that is about half the front width of the pronotum. The ovipositor is as long as or slightly longer than the hind femur. The gray shieldback, *A. dorsalis*, is also smaller; the forewings of males usually project from beneath the pronotum but by less than half the front width of the pronotum. The pronotum has a median ridge on its rear third, and each lateral lobe of the pronotum is marked at the rear with a black triangle with a yellow lower border. The ovipositor is about two-thirds the length of the hind femur.

American shieldback

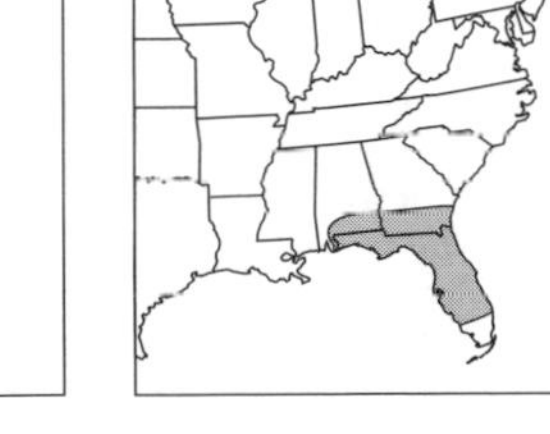

gray shieldback

PROTEAN SHIELDBACK

Atlanticus testaceous (Plate 41)

Distribution: Found in portions of Oklahoma, Arkansas, and Missouri east to the Atlantic coast, excluding New England.

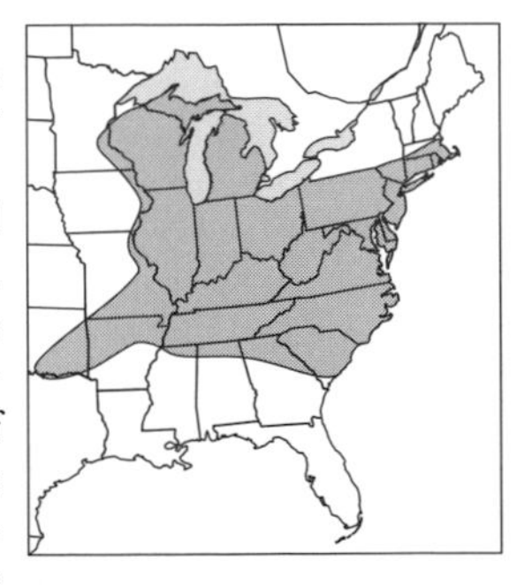

protean shieldback

Identification: The lateral ridges of the pronotum are well defined, and its lateral lobes drop sharply downward. Forewings of the male usually extend from beneath the pronotum by more than half the length of the pronotum. Hind femur is often less than twice the length of the pronotum. Ovipositor is straight. The shieldbacks here treated as a single species vary greatly geographically. This variation needs careful study to determine whether additional species should be recognized. The song is a series of high-pitched rattles lasting 0.5 to >60 s. Pauses between rattles are brief, and the terminal rattle in a series is often much longer than the preceding rattles. Length is 18–24 mm in the North; 22–30 mm in the South.

Ecology: Inhabits woodland borders, open woodland, fencerows, and brushy pastures. It matures earlier than most katydids, with adults occurring as early as late May and as late as September. There is a single, early peak of adult abundance.

Similar species: The male forewings of the least shieldback, *Atlanticus monticola*, extend from beneath the pronotum by a distance of about one-quarter the length of the pronotum, and the hind femur is about twice the length of the pronotum. The ovipositor is usually distinctly curved.

least shieldback

HALDEMAN'S SHIELDBACK

Pediodectes haldemani (Plate 41)

Distribution: Found in the Great Plains from South Dakota south.

Identification: The pronotum is 11–16 mm long and lacks lateral ridges (pronotal disk not defined). The rear two-fifths of the pronotum sometimes is slightly flattened and sometimes has a slight median ridge; it always has two broad, dark brown, dorsal, longitudinal stripes separated by a narrow, buff one that is occasionally poorly defined. Male forewings project beyond the pronotum but by less than the rear width of the pronotum. Male cercus has an inward-projecting tooth that is almost at the end of the cercal shaft. Ovipositor curves gently upward and is about as long as the hind femur. The song is a *zip* lasting less than 0.1 s, repeated every 5–7 s. Length is 32–39 mm.

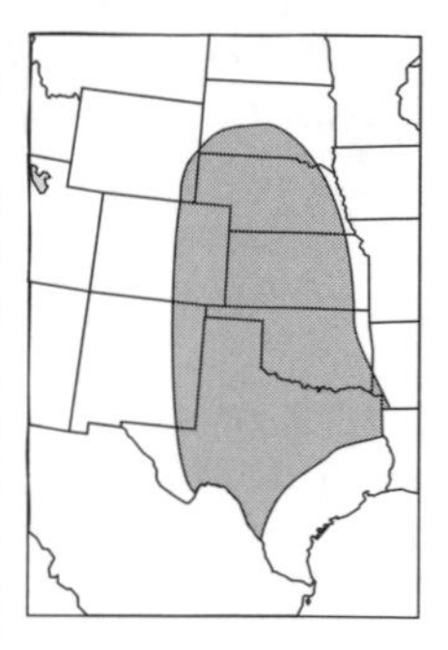

Haldeman's shieldback

Ecology: Occurs in a variety of habitats including old fields, pine-juniper, and mesquite–prickly pear. Specimens have been collected at night, 0.5–2 m up on mesquite, yucca, and low weeds.

Similar species: Because this genus has never been carefully studied, some of the eight other species currently recognized may not be valid and others may yet be discovered. The two species most similar in distribution to Haldeman's shieldback are both smaller. In Stevenson's shieldback, *P. stevensonii*, the male cercus resembles that of Haldeman's shieldback, but the pronotal length is only 5–8 mm. In the black-margined shieldback, *P. nigromarginatus*, the pronotal length is 8–10 mm, and the tooth of the male cercus projects inward from near the middle of the shaft.

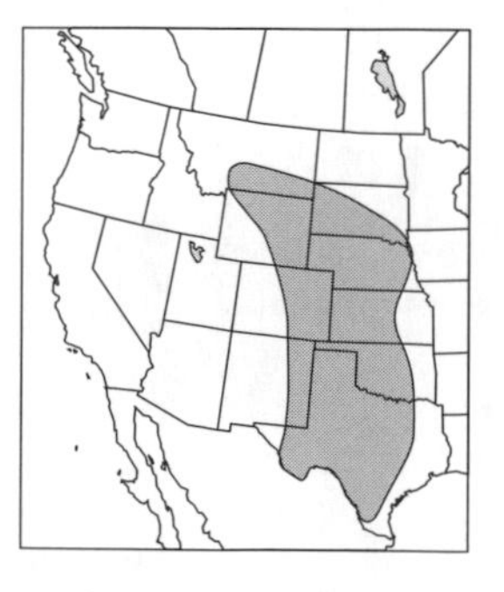

Stevenson's shieldback

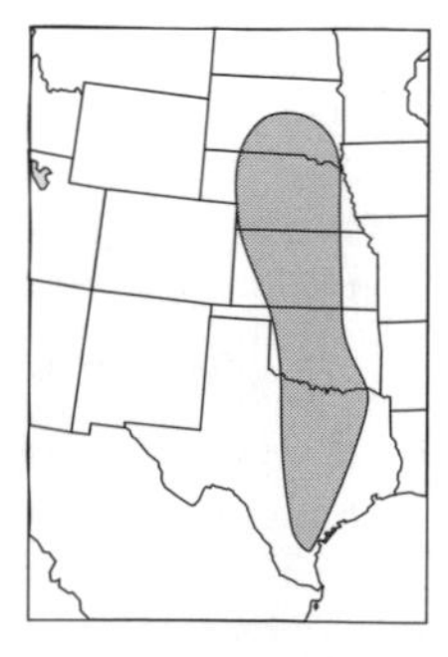

black-margined shieldback

MORMON CRICKET

Anabrus simplex (Plate 41)

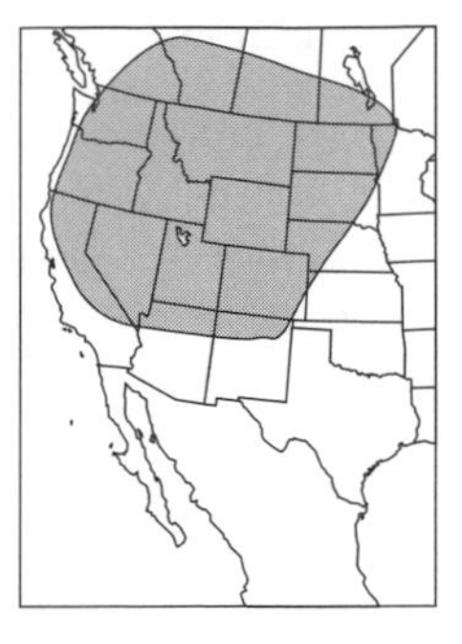
Mormon cricket

Distribution: Found from eastern California to northern New Mexico north to southern British Columbia and Manitoba.

Identification: Large shieldbacks that sometimes travel in bands of thousands. Forewings are short and usually concealed by the pronotum; in some males, they extend rearward from the hind margin of the pronotum but by less than half the rear width of the pronotum. Lateral and median ridges are evident on the rear half of the pronotum but not on the forward half. Hind femur is no more than twice the length of the pronotum. Male cercus has a right-angle bend near its sharp-pointed end and a thumb-like subapical tooth. Female subgenital plate has its rear lateral angles developed into sharp, inward-directed teeth. Ovipositor is equal to or slightly longer than the hind femur and curves upward or is nearly straight. Individuals that travel in bands are generally black or dark brown, whereas solitary ones are green. Length is 28–45 mm.

Ecology: Occurs in a wide variety of open, often scantily vegetated habitats. In early spring, mormon crickets hatch from eggs laid in the soil the previous year. When they are numerous, the nymphs soon gather into bands that move steadily in one direction as they feed. Their appetites become more voracious as they mature, and though they usually eat only low-value broadleaf plants, they may devastate any crops or grazing lands they encounter. They are active during daylight hours, and during the night they rest on the ground or in bushes. Adults are found from June to September.

Similar species: In the long-legged anabrus, *Anabrus longipes*, the hind femur is usually more than twice as long as the pronotum. In the big-tooth anabrus, *A. cerciata*, the subapical cercal tooth of the male is greatly developed and meets the main body of the cercus at a right angle. In coulee crickets, *Peranabrus scabricollis*, which sometimes travel in destructive bands, the upper surface of the pronotum is rough and pebbled rather than smooth.

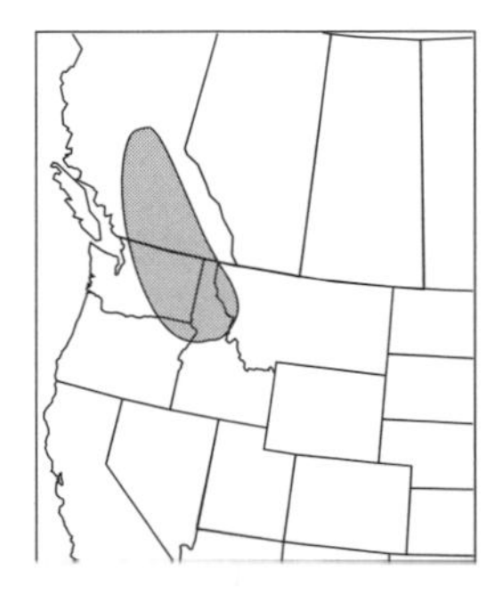
long-legged anabrus

big-tooth anabrus

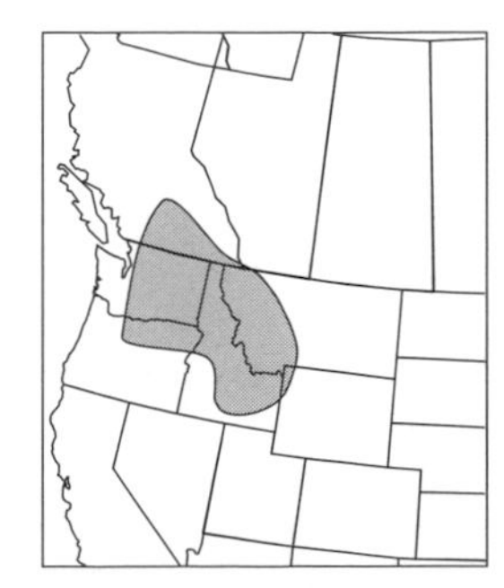
coulee cricket

SPLENDID SHIELDBACK

Aglaothorax ovata (Plate 41)

Distribution: Found in southern California and eastern Nevada.

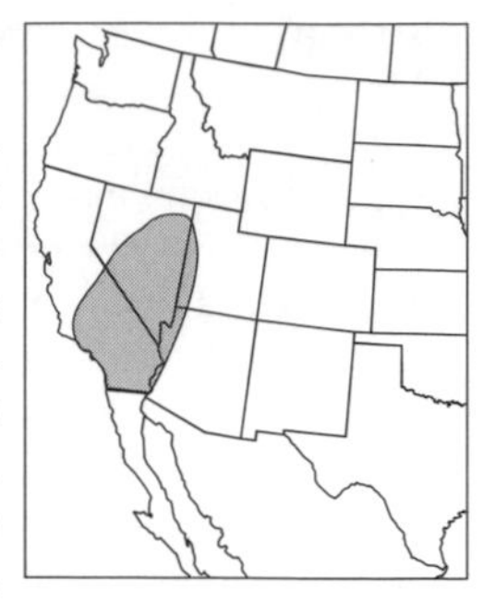

splendid shieldback

Identification: The disk of the pronotum is greatly expanded rearward, with its greatest width at least twice its front width. The dorsum of the abdomen has a median reddish-brown, longitudinal stripe, on either side of which most abdominal segments have white, quadrate spots of varied sizes. Forewings of the male are completely beneath the pronotum or very nearly so. Supra-anal plate has a rounded projection that extends over the bases of the male's "pseudocerci," which are processes that look and function like the true cerci of most other shieldbacks. The true cerci of the spendid shieldback and its close relatives are at the base of the pseudocerci and are short and cone-shaped with pinched tips. Pronotal disk has a series of black marks around its edges. Ovipositor is decidedly curved upward and has many small teeth along the upper and lower edges of the final fourth of its length. This species has six named subspecies that vary in size, male terminal structures, and other features. Length is 18–26 mm.

Ecology: Inhabits the fringes of the Mojave and northern Sonoran deserts. It is found at higher elevations where temperatures are not as extreme as in the hottest parts of these deserts. It probably has a single generation annually, at least in higher areas. Most records are from July to September but a few at lower elevations are as early as April, May, or June.

Similar species: The other four species of this genus are either smaller or colored differently.

BOG KATYDID

Metrioptera sphagnorum (Plate 42)

Distribution: Found in the lower tier of Canadian provinces from eastern British Columbia to western Quebec.

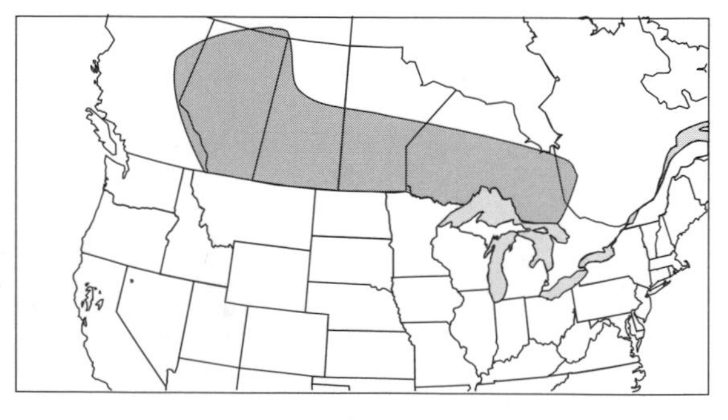

bog katydid

Identification: This species extends farther north than any other North American katydid and is the only Canadian species that has not yet been collected in the United States. Color is primarily brown but with hind tibia, rear margins of abdominal segments, and lower margins of pronotum light green or yellow. Fore tibia has three spines on the upper outer edge. Pronotum is saddle-shaped with the rear width about twice the front width and the lateral lobes bent sharply down. Forewings cover about half the abdomen in most males, but there is a long-winged form in which the forewings extend beyond the tips of the hind femora. Male cerci are nearly covered by the supra-anal plate; each is conical with a sharp, curved tooth near the base. Forewings in females cover about one-fourth of the abdomen; the ovipositor curves upward and is about the length of the hind femur. The calling song is a continuous, very high-pitched trill that alternates regularly between two levels of intensity. Each level is produced for about 0.25 s at a time. Length of short-winged males is 18–20 mm.

Ecology: Occurs only in or near sphagnum bogs in the boreal forest of central and western Canada. Adults are found from mid-July until September. The life cycle may be completed in a year, but this is uncertain because the eggs of some katydids require more than a year to hatch.

Similar species: Roesel's katydid, *Metrioptera roeselii*, is found in grassy fields rather than bogs. The rear width of its pronotum is only slightly greater than the front width; in the short-winged forms, the forewings of the male nearly cover the abdomen and those of the female cover about half of it. This species is native to Europe and was first found in North America at Montreal in 1953. How far it will extend its North American range remains to be seen, but it is worth noting that the long-winged forms, with wings much longer than the abdomen, can fly some distance and are much more commonly collected here than in Europe.

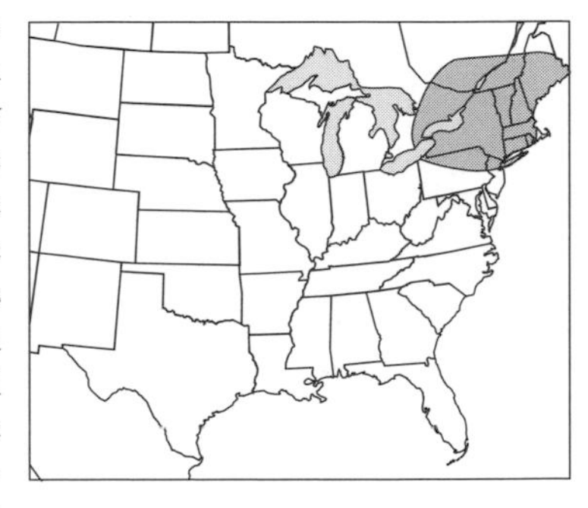

Roesel's katydid

SOOTY-WINGED KATYDID

Capnobotes fuliginosus (Plate 42)

Distribution: Found from southern California north to Nevada and Utah, south to southern New Mexico and east to south-central Texas.

Identification: General color is sooty gray with brown markings. In a few individuals, the brown dominates over the gray. Fore tibia has four spines on the upper outer margin. Forewings are of nearly uniform width throughout, and the hind wings are sooty and barely translucent. Ovipositor curves downward and is nearly as long as the hind femur. The song is a prolonged high-pitched rattle with occasional brief pauses. Males are 61–73 mm long, females 76–82 mm.

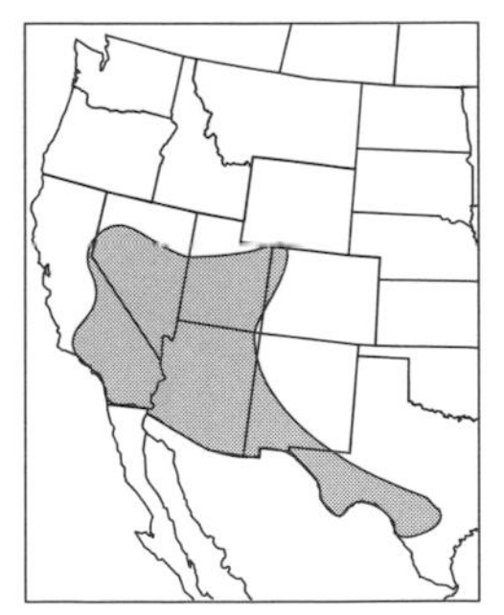

sooty-winged katydid

Ecology: Inhabits the deserts of the Southwest and Great Basin and calls from desert shrubs, such as creosote bush. Adults are surprisingly easy to capture when spotted calling at night because they are reluctant to quit singing or to fly. When severely disturbed, individuals fly clumsily downward rather than flying away. When teased, they display by raising the forewings and spreading the dark hind wings. Adults are found as early as mid-June and as late as November.

Similar species: The seven other species in the genus are similar in general appearance but are smaller, do not have black hind wings, and have forewings that are expanded basally and then taper to the tip. Most have some individuals that are green.

GREATER ARID-LAND KATYDID

Neobarrettia spinosa (Plate 42)

Distribution: Found from central Texas west to southern New Mexico and southeastern Arizona.

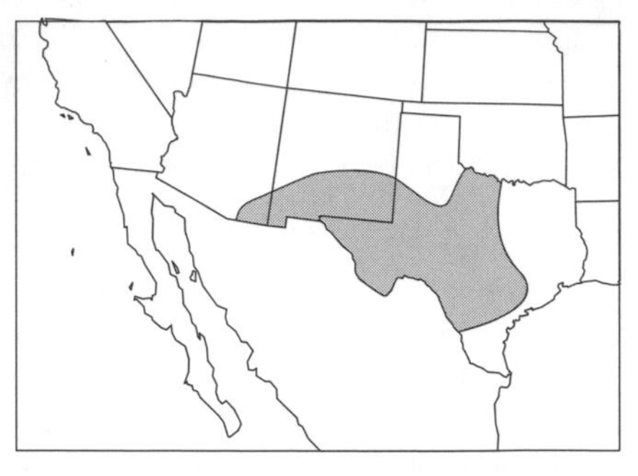

greater arid-land katydid

Identification: A large, formidable species. Individuals do not necessarily retreat when molested and will assume a threatening pose with bright hind wings flared, mandibles opened wide, and spiny forelegs raised high. If given the chance, they may attack, bite, and draw blood! Forelegs are longer than the middle legs and have prominent spines along both lower edges of the femur and tibia. Femoral teeth and the front edge of the pronotum are black. Forewings extend a bit more than halfway toward the end of the abdomen. The center of the hind wings is translucent brown with lighter spots. The song, made at night, is a loud, resonant phrase repeated at intervals greater than 1 s. Males are 34–45 mm long, females 44–52 mm.

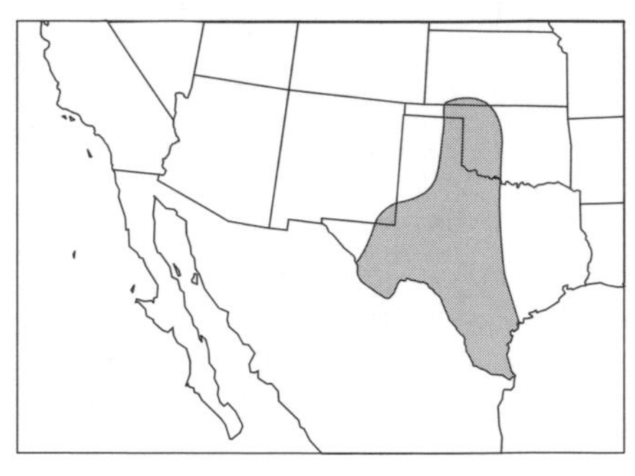

lesser arid-land katydid

Ecology: Inhabits oak-juniper and mesquite open woodland and shrubby desert.

Similar species: Lesser arid-land katydids, *Neobarrettia victoriae*, are smaller (males 25–32 mm; females 31–37 mm) and have a green front edge to the pronotum. Femoral teeth are two-colored, and the center of the hind wings is jet black. The song is a pulsating series of 4- to 5-pulse phrases made mostly at night.

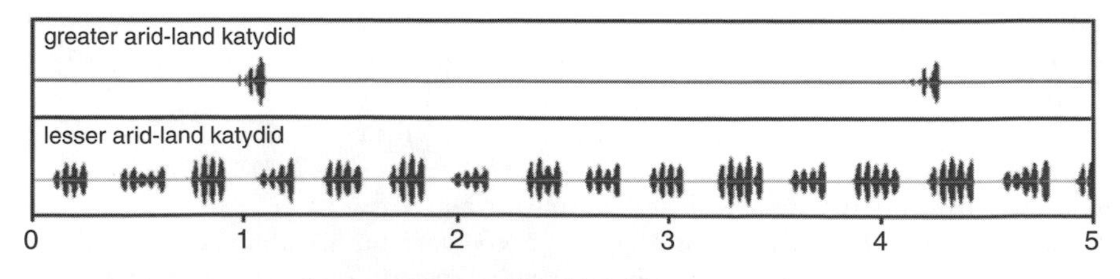

Greater and lesser arid-land katydid song comparison

PINE KATYDID

Hubbellia marginifera (Plate 42)

Distribution: Found from Mississippi east to northern Florida and eastern North Carolina.

Identification: This species is camouflaged for living among pine needles—its forewings are green with a thin, longitudinal light stripe and white margins. A brown dorsal stripe extends from the top of the head to near the tips of the forewings. Forewings are longer than the abdomen. Fore tibia is armed with one or more spines on the dorsal surface. The song consists of high-pitched clicks repeated indefinitely, at night, at intervals of about 2 s. Length is 33–37 mm.

Ecology: Occurs exclusively in the tops of pine trees; therefore, it is rarely seen or collected. Its presence is revealed by the calling of males.

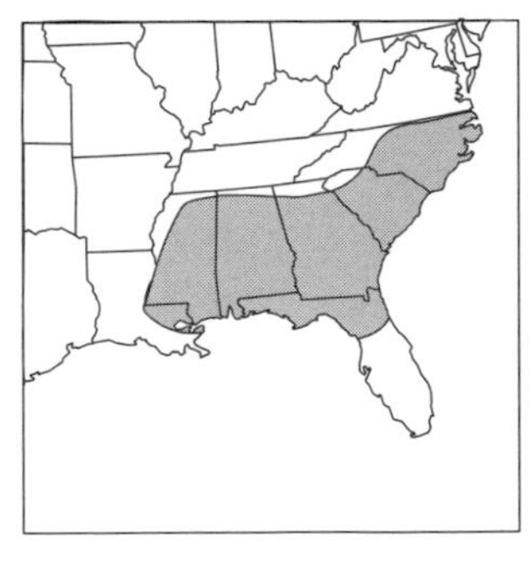
pine katydid

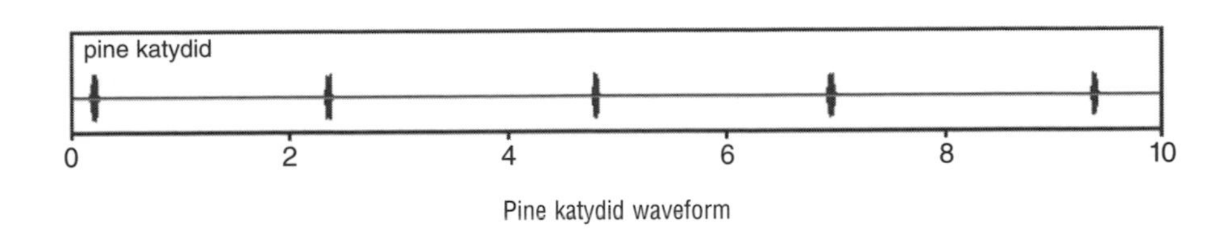

Pine katydid waveform

Similar species: The larger meadow katydids (genus *Orchelimum*) are similar in form and color to the pine katydid but lack the white margins on the wings. The lesser pine katydid, *Orchelimum minor*, is the only species of its genus that lives in pine trees; its length is less than 27 mm.

Hump-winged Grigs

Family Prophalangopsidae

This remarkable group is known mostly from fossils more than 135 million years old. Of the three surviving genera, one is North American and two are Asian. The term *grig*, a little-used English word for all jumping orthopterans, is applied here because the fossil species have characteristics somewhat intermediate between crickets and katydids. The modern species, however, definitely belong with the katydids (superfamily Tettigonioidea).

GREAT GRIG

Cyphoderris monstrosa (Plate 41)

Distribution: Found in southwestern Canada, along the Coastal Range in Washington and Oregon, and in the Rocky Mountains region of Idaho and Montana.

Identification: In New-World hump-winged grigs (genus *Cyphoderris*) the antennal sockets are low on the head—about midway between the top of the head and the transverse suture that is above the mouthparts. Hind femora are short, extending no more than 3 mm beyond the end of the abdomen. Forewings of the male cover half or

great grig

more of the abdomen; those of the female are tiny. Unlike most katydids, the forewings of the male are mirror images; either wing may be upmost at rest, and either file may be used in stridulation. Tympana are exposed, as in the false katydids (subfamily Phaneropterinae). In male great grigs, the subgenital plate has a ventrally directed process shaped like the nail-pulling claw of a hammer. The song is a shrill, metallic, short trill repeated every few seconds. The carrier frequency is about 13 kHz. Length is 20–30 mm.

Ecology: Inhabits forests of lodgepole pine, Englemann spruce, and mountain hemlock. Spends the daylight hours in burrows and feeds, calls, and mates at night. Food includes the staminate cones of lodgepole pine. Males, females, and nymphs have been observed climbing conifers at dusk, presumably to feed on the cones. Males sometimes call from high in trees. During mating, the females feed on the fleshy hind wings of the male.

Similar species: Sagebrush grigs, *Cyphoderris strepitans*, have no cleft in the ventrally directed process of the male subgenital plate. Buckell's grigs, *Cyphoderris buckelli*, have no ventrally directed process. Sagebrush grigs inhabit high altitude sagebrush prairie as well as open coniferous forests.

sagebrush grig

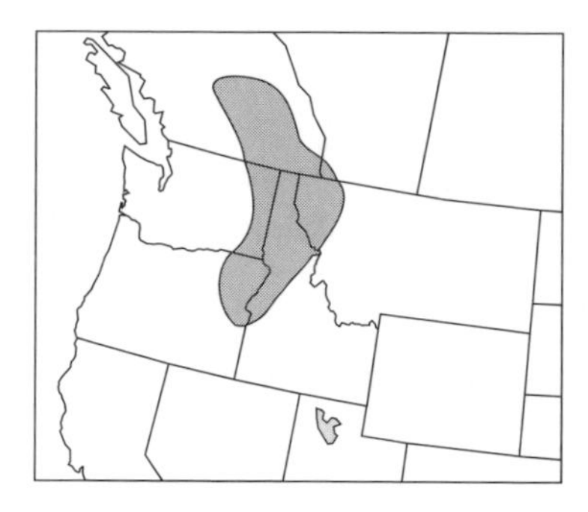

Buckell's grig

Crickets

SUPERFAMILY GRYLLOIDEA

Crickets and katydids are alike in that the antennae are at least as long as the body (except in mole crickets), the ears are at the bases of the fore tibiae, and males make species-specific calling songs by rubbing together structures at the bases of the forewings. They differ in that crickets have tarsi with three segments, rather than four, and that cricket calling songs are musical, rather than raspy or buzzy. Furthermore, the right forewing of male crickets is generally over the left, whereas in katydids the left forewing nearly always overlaps the right. In crickets, except for the sword-tail subfamily, the ovipositor is needle-like rather than blade-like.

Some Orthoptera that are not "true" crickets (superfamily Grylloidea) are nonetheless called crickets, for example, camel crickets, Mormon crickets, and Jerusalem crickets.

In cases where songs are important, in addition to our pictures and descriptions of songs, to resolve identification you can listen to samples of the songs posted on the Internet at http://buzz.ifas.ufl.edu/. This is a site where you also can learn about crickets not dealt with in this book and more about those that are.

Field Crickets

Subfamily Gryllinae

Crickets in this subfamily are large (length 13–34 mm) and occur almost exclusively on the ground or in burrows. They have stout spines along the upper margins of the hind tibiae, and no small teeth between the spines. Their second tarsal segments are not bilobed with flattened pads. Most North American species belong to the genus *Gryllus.* The species of this genus are primarily black or dark brown and so similar in appearance that until 1957 all North American populations were assigned to a single species. By that time, studies of their songs and life cycles had begun to reveal that there were many species. Now, ten species are known from the eastern United States and at least twice that many, mostly still undescribed, are known from the western states.

Because field crickets are diverse and easily collected and cultured, they are attractive for studies in schoolrooms and laboratories.

SOUTHEASTERN FIELD CRICKET

Gryllus rubens (Plate 43)

Distribution: Found in the southeastern United States and as far north as central Missouri and Delaware.

Identification: Head and pronotum are mostly black, sometimes with pale markings on the lateral lobes of the pronotum. Forewings are light brown with a longitudinal dark streak at the junction of the dorsal and lateral fields. Hind-wing length is dimorphic: in the short-winged form (which cannot fly) the hind wings are concealed by the forewings, whereas in the long-winged form (some of which fly) they protrude well beyond. As in other species of the genus, the ocelli form a triangle, and the maxillary palps are not white. The calling song consists of prolonged trills briefly interrupted at irregular intervals. The pulse rate within the trills is about 55 per s at 77 °F. Length is 16–22 mm.

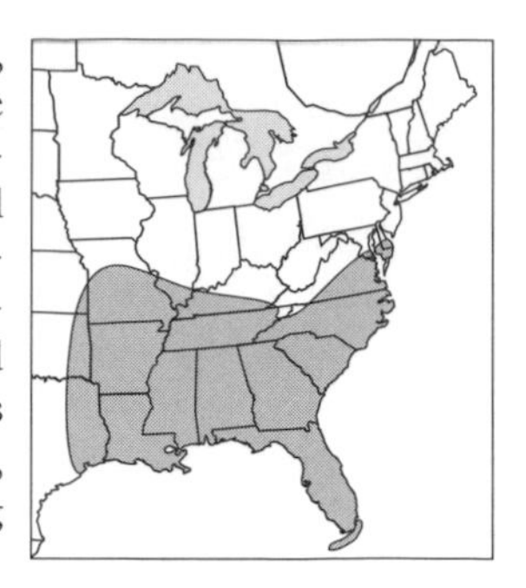

southeastern field cricket

Ecology: Occurs in abundance in lawns, pastures, and grassy roadsides. It sometimes flies to lights, especially in autumn. Even at the northern extremes of its range, it has two annual generations, with peaks of adult abundance in spring and fall. Midsized juveniles are the chief overwintering stage.

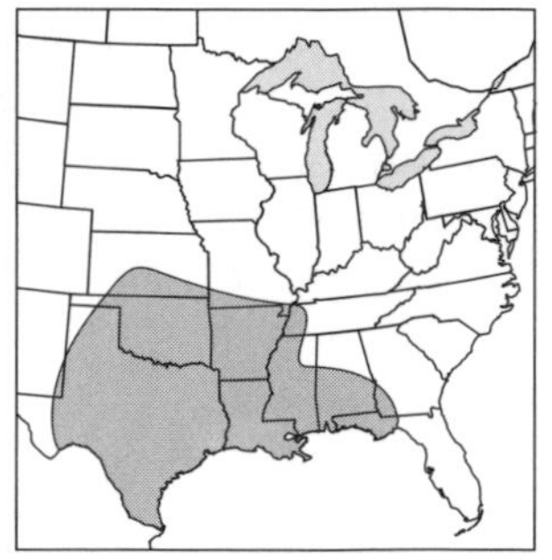

Texas field cricket

Similar species: Of the nine other species of *Gryllus* that occur in the eastern United States, the most similar is the Texas field cricket, *G. texensis.* It is practically identical morphologically to the southeastern field cricket and is the only other species with a trilling calling song. However, its calling song has a faster pulse rate (about 75 per s at 77 °F), the trills are shorter, and the interruptions are more regular and distinct. Of the remaining eight species, four resemble the southeastern field cricket in being characteristic of pastures, lawns, and other open areas. The calling songs of three of these, the sand field cricket, *G. firmus*, the spring field cricket, *G. veletis*, and the fall field cricket, *G. pennsylvanicus*, consist of chirps, mostly with 4 pulses, at a rate of 2 to 3 per s. The pulse rate within the chirps is less than 25 per s at 77 °F. The spring field cricket and the fall field cricket each have one generation annually, but they overwinter in different stages. The spring field cricket overwinters as a midsized juvenile and matures in spring. Its adults are dying out by the time the fall field cricket, which overwinters in the egg stage, begins to mature and call. The sand field cricket produces chirps with a slower pulse rate than the spring and fall field crickets (about 17 as opposed to 23 per s at 77 °F). It occurs year-round in Florida and mostly in late summer and fall farther north.

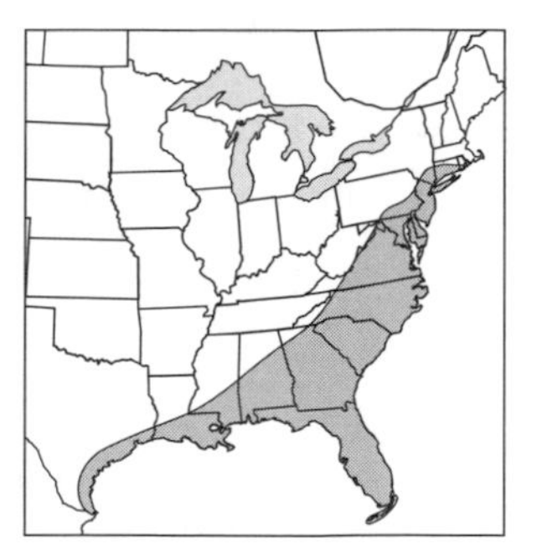

sand field cricket

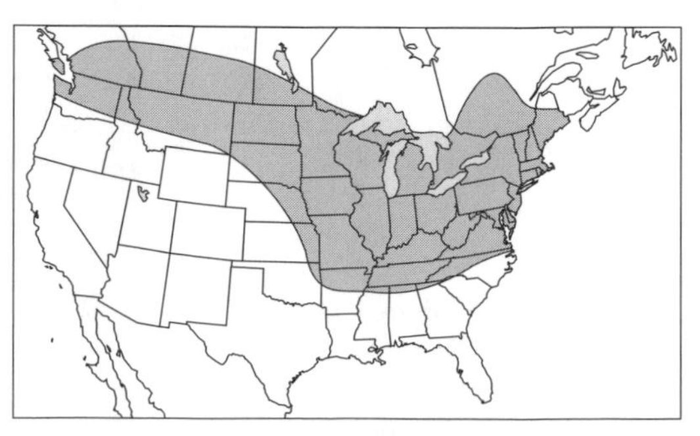

spring field cricket

The Jamaican field cricket, *G. assimilis*, differs from the other open-ground species in having brown pubescence on the pronotum and a calling song that consists of chirps produced at the slow rate of about 1 per s. Each chirp consists of 6 to 10 pulses with the initial pulses produced more quickly than the terminal ones. All Jamaican field crickets are long-winged, whereas the other open-ground species produce both long- and short-winged forms.

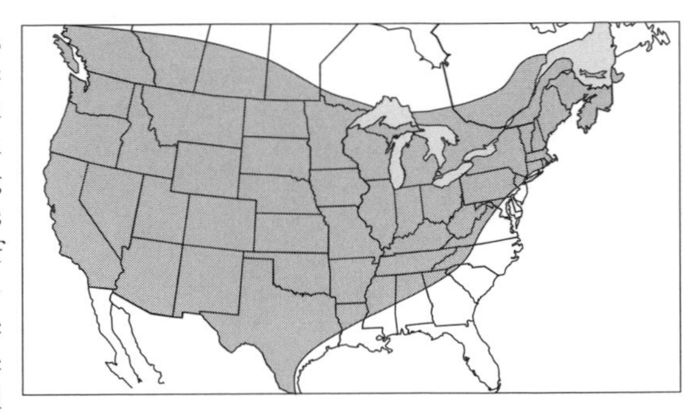

fall field cricket

Jamaican field cricket

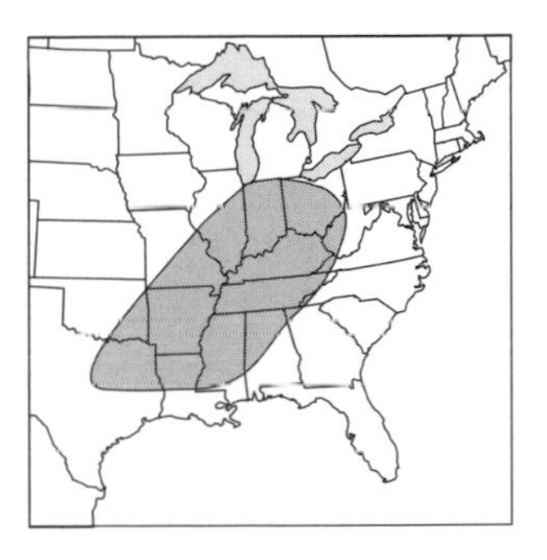

northern wood cricket

southern wood cricket

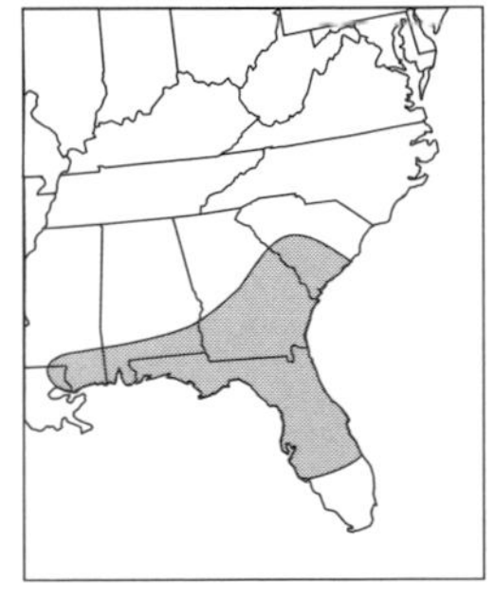

taciturn wood cricket

Keys wood cricket

The remaining four species of eastern *Gryllus* are largely confined to wooded habitats and are always short-winged. The northern wood cricket, *G. vernalis*, is a spring species that is jet black, small, and nearly always in hardwood leaf litter. The southern wood cricket, *G. fultoni*, is likewise a spring species, although in Florida there is a partial second generation that matures in August. Unlike the northern wood cricket, the southern wood cricket has light-brown forewings and femora, and reddish tibiae and tarsi. Where the two species overlap geographically, the southern wood cricket is characteristic of forest edges and of old fields dominated by beard-grass (*Andropogon*). Elsewhere, it occurs in hardwood forests and open pine woods. Both of these wood crickets have calling songs that consist of sequences of brief, fast-pulsed chirps, usually with 3 pulses per chirp. The chirp rates are between 2 and 6 per s, and the pulse rates within the chirps are higher in the southern wood cricket than in the northern wood cricket. In both species, pulse rates are higher than those for the open-ground *Gryllus* that occur in the same regions. The taciturn wood cricket, *G. ovisopis*, is an egg-overwintering, fall species whose adults first appear in September. It is the largest of the eastern *Gryllus* and has no calling song. The Keys wood cricket, *G. cayensis*, was apparently extirpated from the Florida Keys by aerial mosquito spraying but still thrives in the pineland of Everglades National Park. It has no regular calling song, but some solitary males occasionally produce sequences of soft, high-pitched chirps.

southeastern field cricket 4.8 kHz
Texas field cricket 4.8 kHz
fall field cricket 4.7 kHz
Jamaican field cricket 3.6 kHz
northern wood cricket 4.7 kHz
southern wood cricket 4.5 kHz
0 1 2

Field cricket (genus *Gryllus*) song comparisons

HOUSE CRICKET

Acheta domesticus (Plate 43)

house cricket

Distribution: Because live house crickets are sold as fish bait and pet food practically everywhere, and sometimes escape or are released, individuals are often encountered outdoors at sites where feral populations do not occur. In the eastern United States, feral populations are sporadic and seem to occur more frequently in the north than in the south. None are known in peninsular Florida. A few urban areas in southern California support feral populations. House crickets are native to the Old World, perhaps East Africa or India.

Identification: A straw-colored cricket with dark-brown markings. From above, the most conspicuous markings are two bars across the head, one behind the eyes and the other between the eyes. Pronotum is symmetrically blotched. Distance between the antennal sockets is more than twice the width of the basal antennal segment. Forewings are as long as the abdomen or nearly so. Hind wings exceed the abdomen except when they have been shed by breaking off at their bases—house crickets never have short hind wings. The song consists of irregularly delivered brief chirps at an average rate of 1 or 2 per s. Each chirp has 2 or 3 pulses produced at a rate of about 20 pulses per s at 77 °F. Length is 16–21 mm.

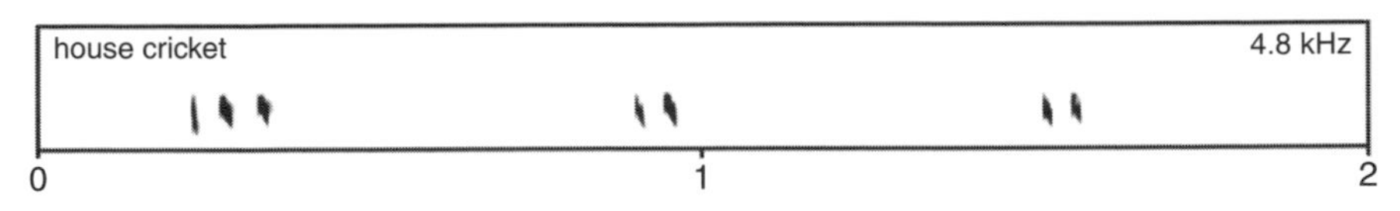

House cricket spectrogram

Ecology: Dwells in and around buildings, in rubbish dumps, and along the edges of lakes and streams—where individuals have escaped from fishermen who bought them for bait.

Similar species: In the tropical house cricket, *Gryllodes sigillatus* (Plate 43), the distance between the antennal sockets is about the width of the basal antennal segment. Male forewings are about half as long as the abdomen, and the female forewings are reduced to non-overlapping flaps.

TROPICAL HOUSE CRICKET

Gryllodes sigillatus (Plate 43)

Distribution: The homeland of the tropical house cricket is unknown, but the species now occurs in tropical areas around the world. In the United States, it is broadly distributed in the lowest latitudes and has been reported as far north as the tip of Nevada.

tropical house cricket

Identification: A light yellowish-brown cricket with dark markings. It is somewhat flattened and roach-like in behavior. The space between the antennal sockets is narrow—about the width of the basal segment of either antenna. There is a single, dark transverse band between the eyes, a narrow one at the front margin of the pronotum, and a broader band at its rear. Females and juveniles have a broad, transverse, dorsal dark band near the base of the abdomen. In males, the forewings only half cover the abdomen; females are practically wingless. The song consists of brief chirps produced at a rate of about 5 per s. Each chirp has 3 or 4 pulses made at a rate of about 50 per s at 77 °F. Length is 13–18 mm.

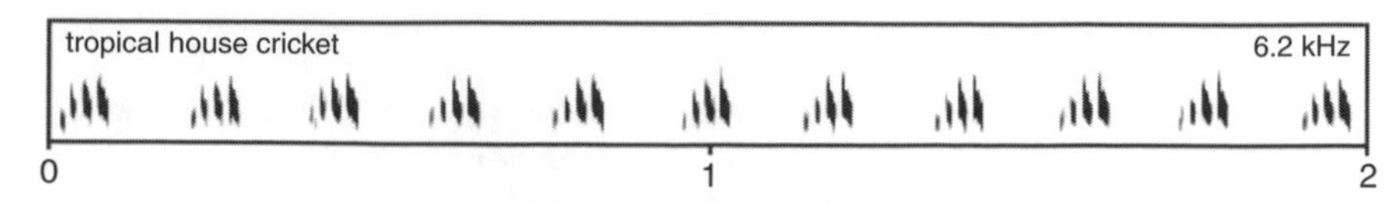

Tropical house cricket spectrogram

Ecology: Usually inhabits urban areas, where it may harbor in storm sewers, piles of rock, or cracks in masonry structures. Adults are found at all times of year.

Similar species: The house cricket, *Acheta domesticus* (Plate 43), has forewings that cover the abdomen or nearly so, and the antennal sockets are separated by more than twice the width of a basal antennal segment.

COMMON SHORT-TAILED CRICKET

Anurogryllus arboreus (Plate 43)

Distribution: Found in the southeastern United States north to New Jersey and southern Illinois.

common short-tailed cricket

Identification: Both the generic name (Latin for "no-tail cricket") and the common name refer to the fact that females have no easily visible ovipositor. (Like mole crickets, which have no ovipositors at all, they lay their clutches of eggs in underground chambers.) Both sexes are light brown with no conspicuous dark markings. Ocelli are in a transverse row rather than in a triangle. Forewings cover all or nearly all of the abdomen in the male and more than half the abdomen in the female. The calling song is a loud continuous trill, usually made from the entrance to a burrow or from an above-ground perch—often a meter or more up a tree trunk. At 77 °F, the pulse rate is about 75 per s. Length is 16–19 mm.

Ecology: For most of its life, it inhabits burrows that were dug by carrying mouthfuls of soil to the surface. The only time these crickets are likely to be seen, even when they occur in large numbers on a lawn, is in spring, when the adults mature from overwintering nymphs. Nearly all the males in a population call each evening for about 45 min, starting at sunset. Males initially call from the burrows where they spent the winter, but after a few days abandon them in favor of calling from perches at dusk and then spending most of the night searching for females in their burrows. After mating, the female lays a clutch of eggs in her burrow and cares for the young after they hatch by bringing food into the burrow. By late summer, the female dies and the young disperse to dig burrows of their own. There is a single generation annually.

Similar species: House crickets, *Acheta domesticus*, have conspicuous dark markings.

JAPANESE BURROWING CRICKET

Velarifictorus micado (Plate 43)

Distribution: Found in localities scattered throughout the southeastern states. They probably entered the United States at Mobile, Alabama. Their ultimate North American distribution is uncertain.

Identification: Dark brown with seven or fewer ragged pale longitudinal stripes extending forward on the top of the head from the front margin of the pronotum. A pale transverse band connects the lateral ocelli, and the pronotal disk has pale spots or blotches. Maxillary palps are white. The calling song is a regular sequence of loud 5- to 7-pulsed chirps delivered at a rate of about 4 per s. Length is 13–19 mm.

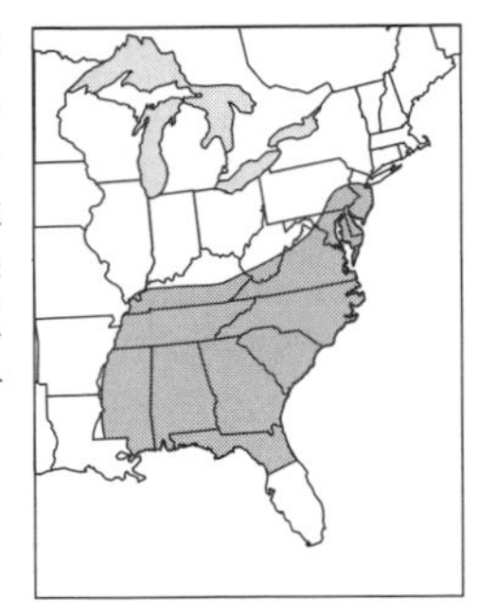

Japanese burrowing cricket

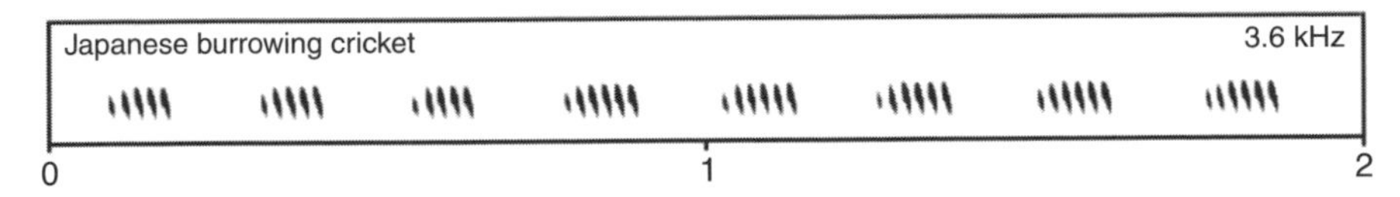

Japanese burrowing cricket spectrogram

Ecology: Partial to wet and mesic habitats. As the name implies, this species is native to Japan and sometimes makes burrows. It was first discovered in North America in 1959 in Alabama and now occurs widely but spottily throughout the Southeast. Its loud, distinctive song has made its spread easy to study. Its largely suburban distribution suggests that it may have spread primarily from overwintering eggs in soil shipped with ornamentals from infested nurseries. Very few individuals have the long hind wings required for flight, but such individuals could become common if dispersal by flight becomes a major means of spread. When these crickets make burrows, they move the dirt a mouthful at a time, in the manner of short-tailed crickets. Their reliance on burrows for protection is less than in short-tailed crickets, as demonstrated by their occurrence beneath logs, rocks, and trash. Some males, when located by their calls, are on the ground but away from any detectable burrow. Others are found at the entrance to a burrow. The entrance may be cricket-sized or larger and may have a hood constructed of soil particles. There is a single generation annually, with adults present from August through October.

Similar species: Other eastern field crickets (subfamily Gryllinae) of similar size have dingy palps and lack the stripes and band on the head. The calling songs of other eastern field crickets that chirp several times per s have a less regular chirp rate and fewer pulses per chirp.

Ground Crickets

Subfamily Nemobiinae

In general form, ground crickets resemble field crickets (subfamily Gryllinae), but they are smaller (length less than 14 mm), the spines on the hind tibia are not as stout (length of the median spine is more than twice the width of the hind tibia at midpoint), and the last segment of the maxillary palp is nearly twice the length of the preceding segment or longer. Like field crickets, they have no small teeth between the spines on the hind tibia, and the second tarsal segment is not heart shaped.

Like field crickets, ground crickets are diverse and easily collected and cultured. They are well suited for studies in schoolrooms and laboratories. However, being smaller, their identifying structures are more difficult to see and, unlike field crickets, few species occur in the West or Southwest.

SOUTHERN GROUND CRICKET

Allonemobius socius (Plate 44)

Distribution: Found in the southeastern United States north to central Illinois and New Jersey. Its occurrence in California's Central Valley is probably a result of human transport.

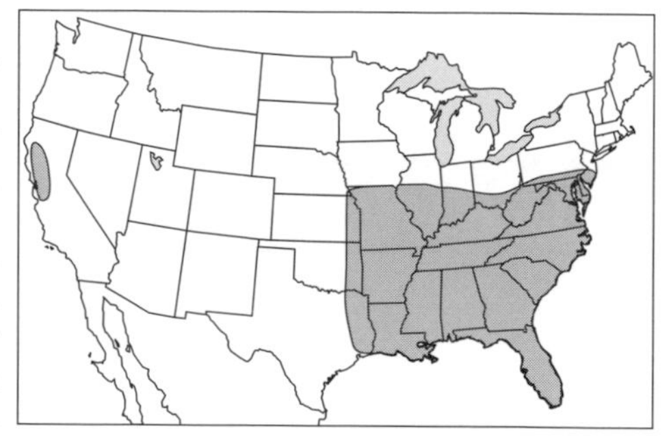

southern ground cricket

Identification: Top of the head has seven to eleven alternating light and dark longitudinal stripes. Ovipositor is straight, or nearly so, and at least three-fourths the length of the hind femur. Stridulatory file has 100 to 150 teeth. Both long- and short-winged individuals occur. The calling song is a steady series of short, buzzy chirps produced at rates of 3 to 5 chirps per s at 77 °F. Length is 9–12 mm.

Ecology: Inhabits open grassy areas such as pastures and lawns, especially those that are moist or wet. It has one generation annually in the northern extreme of its distribution and two or more farther south. In Ohio, adults are found as early as July and as late as November. In peninsular Florida, they occur year-round.

Similar species: The striped ground cricket, *Allonemobius fasciatus*, is the northern counterpart of the southern ground cricket—so much so that the two cannot be reliably distinguished by habitat, morphology, or calling song. In the region of geographic overlap, they occur in mixed populations and are currently distinguished only by electrophoretic analysis of their enzymes. (Because they are so similar yet maintain their status as distinct species even where they occur intermingled, they are the most studied of any pair of cricket species.) Allard's ground cricket, *A. allardi*, and the tinkling ground cricket, *A. tinnulus*, have more than 160 teeth in their stridulatory files. The songs of both species are slow-pulsed trills. At 77 °F, Allard's ground cricket has a pulse rate of about 15 per s, whereas the tinkling ground cricket has a pulse rate of about 8 per s. The

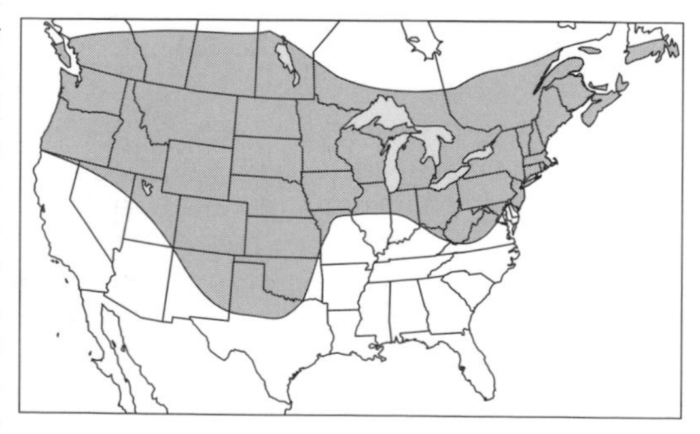

striped ground cricket

former species inhabits open grassy areas, whereas the latter inhibits xeric woodland and woodland borders. The former is red-brown to black, usually with dorsal head striping faintly visible. The latter is pale and reddish with no head striping.

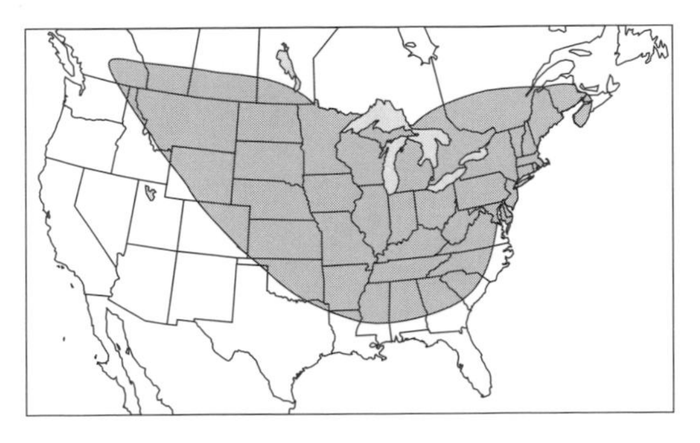

Allard's ground cricket

tinkling ground cricket

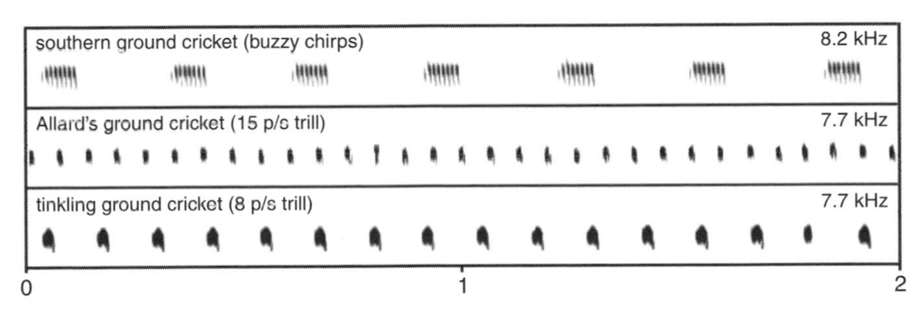

Song comparisons of three ground crickets

Sword-tail Crickets

Subfamily Trigonidiinae

Sword-tail crickets ("trigs") are tiny, active crickets that are found on vegetation. The heads are wide with large bulging eyes. The first segment of the tarsus is longer than the other two combined. The second segment is short and densely pubescent beneath; the third is very slender. Females have short, up-curved, sharp-pointed, blade-like ovipositors.

COLUMBIAN TRIG

Cyrtoxipha columbiana (Plate 44)

Distribution: Found in the southeastern United States and as far north as southern Ohio and New Jersey.

Identification: Tiny, fast-moving, pale-green crickets. The area between the eyes is flattened. Hind wings project well beyond the forewings. Stridulatory file has fewer than 175 teeth. Ovipositor, measured on a straight line from the posterior of the dorsal base to the tip, is more than 2.3 mm long. The calling song is a tinkling chirp, with the 4 to 6 pulses within a chirp produced at a rate of about 39 per s at 77 °F. The chirp rate is about 3 per s. When many males sing in close proximity, they synchronize their chirps. Length is 7–9 mm.

Columbian trig

Ecology: Found on the foliage of broadleaf trees and shrubs. They are usually out of reach and, when spotted, are so quick as to be hard to catch. Overwintering is in the egg stage. North of peninsular Florida there is a single generation annually, with the earliest adults maturing in June or July.

Similar species: Gundlach's trig, *Cyrtoxipha gundlachi*, has more than 180 teeth in the stridulatory file and an ovipositor less than 1.6 mm long. Its song has 2 to 3 pulses per chirp produced at a rate of about 19 per s at 77 °F. Brown trigs (genus *Anaxipha*) are brown rather than green, the area between the eyes is not flattened, and most species are never found in the crowns of trees.

Gundlach's trig

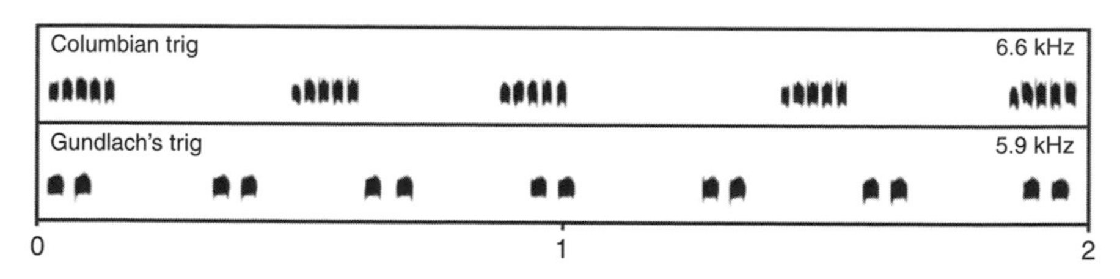

Trig song comparison

HANDSOME TRIG

Phyllopalpus pulchellus (Plate 44)

Distribution: Found in the southeastern United States and as far north as Illinois and New Jersey, but not in south Florida.

Identification: An easily recognized species with red head and thorax, pale yellowish-white legs, and dark blue-black forewings. Last segment of the maxillary palp is flattened, oval, and black. Basal segments of the antennae are much darker than the remaining segments. Hind wings are always fully concealed by the forewings. Forewings in females are convex and beetle-like. Males call both day and night with a con-

handsome trig

tinuous, jerky trill. The jerkiness comes from the caller stopping the trill momentarily every several pulses. The pulse rate within any group of pulses produced without pause is about 60 per s at 77 °F. Length is 7–9 mm.

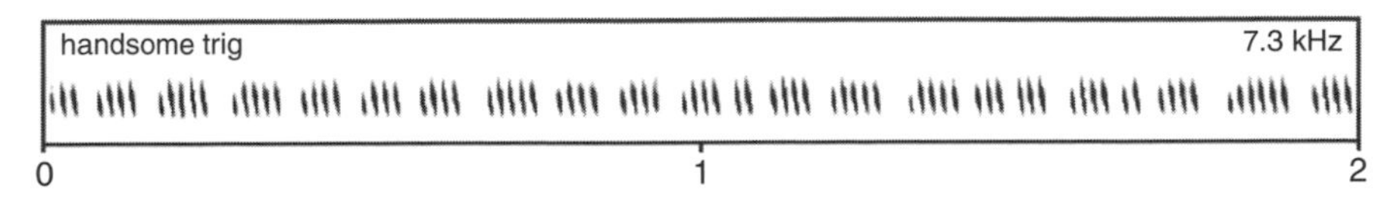

Handsome trig spectrogram

Ecology: Found a meter or so above the ground in vegetation around streams and marshes and sometimes in similar brushy and weedy habitats elsewhere. There is a single generation annually, with adults first appearing in July or early August.
Similar species: None: no other cricket is nearly so colorful.

Bush Crickets

Subfamily Eneopterinae

These medium-sized crickets live not only in bushes but may, depending on the species, be found in the canopies of trees and on herbs. The hind tibia is armed on each upper margin with five to eight spines, between which are small teeth. The second segment of all tarsi is much shorter than the other two segments and is bilobed, with two flattened pads.

RESTLESS BUSH CRICKET

Hapithus agitator (Plate 45)

Distribution: Found in the southeastern United States, west to central Texas, and north to Missouri, Ohio, and New Jersey.

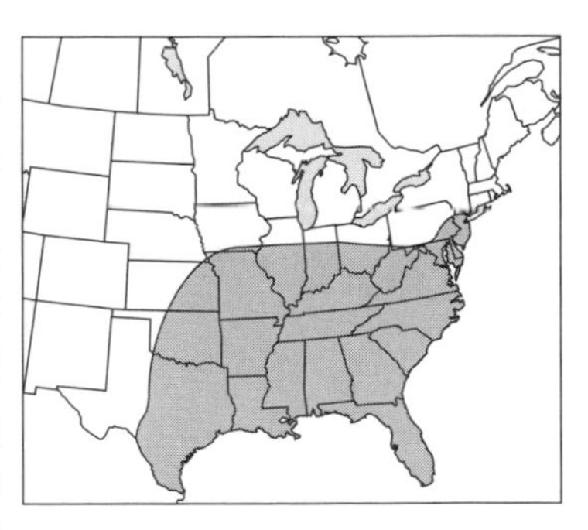

restless bush cricket

Identification: Forewings cover three-fourths or more of the abdomen. Hind wings are shorter than the forewings. Fore tibia has only an anterior tympanum. Northern populations have no calling song, but in peninsular Florida and eastern Texas, males produce a loud distinctive call consisting of 5 to 20 buzzy chirps at a rate of 1 or 2 per s. Length is 9–14 mm.

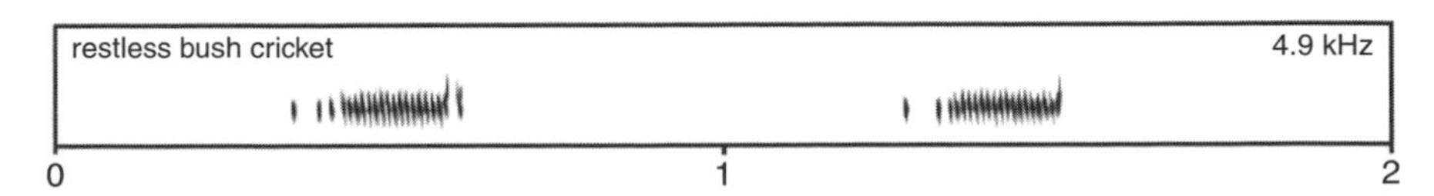

Restless bush cricket spectrogram

Ecology: Inhabits undergrowth in moist or wet wooded areas and roadside weeds. In the region where males do not call, males are often found with their forewings badly mutilated or even missing. Their damaged wings are not a result of predation but a badge of sexual success. During copulation, the male allows the female to feed on his wings while the bag of sperm (spermatophore) he has attached to the female's genitalia empties. During the 6 to 12 minutes that the female is allowed to feed, she may consume a quarter of the male's forewings. Except in south Florida, there is one generation annually, and overwintering is in the egg stage. In areas with one generation per year, adults first appear in July or August.

short-winged bush cricket

Similar species: In short-winged bush crickets, *Hapithus brevipennis*, the forewings cover less than two-thirds of the abdomen.

JUMPING BUSH CRICKET

Orocharis saltator (Plate 45)

Distribution: Found in the southeastern United States north to Missouri, Ohio, and New Jersey but not in peninsular Florida or southeast Georgia.

jumping bush cricket

Identification: Forewings cover the abdomen and hind wings project from beneath the forewings. Fore tibia has anterior and posterior tympana, with both fully exposed, although the posterior one is small and sometimes scarcely evident. Most specimens are easily categorized as belonging to one of two color forms. The light form is light brown marked with dark flecks especially on the lighter face. The dark form is brownish gray heavily marked with brownish black, including a stripe that extends from the eye to the rear of the lateral lobe of the pronotum. In both the light and dark forms, the veins that lie along the juncture of the lateral and dorsal fields of the forewings are light yellow. The song is a loud, clear, hard-to-locate chirp repeated at irregular, 1.5–3 s intervals. Chirps have 10 to 18 pulses produced at a rate of 55 per s at 77 °F. Length is 15–20 mm.

Ecology: Found in broadleaf trees and occasionally in herbaceous undergrowth, shrubs, and pine trees. It has one generation annually and overwinters as eggs. Adults are found as early as July or August.

Similar species: False jumping bush crickets, *Orocharis luteolira*, cannot be distinguished morphologically from jumping bush crickets, but their calling songs have chirps with fewer pulses (4 to 9) produced at a faster rate (71 per s at 77 °F), making their chirps noticeably briefer. The false jumping bush cricket has at least two generations annually and, where the two

false jumping bush cricket

species occur together, begins to call about a month earlier. In peninsular Florida it calls year-round.

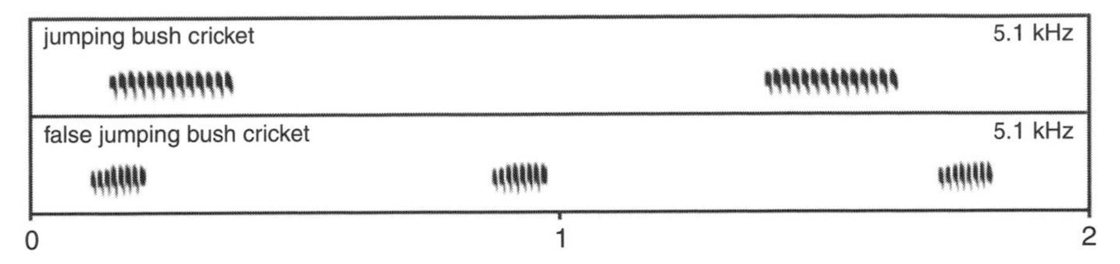

Jumping bush cricket and false jumping bush cricket song comparison

Tree Crickets
Subfamily Oecanthinae

Tree crickets are delicate and usually pale green. They have forward-directed mouthparts, thin hind femora, and two-lobed tarsal claws. The forewings of males are often much wider than the abdomen. Most North American species can be identified by markings on the ventral surface of the first two antennal segments (Fig. 61). The songs are unusually melodious and are lower in pitch

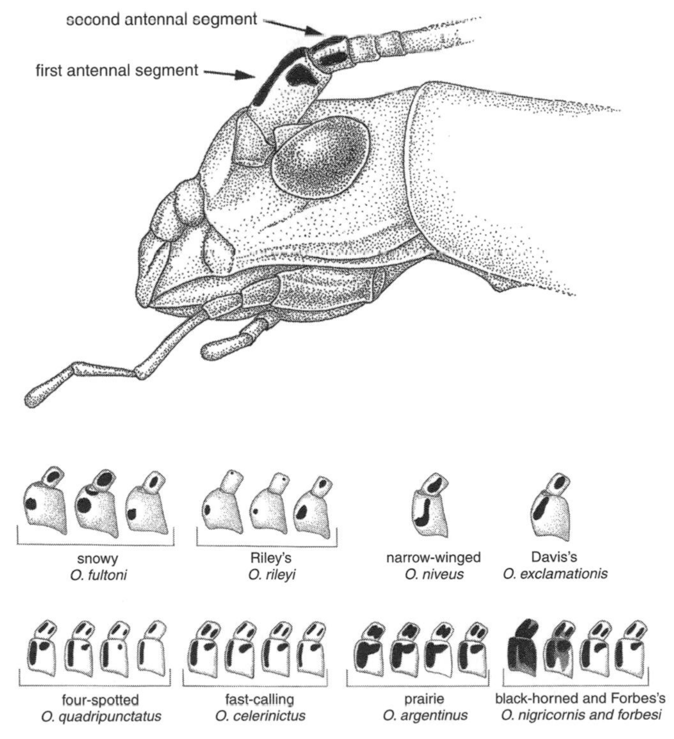

Figure 61. Antennal markings of tree crickets, genus *Oecanthus*. Above, lateral view of head. Below, markings on the first and second antennal segments of nine species of tree crickets.

than those of most other crickets. Although some species live in trees, others are characteristic of shrubs and of herbaceous vegetation. All species overwinter as eggs laid in the stems of their host plants.

SNOWY TREE CRICKET

Oecanthus fultoni (Plate 46, Fig. 61)

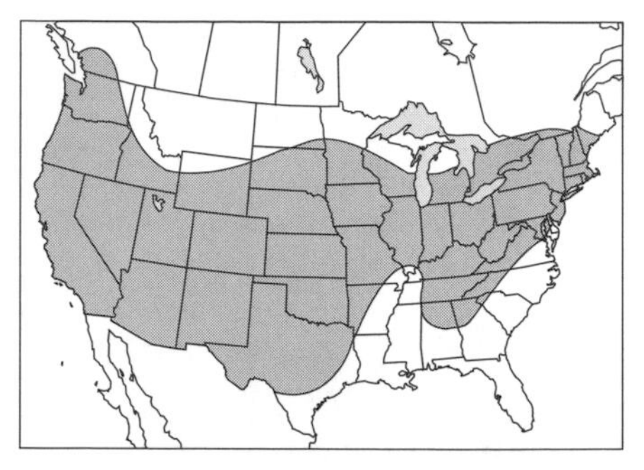

snowy tree cricket

Distribution: Found in southwest British Columbia and southern Ontario and throughout the United States except in the Southeast and in northern areas with very cold winters.

Identification: The inner edge of the ventral surface of the first antennal segment has a pale swelling that bears a round or oval black mark. The second segment has a similar but more elongated mark. The greatest width of the dorsal field of the forewings of the male is more than four-tenths its length. The song is a continuous sequence of low-pitched, 5- to 8-pulse chirps produced at a regular rate of 165 to 195 per min at 77 °F. Neighboring individuals synchronize their chirps. Length is 15–18 mm.

Ecology: Found in dooryard shrubbery and vines, in unsprayed orchards and blackberries, and in fencerows and forest edges. In less-disturbed habitats, it occurs in shrubby trees and undergrowth but seldom in the crowns of tall trees. It has a single generation annually. In most areas, the first adults are heard in July or August.

The snowy tree cricket is famous as the cricket whose song can be used to tell the temperature. As explained in the Introduction, nearly all crickets and katydids make songs from which temperature can be calculated. However, in most cases, the elements that are linearly related to temperature are too rapid to be quantified by ear. The fame of the snowy tree cricket as a thermometer is based on the fact that it is a widespread, dooryard species that produces a countable chirp at a very regular rate. This allows the temperature at the calling cricket's location to be easily estimated from the chirp rate. In the eastern United States, a simple formula for determining Fahrenheit temperature is to count the number of chirps in 13 s and add 41. Snowy tree crickets in the West call with a somewhat higher chirp rate. In Oregon, for example, count the number of chirps in 12 s and add 38. To use the closely related Riley's tree cricket as a thermometer, count the number of chirps in 21 s and add 38.

The song of the snowy tree cricket should be familiar to moviegoers because directors often use it to set the mood when the action moves outside on a warm suburban or country night.

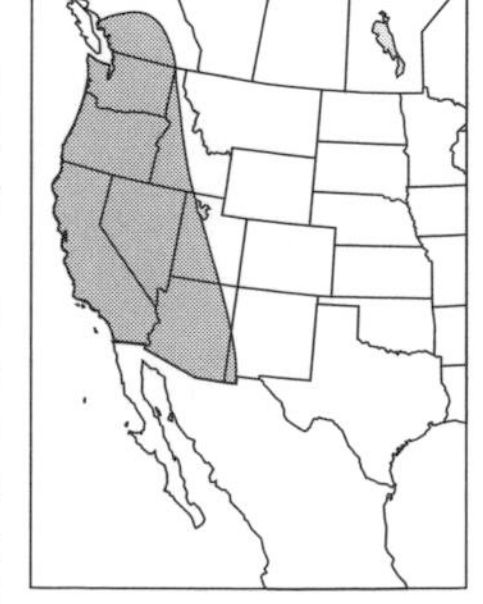

Riley's tree cricket

Similar species: Riley's tree cricket, *Oecanthus rileyi*, has a chirp rate of about 110 per min at 77 °F, and the chirps have 8 to 14 pulses. The length of the dark mark on the second antennal segment is usually much less than half the length of the segment, and the center of the mark is near the distal border of the segment (see Fig. 61). Males of the narrow-winged tree cricket, *O. niveus*, and Davis's tree cricket, *O. exclama-*

tionis, have narrower wings and call with trills rather than with chirps. The former species has a J-shaped black mark on the pale swelling of the first antennal segment, and the swelling of the latter species has a straight, long mark, slightly expanded proximally (see Fig. 61).

narrow-winged tree cricket

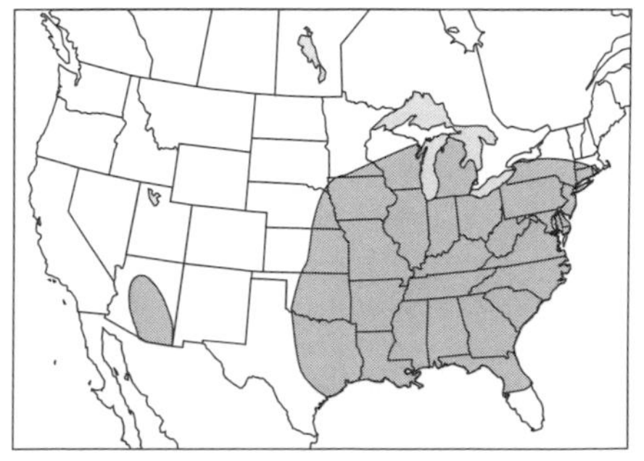

Davis's tree cricket

snowy tree cricket 2.9 kHz

Riley's tree cricket 2.3 kHz

0 1 2

Snowy tree cricket and Riley's tree cricket song comparison

BROAD-WINGED TREE CRICKET

Oecanthus latipennis (Plate 46)

Distribution: Found in the southeastern United States, except peninsular Florida, and north to Iowa, Ohio, and Connecticut.

broad-winged tree cricket

Identification: This, the largest tree cricket in the United States and Canada, has the top of the head and basal segments of the antennae tinged with a raspberry color. The first antennal segment has no pale swelling and no dark marks. The third antennal segment is seldom darker than the second. Width of the dorsal field of the male's forewings is about half its length. The song is a loud, bell-like continuous trill. At 77 °F, the pulse rate is about 55 per s. Length is 17–22 mm.

Ecology: Found in the vines, brambles, and coarse weeds of forest edges and fencerows. It also occurs on coarse weeds in old fields, and on shrubby trees, usually oaks, in dry open woods. It has one generation annually, and, as befits its size, it is the latest maturing of the eastern tree crickets. Even in the southern portion of its range, adults are seldom encountered until August.

Similar species: Western tree crickets, *Oecanthus californicus*, usually have the third segment of the antennae much darker than the second and often have a narrow dark line along the inner edge of the first and second antennal segments.

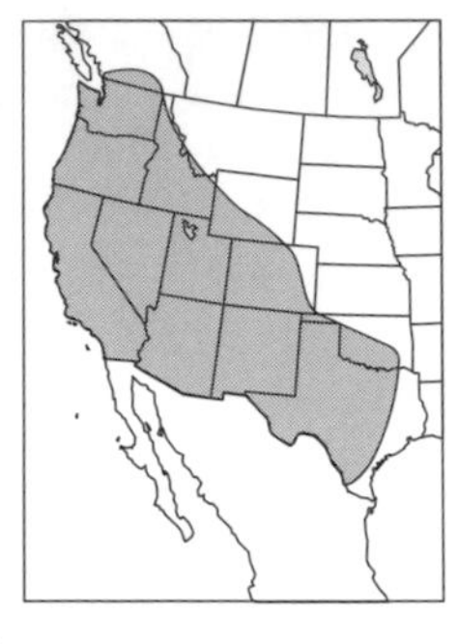

western tree cricket

FOUR-SPOTTED TREE CRICKET

Oecanthus quadripunctatus (Plate 46, Fig. 61)

Distribution: Found throughout the United States and southern Canada.

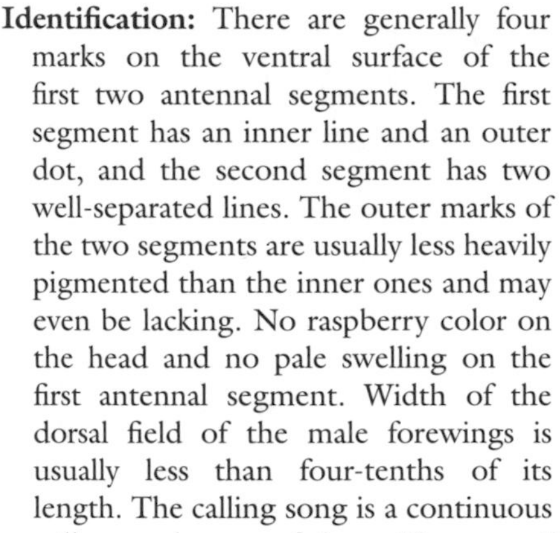

Identification: There are generally four marks on the ventral surface of the first two antennal segments. The first segment has an inner line and an outer dot, and the second segment has two well-separated lines. The outer marks of the two segments are usually less heavily pigmented than the inner ones and may even be lacking. No raspberry color on the head and no pale swelling on the first antennal segment. Width of the dorsal field of the male forewings is usually less than four-tenths of its length. The calling song is a continuous trill at a pulse rate of about 40 per s at 77 °F. Length is 13–17 mm except in a dwarf form that occurs on tarweed in the Bay Area of California and measures 10–14 mm.

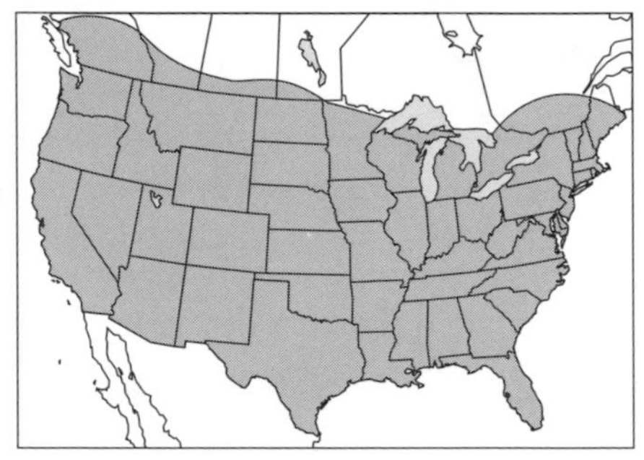

four-spotted tree cricket

Ecology: Seldom found on trees but is common on herbaceous plants of many types. Males often call from a single species of plant among the mixtures that occur, for example, in old fields. Males sometimes call from the lower branches of small trees in open situations, and at some western sites, they call almost exclusively from sagebrush and from tarbush. There are two generations annually in the southern portion of the range and one in the north. In the two-generation area, the first adults are found in May or June. In the one-generation area, the first adults are found in July or August.

Similar species: Fast-calling tree crickets, *Oecanthus celerinictus*, call at about 65 pulses per s at 77 °F. The outer marks of the antennae are never paler than the inner marks, and the outer mark of the first segment is never dot-like (see Fig. 61). In prairie tree crickets, *O. argentinus*, the marks on the antennal segments are broad and confluent or very narrowly separated. The calling song has a pulse rate of about 52 per s. The pronotum of black-horned and Forbes's tree crickets, *O. nigricornis* and *O. forbesi*, is usually marked with one, two, or three black or dusky longitudinal stripes, and the venter and distal portions of the legs and antennae are black or dusky. The basal two segments of the antennae may be suffused with

fast-calling tree cricket

black, and the inner and outer marks of the first antennal segment are usually separated by less than the width of the inner mark. The pulse rates of the calling songs of the two species are about 52 and 70 per s at 77 °F. Pine tree crickets, *O. pini*, occur exclusively on pine trees and are easily recognized by their brown head, pronotum, and legs, which contrast with the light green of their forewings. This color pattern makes them hard to see when their heads are next to the needle-bundle sheaths and their forewings are next to the needles.

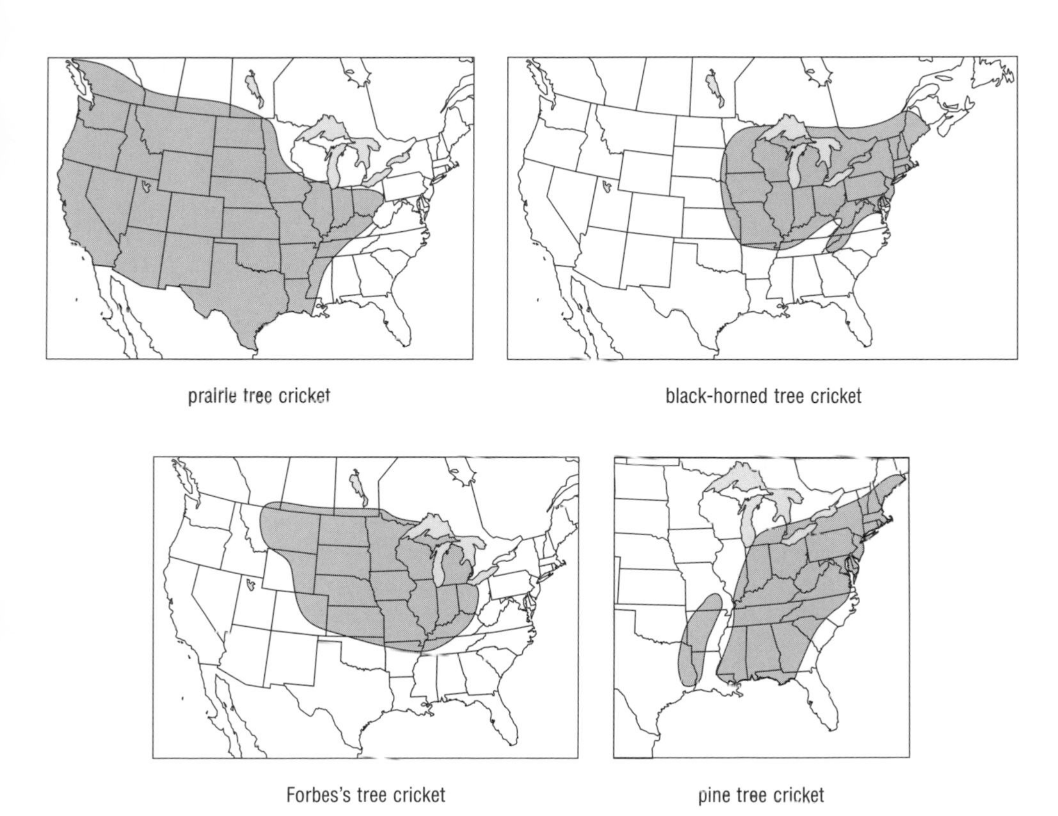

prairie tree cricket

black-horned tree cricket

Forbes's tree cricket

pine tree cricket

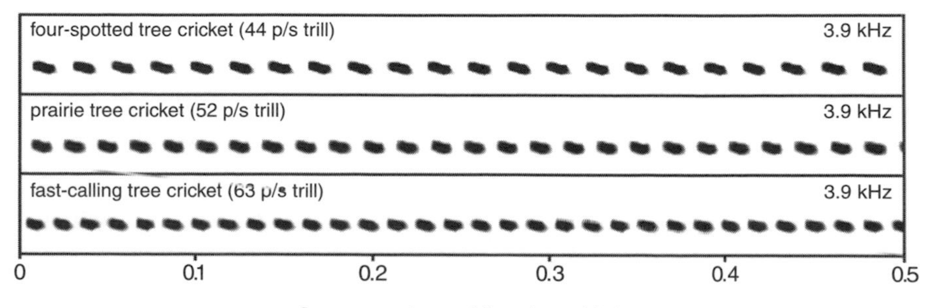

Song comparisons of three tree crickets

Scaly Crickets

Subfamily Mogoplistinae

These peculiar crickets are tiny, somewhat flattened, and covered with scales. Females are wingless and males have no hind wings. The forewings of males do not cover the abdomen and in most species are quite short and largely covered by a rearward extension of the pronotum. In several respects they resemble silverfish rather than other crickets.

There are 20 species of scaly crickets in the United States and Canada. All but three are in the genus *Cycloptilum*, and many occur no farther north than peninsular Florida.

SLOSSON'S SCALY CRICKET

Cycloptilum slossoni (Plate 45)

Distribution: Found in Florida, west to eastern Texas and north to southeastern North Carolina.

Slosson's scaly cricket

Identification: Terminal segment of the maxillary palp is weakly expanded distally, and its shortest side is much longer than the diameter at the tip. The calling song is a sequence of buzzy high-pitched chirps, each lasting about 0.2 s and containing 15 to 25 pulses. After the first pulse, the remaining pulses are produced in pairs. The pulses within a pair, however, are hardly closer to each other than they are to the nearest pulses in the adjacent pairs. Pairs are produced at a rate of about 46 per s at 77 °F. Length is 8–10 mm.

Ecology: Found on shrubs and low trees in woodland and scrub habitats. In urban areas, it is common on shrubby ornamentals.

Similar species: The forest scaly cricket, *Cycloptilum trigonipalpum*, has the terminal segment of the maxillary palp strongly expanded distally and obliquely truncated. As a result, the short side of the segment is about half the length of the long side and less than the greatest diameter at the end. Its song consists of unpaired pulses in chirps that last about twice as long as those of Slosson's scaly cricket.

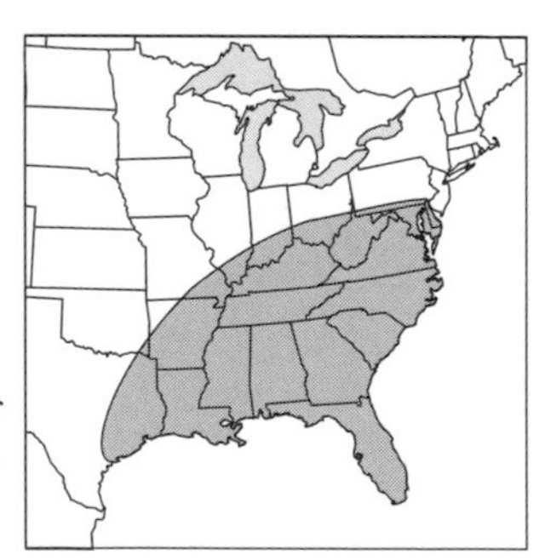

forest scaly cricket

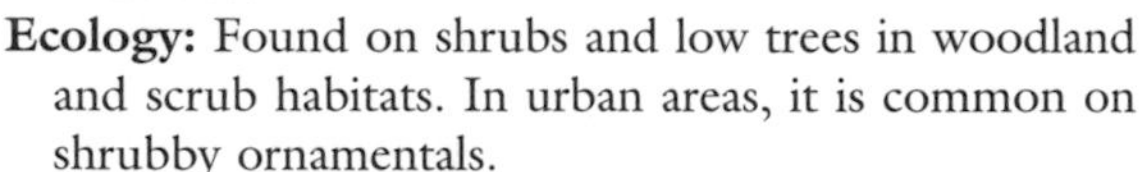

Slosson's scaly cricket and forest scaly cricket song comparison

Ant Crickets

Subfamily Myrmecophilinae

These tiny, wingless crickets live in ant nests and are seldom found anywhere else. The compound eyes are reduced to a cluster of large facets, and the fore tibiae have no tympana. The length of the hind femur is less than twice its width.

Ant crickets are unwelcome guests in ant nests but do little if any damage. They have been observed to mouth the legs and bodies of their hosts, apparently feeding on the ants' oily secretions. They also nibble at the walls of ant galleries, apparently getting grease that has rubbed from the ants. The ants do not accept being attended by crickets but move away or even lunge at the crickets with open mandibles. The crickets are usually agile enough to escape, but crickets that are caught are quickly eaten.

Ant crickets are not host specific. Each species is found with many species of ants. Many crickets are often found in a single nest, with females generally outnumbering males. Within each species of ant cricket, the largest adults are about twice the length of the smallest. To some extent, variation in size of the cricket is related to the size of the host ant—the smallest crickets are found in the nests of small ants. The smallest adult ant crickets are less than 1.5 mm long. No other adult orthopterans are as diminutive.

EASTERN ANT CRICKET

Myrmecophilus pergandei (Plate 47)

eastern ant cricket

Distribution: Found in the eastern United States north to Illinois and Maryland.

Identification: Dark brown in general coloration, with abdominal segments that usually do not appear banded. Upper internal margin of the hind tibia is armed with four spines that alternate in length. Spurs of the basal segment of the hind tarsus are usually as long as the terminal tarsal segment. Length is 2.2–4.7 mm.

Ecology: Eastern ant crickets have been collected in association with ants of eleven species in six genera and two subfamilies. Both adults and juveniles have been collected in most months of the year.

Similar species: The other three species of ant cricket found north of Mexico are western. The Oregon ant cricket, *M. oregonensis*, is dark like the eastern ant cricket, but the spurs of the basal segment of its hind tarsus are usually slightly longer than the terminal tarsal segment. Mann's ant cricket, *M. manni*, and the Nebraska ant cricket, *M. nebrascensis*, are both pale yellowish brown, with some individuals having darker rear margins to their abdominal segments, giving their abdomens a banded appearance. Along the upper inner margin of the hind tibia, Mann's ant cricket usually has four spines, which alternate in length, whereas the Nebraska ant cricket always has three, which increase in length distally.

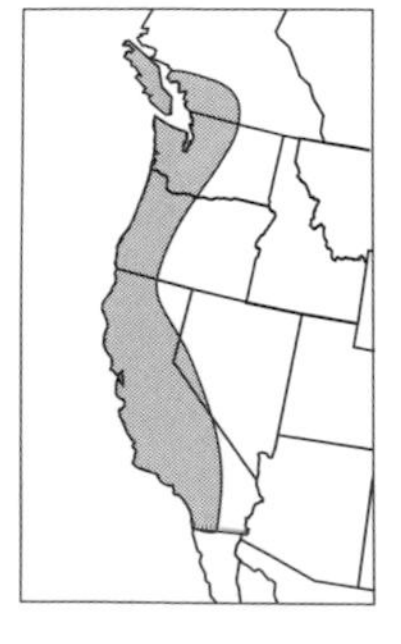

Oregon ant cricket

Mann's ant cricket

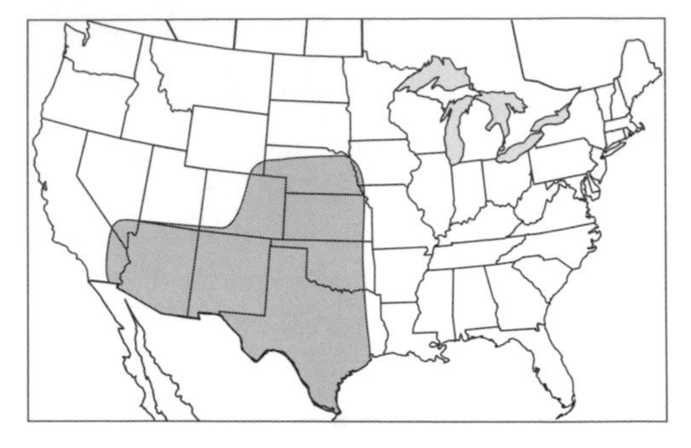
Nebraska ant cricket

Mole Crickets
Family Gryllotalpidae

With their cylindrical bodies and stubby, digging forelimbs, mole crickets look, at first glance, like tiny versions of mammalian moles. However, closer inspection reveals six legs, dense pubescence rather than fur, and well-developed eyes and wings. Mole crickets have two or four strong blade-like projections on the fore tibia, and the first two segments of the fore tarsus are blade-like as well. The hind femur does not reach the tip of the abdomen. Females have no ovipositor. Mole crickets spend nearly all their lives underground and are powerful tunnelers. When dug up, they do not leap away as other crickets do but quickly dig their way underground with powerful strokes of their forelegs. They frequently tunnel immediately below the soil surface and leave trails of pushed-up soil that resemble, in miniature, the surface tunnels of mammalian moles.

NORTHERN MOLE CRICKET

Neocurtilla hexadactyla (Plate 47)

Distribution: Found in the eastern United States west to Nebraska and central Texas.

northern mole cricket

Identification: There are six digging claws (dactyls) on each foreleg. Four project from the tibia and the other two project from the first and second segments of the tarsus. Fore femur has no blade-like basal projection. Hind tibia has no spines except at its tip, where there are four on the inside and four on the outside. The calling song consists of low-pitched chirps that issue from the ground during the afternoon as well as at night. The chirps have 8 to 16 pulses, and at 77 °F, the chirp rate is 2 to 3 per s. Length is 19–33 mm.

Ecology: Inhabits low, mucky ground and the edges of lakes and streams. In central Florida, they have a one-year life cycle, overwintering as adults, and calling primarily from January to August. In North and South Carolina, the life cycle is two years, with the first winter passed as mid-juveniles and the second as adults. Calling is from July to November. The life cycle farther north, has not been studied but it must require at least two years.

Similar species: The prairie mole cricket, *Gryllotalpa major*, is larger (length 35–50 mm) and louder (song easily heard from more than 30 m away), and its fore femur has a blade-like basal projection.

prairie mole cricket

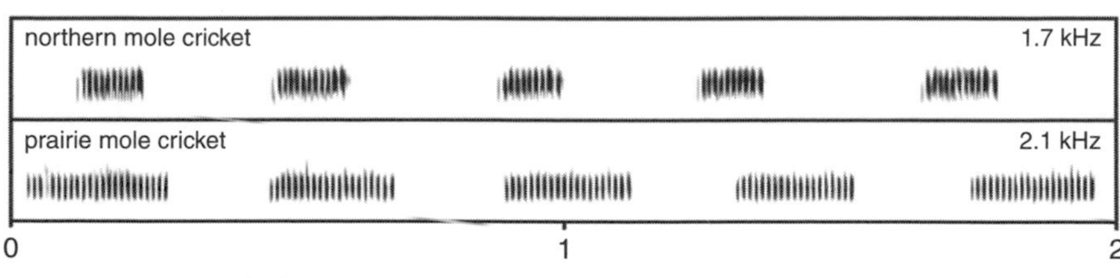

Northern mole cricket and prairie mole cricket song comparison

TAWNY MOLE CRICKET

Scapteriscus vicinus (Plate 47)

Distribution: Native to temperate South America and first reached North America at Brunswick, Georgia, in about 1900. It has since spread north to southern North Carolina, south to southern Florida, and west to eastern Texas.

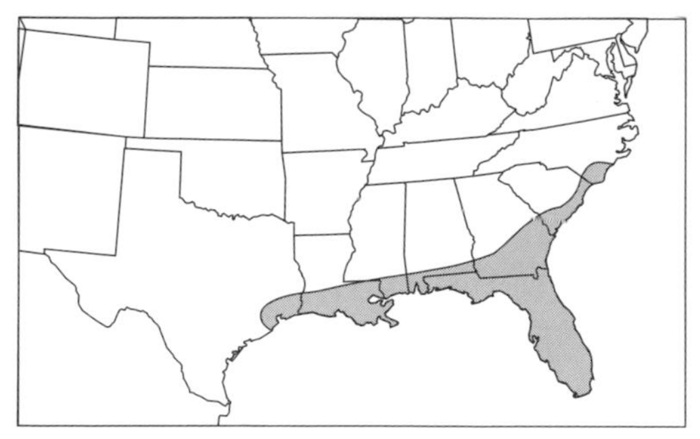

tawny mole cricket

Identification: There are four digging claws (dactyls) on each foreleg. Two project from the tibia and the other two project from the first and second segments of the tarsus. The gap between the tibial dactyls is less than half the basal width of either dactyl—that is, they nearly touch at the base. Pronotal pattern is complex. Forewings are longer than the pronotum, and the hind wings are longer than the abdomen. The lower abdominal surface is tawny in color. Males make a horn-shaped opening in the ground and call from it during the first 90 minutes after sunset. The song is a loud, nasal trill with a very high pulse rate: about 135 per s at 77 °F. Length is 24–33 mm.

Ecology: Underground inhabitant of open, grassy areas and a major pest of pastures, lawns, and golf courses. This species damages grasses by feeding on the blades and by cutting the roots as it tunnels near the surface. On the first warm evenings in spring, tawny mole crickets fly from their burrows to disperse and find mates. Before their populations were reduced by classical biological control (see Introduction), these flying crickets were sometimes attracted to lights in such enormous numbers that they stopped lighted sports events and once closed a major theme park for an evening. There is a single generation annually with eggs laid in the spring. Most individuals overwinter as adults, but some overwinter as large nymphs. Calling is primarily in spring, but because adults are long-lived, some occurs throughout the year.

Similar species: In the southern mole cricket, *Scapteriscus borellii*, and the short-winged mole cricket, *S. abbreviatus*, the tibial dactyls are separated at the base by a gap equal to at least half the basal width of either dactyl. The abdomen of the southern mole cricket is gray beneath, and its dark-gray pronotum usually has four dots of light gray. Its call is a low-pitched, ringing trill at a pulse rate of about 55 per s at 77 °F. The short-winged mole cricket has the forewings shorter than the pronotum and the hind wings concealed by the forewings. It has no calling song.

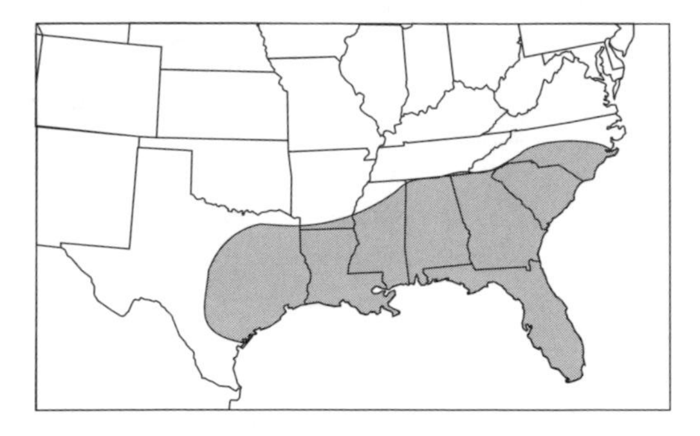

southern mole cricket

short-winged mole cricket

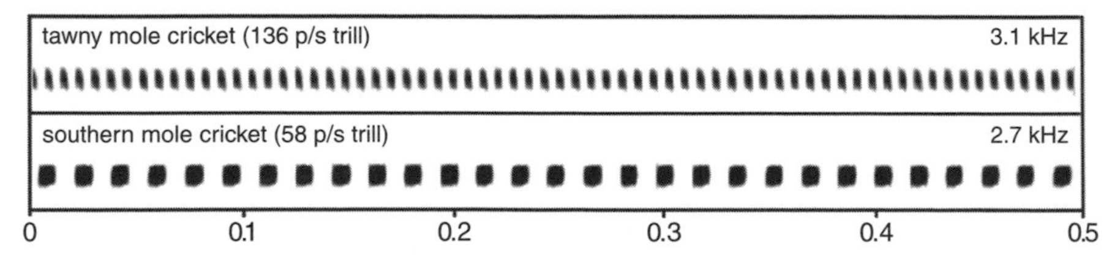

Tawny mole cricket and southern mole cricket song comparison

Gryllacridoids

SUPERFAMILY GRYLLACRIDOIDEA

Gryllacridoids resemble katydids in having four-segmented tarsi and blade-like ovipositors, but none from north of Mexico have fore-tibial ears or wings. They are usually brownish or grayish, sometimes patterned, but never green. North American species belong to four families, three of which are treated here. Members of the remaining family, Anostostomatidae (silk-spinning sand-crickets), resemble camel crickets but have the antennae more widely separated at the base and, like Carolina leaf-rollers, spin silk from the mouth. They are known only from mainland southern California and California's Channel Islands.

Camel Crickets

Family Rhaphidophoridae

Camel crickets have antennae that touch or nearly touch at the base. The tarsi are laterally compressed and lack ventral pads. The hind femur extends beyond the tip of the abdomen. Most are somewhat humpbacked. Length is 9–35 mm.

Camel crickets are active at night. During daylight, they remain in many types of shelters, including basements, caves, animal burrows, tree holes, and burrows that they make themselves; they also rest under logs, stones, and wooden bridges.

The approximately 150 species of camel crickets in the United States and Canada are distributed among 21 genera. Some 89 of the species, including the ones that are most frequently encountered, belong to the genus *Ceuthophilus*.

WALKER'S CAMEL CRICKET

Ceuthophilus walkeri (Plate 48)

Distribution: Found from central Florida northward in the Atlantic coastal plain to North Carolina.

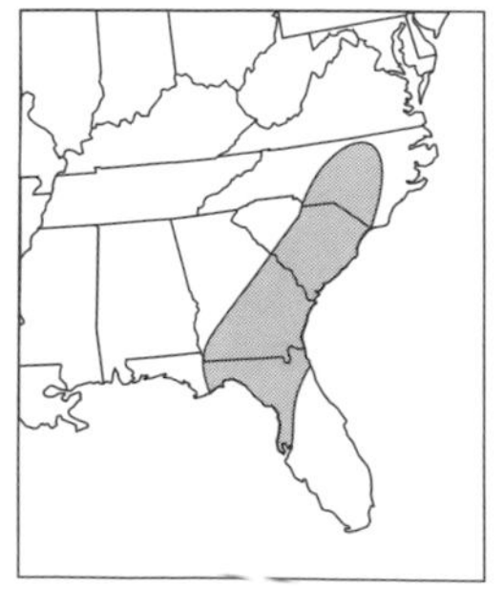

Walker's camel cricket

Identification: A reddish-brown species that has only one non-terminal spine on the ventral surface of the extended hind tibia. Lower blades of the ovipositor end with four ventral teeth and a terminal hook.

Ecology: Partial to habitats with dry, sandy, well-drained soils, such as scrub oak and upland longleaf pine. Shelters include both occupied and unoccupied burrows of the gopher tortoise. It matures in fall and overwinters as an adult. No adults have been taken from mid-April through late October.

Similar species: DeFuniak camel crickets, *Ceuthophilus armatipes*, and spinose camel crickets, *C. spinosus*, although morphologically similar to Walker's camel cricket, occur in localities west of the range of this species.

DeFuniak camel cricket

spinose camel cricket

Jerusalem Crickets

Family Stenopelmatidae

In Jerusalem crickets, the antennae are widely separated at the base, the head is large, and the front of the pronotum is wide. The tibiae are stout with strong spines for digging, and the tarsal segments have pads underneath. The hind femora do not reach beyond the tip of the abdomen. The ovipositor is short and inconspicuous. Length is 21–69 mm.

Jerusalem crickets live in burrows and under rocks and logs. At night they may come to the surface and wander. Adults, and sometimes nymphs, produce audible sequences of substrate vibrations by rhythmically striking the ground with the abdomen. Drumming patterns are species specific and, in adults, function in sexual pair formation.

D. B. Weissman is preparing a taxonomic revision of Jerusalem crickets and has concluded that at least 60 species of Jerusalem crickets occur in the United States, although only 14 have so far been described (Field 2001, p. 57). Thus, the dark Jerusalem cricket described below may prove to be not a single species but a complex of species.

DARK JERUSALEM CRICKET

Stenopelmatus fuscus (Plate 48)

Distribution: Found in the western United States west of the Central Plains.

Identification: In addition to the features listed above for the family, this species has a dark-brown abdomen with paler, buff-colored rear margins to the segments. The fore tibia has three narrow spines on its ventral surface in near-linear sequence as well as five broad-based spines at or near its tip. In addition to broad-based spines, the hind tibia has two side-by-side, narrow spines on its

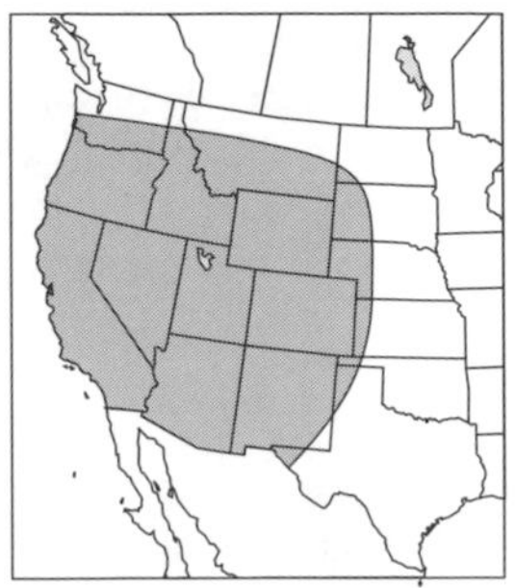

dark Jerusalem cricket

ventral surface near the tip. Adults and nymphs are difficult to distinguish, but adult males have a pair of black, incurved hooks on the sides of the supra-anal plate near the bases of the cerci, and adult females have the tips and ventral surfaces of the blades of the ovipositor dark and hard. Length is 35–50 mm.

Ecology: Occurs in a wide variety of habitats but not sand dunes or coastal areas. In the southern portion of its range, the life cycle requires at least two years; in the northern portion it probably requires at least three.

Similar species: Distinguishing dark Jerusalem crickets from similar species must await the publication of a taxonomic revision of the genus *Stenopelmatus.*

Raspy Crickets

Family Gryllacrididae

There is a single representative of this family in the United States, but more than 600 species occur worldwide. The vernacular name "raspy cricket" comes from Australia, which is home to about 200 species.

CAROLINA LEAF-ROLLER

Camptonotus carolinensis (Plate 48)

Carolina leaf-roller

Distribution: Found from New Jersey south to northern Florida and west to Louisiana and Missouri.

Identification: Small, smooth, yellowish-brown, rapidly running orthopteran with antennae that are at least five times the length of the body. The hind femur extends beyond the tip of the abdomen, and all but the terminal tarsal segments have paired pads underneath. Ovipositor is up-curved and carried over the back rather than directly behind. Length is 12–15 mm.

Ecology: Found on foliage in deciduous forests. During the day, individuals shelter in leaf rolls made by cutting into the edge of a leaf, folding over a flap, and holding the surfaces together with silk spun from the mouth. Somehow, the long antennae are completely contained within the shelter.

Pronunciation of Scientific Names

Scientists have agreed on a system to ensure that each species of animal and plant has a unique, two-part scientific name (genus and species), but they have not agreed on a system that specifies how these names (or those of higher taxonomic groups) should be pronounced. The challenge starts with the fact that scientific names are Latin words or words from other languages that are treated as Latin. It continues with the fact that there have been two widely taught systems for pronouncing Latin words. The earlier system was in vogue during the first two-thirds of the development of scientific nomenclature, when nearly all scientists took courses in Latin. By the time the current system came into style in the mid-twentieth century in English-speaking countries, few scientists were trained in Latin. As a result, there has been little replacement of the old-style pronunciations with the new; indeed, few scientists know either system.

Under the circumstances, it is not surprising that when scientists pronounce scientific names they generally apply neither system but pronounce the names as they have heard their teachers and colleagues pronounce them (who may have used the old system, the new system, or, most likely, neither system). Thus, you need not worry much about pronouncing scientific names "correctly." In most cases, the listener will figure out your intent, and likely will not embarrass you by suggesting a correction. Nonetheless, at the suggestion of our editors, we indicate below how U.S. orthopterists tend to pronounce the orthopteran names that apply to the genus level and higher. For those who wish to apply a system instead of relying on our suggested pronunciations, any modern textbook of Latin will explain the newer system, and a widely used entomology textbook by Borror et al. (1989) will explain the older one.

To pronounce a scientific name, you must know what syllable is stressed and how the consonants and vowels are sounded. In the suggested pronunciations below, we have capitalized the stressed syllable and used unambiguous consonants. Vowels are pronounced as follows:

a	pronounced as in mad, except at the end of a word, when it is pronounced as in idea
ā	pronounced as in made
au	pronounced as in August
e	pronounced as in send
ē	pronounced as in seek
i	pronounced as in pick
ī	pronounced as in pike
o	pronounced as in hop
ō	pronounced as in hope
oi	pronounced as in oil
u	pronounced as in fun
ū	pronounced as in puny

Names of Taxonomic Groups above the Genus Level

Acrididae	a KRID i dē
Acridinae	a krid Ī nē
Acridoidea	a kri DOI dē a
Caelifera	sē LIF e ra
Conocephalinae	kon ō sef a LĪ nē
Copiphorinae	kop i fo RĪ nē
Cyrtacanthacridinae	sir ta kan tha kri DĪ nē
Eneopterinae	ē nē op te RĪ nē
Ensifera	en SIF e ra
Gomphocerinae	gom fō se RĪ nē
Gryllacrididae	gril a KRID i dē
Gryllacridoidea	gri lak ri DOI de a
Gryllidae	GRIL i dē
Gryllinae	gri LĪ nē
Grylloidea	gri LOI dē a
Gryllotalpidae	gri lō TALP i dē
Meconematinae	mē cō nē ma TĪ nē
Mogoplistinae	mō gō plis TĪ nē
Myrmecophilinae	mir mē kof i LĪ nē
Nemobiinae	nē mō bi Ī nē
Oecanthinae	ē kan THĪ nē
Oedopodinae	ē dō pō DĪ nē
Orthoptera	or THOP ter a
Phaneropterinae	fan er op te RĪ nē
Prophalangopsidae	prō fā lan GOPS i dē
Pseudophyllinae	sū dō fi LĪ nē
Rhaphidophoridae	ra fid ō FOR i dē
Romaleinae	rō mā lē Ī nē
Stenopelmatidae	sten ō pel MAT i dē
Tetrigidae	te TRIG i dē
Tettigoniidae	tet ti gō NĪ i dē
Tettigoniinae	tet ti gō ni Ī nē
Tettigonioidea	tet ti gōn i OI dē a
Tridactylidae	trī dak TIL i dē
Trigonidiinae	trig ō nid i Ī nē

Names of Genera

Acheta	a KĒ ta
Achurum	a KŪ rum
Acrolophitus	a krō lō FĪT us
Aeoloplides	ē ō LOP li dēz
Aeropedellus	ēr ō PED el us
Ageneotettix	a gēn ē ō TET iks
Aglaothorax	ag lā ō THŌ raks
Allonemobius	al lō nē MŌ bi us
Amblycorypha	am bli CŌR i fa
Amblytropidia	am bli trō PID i a
Amphitornus	am fi TOR nus
Anabrus	a NAB rus
Anaxipha	a NAKS i fa
Anurogryllus	a nūr ō GRIL us
Aptenopedes	ap ten ō PED es
Arethaea	ar ē THĒ a
Arphia	ARF i a
Atlanticus	at LANT i kus
Aulocara	au LOK a ra
Belocephalus	bel ō SEF a lus
Boopedon	bō ŌP e don
Bootettix	bō ō TET iks
Brachystola	brak i STŌ la
Bradynotes	brad i NŌ tēz
Bucrates	bū KRĀ tēz
Camnula	kam NŪ la
Camptonotus	kam tō NŌ tus
Campylacantha	kam pē la KAN tha
Capnobotes	cap nō BŌ tēz
Ceuthophilus	sū THOF i lus
Chimarocephala	ki mar ō SEF a la
Chloealtis	klē AL tis
Chorthippus	kor THIP us
Chortophaga	kor TOF a ga
Circotettix	sir kō TET iks
Conocephalus	con ō SEF a lus
Conozoa	kon o ZŌ a
Cordillacris	kor di LAK ris
Cratypedes	kra ti PĒ dēz
Cycloptilum	sī KLOP ti lum
Cyphoderris	sī fō DĒR ris
Cyrtoxipha	sir TOKS i fa
Dactylotum	dak ti LŌ tum
Derotmema	der ot MĒ ma
Dichopetala	dī kō PET ā la
Dichromorpha	dī krō MOR fa
Dissosteira	dis ō STĪR a
Dracotettix	drak ō TET iks
Ellipes	EL li pēz
Encoptolophus	en cop tō LŌ fus
Eotettix	ē ō TET iks
Eritettix	er i TET iks
Gryllodes	gri LŌ dēz
Gryllotalpa	gri lō TAL pa
Gryllus	GRIL us
Gymnoscirtetes	jim no SKIR tē tēz
Hadrotettix	ha drō TET iks
Hapithus	ha PĪ thus
Heliastus	hē li AS tus
Heliaula	hē li AU la

Hesperotettix	hes per ō TET iks
Hippiscus	hi PIS kus
Hippopedon	hi PŌ pe don
Hubbellia	hu BEL i a
Hypochlora	hī pō KLOR a
Insara	in SĀ ra
Inscudderia	in sku DĒR i a
Lea	LĒ a
Leprus	LEP rus
Leptysma	lep TIS ma
Ligurotettix	li gur ō TET iks
Meconema	mē cō NĒ ma
Melanoplus	me LAN ō plus
Mermiria	mer MIR i a
Mestobregma	mes tō BREG ma
Metaleptea	met a LEP te a
Metator	me TĀ tor
Metrioptera	me tri OP ter a
Microcentrum	mī krō SEN trum
Montezumina	mon te ZŪM i na
Myrmecophilus	mir mē KOF i lus
Neobarrettia	nē ō ba RET i a
Neoconocephalus	nē ō kon ō SEF ā lus
Neocurtilla	nē ō kur TIL la
Odontoxiphidium	ō don tō zi FID i um
Oecanthus	ē KAN thus
Oedaleonotus	ēd a lē ō NŌ tus
Opeia	ō PĪ a
Orchelimum	or KEL i mum
Orocharis	ō ROK a ris
Orphulella	or fū LEL a
Paracyrtophyllus	par a sir TOF i lus
Parapomala	par a POM a la
Paratettix	par a TET iks
Pardalophora	par da LOF for a
Paroxya	pa ROKS i a
Pediodectes	pē di ō DEK tēz
Peranabrus	per a NAB rus
Phlibostroma	fli BOS trō ma
Phoetaliotes	fē TAL i ō tēz
Phrynotettix	fri nō TET iks
Phyllopalpus	fil ō PAL pus
Platylyra	plat i LĪR a
Poecilotettix	pē sil ō TET iks
Pseudopomala	sū dō POM a la
Psinidia	si NID i a
Psoloessa	sō LĒS a
Pterophylla	ter ō FIL a
Pyrgocorypha	pir gō COR i fa
Pyrotettix	pī rō TET iks
Romalea	rō MĀL e a
Scapteriscus	skap te RIS kus
Schistocerca	shis tō SIR ka
Scudderia	sku DĒR i a
Spharagemon	sfa RAG e mon
Stenacris	sten AK ris
Stenopelmatus	sten ō PEL ma tus
Stethophyma	steth ō FĪ ma
Stilpnochlora	stilp nō KLŌR a
Syrbula	SIR bu la
Taeniopoda	tē nē OP a da
Tettigidea	te tig i DĒ a
Trachyrhachys	trak i RAK is
Trepidulus	tre PID ū lus
Trimerotropis	trī mer ō TRŌ pis
Tropidolophus	trō pid ō LŌ fus
Turpilia	tur PIL i a
Velarifictorus	ve lar i fik TŌR us
Xanthippus	zan THIP pus

Glossary

abdomen The third or posterior division of the insect body, the other divisions being the head and thorax.

acridoid An insect belonging to the superfamily Acridoidea.

Acridoidea The superfamily of Orthoptera, in the suborder Caelifera, that contains the grasshoppers.

acute angle An angle of less than 90 degrees.

annual An herb that lives for only one growing season.

antenna (*pl*, antennae) Paired, elongate sensory structures located on the head.

antennal socket A recessed area at the point of attachment of the antenna to the head.

anterior Referring to the front or forward position.

apical Referring to the tip or distal portion.

band A narrow or broad line, of contrasting color, oriented across the length of the body.

basal Referring to the base or point of origin.

broadleaf Having a flat, wide leaf structure.

Caelifera The suborder of Orthoptera that contains the grasshoppers.

caeliferan An insect belonging to the suborder Caelifera.

calling song A sound made prior to close contact that attracts the other sex of the same species.

carina (*pl*, carinae) An elevated, longitudinal ridge on the pronotum.

carrier frequency The sound frequency (pitch) of maximum power in a cricket or katydid song.

cercus (*pl*, cerci) One of a pair of appendages near the posterior end of the abdomen.

conspecific Belonging to the same species.

crepitation Crackling sound produced during flight.

dactyl A digging claw on the fore tibia or fore tarsus of a mole cricket.

distal Pertaining to the portion away from the base or point of origin.

dorsal Pertaining to the "back," which in orthopterans is the upper surface.

Ensifera The suborder of Orthoptera that includes katydids, crickets, and their close relatives.

ensiferan An insect belonging to the suborder Ensifera.

femur (*pl*, femora) The third and most stout segment of the leg; the "thigh."

forb A broadleaf herb.

forewings The front pair of wings, closest to the head, and attached to the mesothoracic body segment. In Orthoptera, the forewing is often called the tegmen (*pl*, tegmina). The forewings are on top of the hind wings when the insect is not in flight; in some katydids and crickets they are shorter than the hind wings.

frontal ridge A broad, flat ridge or elevated region on the front of the head, extending from the eyes to above the mouth.

furcula A forked process at the posterior of male grasshoppers that overlies the supra-anal plate; only the tips of the forks are visible, so it appears to be a paired structure.

grass Common name for plants in the monocotyledenous family Graminae.

gryllacridoid An insect belonging to the superfamily Gryllacridoidea.

Gryllacridoidea The superfamily of Orthoptera, in the suborder Ensifera, that contains close relatives of katydids and crickets.

grylloid An insect belonging to the superfamily Grylloidea.

Grylloidea The superfamily of Orthoptera, in the suborder Ensifera, that contains the crickets.

head The forward or anterior division of the insect body, followed by the thorax and abdomen.

herb A nonwoody, short-lived plant.

herbivore A plant-feeding animal.

herbivorous Feeding on plants.

hind wings The second pair or posterior wings, attached to the metathoracic body segment; in orthopterans capable of flight, these are the larger wings, but they fold and may be hidden beneath the forewings when the insect is at rest.

horizontal Oriented in a plane parallel to the horizon, or along the length of the body.

instar The stage of the insect between molts; this term is applied to nymphal orthopterans, usually in combination with a number to indicate whether it is early in development (e.g., 1st) or late (e.g., 5th).

lateral Relating to the side.

lateral lobes (of pronotum) The portions of the pronotum that extend down either side from the pronotal disk (or top of the pronotum).

leg Appendage associated with the thorax that is used for terrestrial locomotion.

length In grasshoppers with wings extending beyond the tip of the abdomen, the distance from the front of the head to the tip of the wings; in grasshoppers with abbreviated wings, the distance from the front of the head to the tips of the hind femora; in katydids and crickets where wings do not extend beyond the tip of the abdomen, the maximum measurement from the head to the tip of the abdomen, excluding the ovipositor; in katydids (except false katydids) and crickets where the wings extend beyond the tip of the abdomen, the distance from the head to the tip of the forewings (positioned at rest); in false katydids (Phaneropterinae), the distance from the head to the tips of the hind wings. (Some authors measure length differently; see Introduction.)

long-winged In grasshoppers, wings extending to about the tip of the abdomen or beyond; in crickets, having hind wings that project from beneath the forewings at rest.

mandible One of a pair of stout jaw-like structures used for chewing, and which open and close from side to side in orthopterans.

maxillary palp One of a pair of feeler-like appendages that attach to mouth parts immediately behind the mandibles.

mesic Moist but not wet.

metamorphosis Change in body form (e.g., from a caterpillar to a moth, or, in orthopterans, from a nymph to an adult).

molt To cast off the outgrown exterior body covering; a process that occurs between instars and between nymph and adult stages.

morph A morphological variant; sometimes with reference to different color forms.

morphology The form and structure of organisms.

multivoltine Having more than one generation (complete life cycle) per year.

nymph The young of a hemimetabolous insect (one not undergoing radical change in body form), such as any orthopteran; an immature insect that resembles the adult in body form, differing principally in wing development and reproductive capabilities.

omnivorous Feeding on many kinds of food, including both plant and animal matter.

Orthoptera The order of insects consisting of grasshoppers, katydids, crickets, and their relatives.

orthopteran A member of the order Orthoptera.

ovipositor The structures located at the tip of the abdomen in females and used to deposit eggs.

pallium An erect conical structure near the tip of the abdomen.

perennial A plant living several years, not dying soon after reproduction.

posterior Referring to the back or rear position.

pronotal disk The upper or dorsal part of the pronotum, often set off from the sides or lateral lobes of the pronotum by a longitudinal ridge (carina) on either side.

pronotum The upper or dorsal part of the prothorax, the first of the three thoracic segments; the pronotum is often divided by ridges into a dorsal pronotal disk and side pieces called lateral lobes; in orthopterans, the pronotum largely hides the other upper parts of the thorax.

prosternal spine A small spur or spine located ventrally on the first thoracic segment; a spine protruding from between the front legs, especially in the subfamilies Cyrtacanthacridinae and Romaleinae.

prothorax The first or anterior segment of the three thoracic segments.

pulse A physical unit of sound; each pulse in the calling songs of crickets is a brief, nearly pure tone made as the scraper and file engage on the closing stroke of an opening and closing cycle of the forewings (no sound is made on the opening stroke).

scraper The hardened edge of a forewing that engages the stridulatory file on the opposite wing during stridulation (sometimes called the plectrum).

short-winged in grasshoppers, wings not extending to the tip of the abdomen or beyond; in crickets, having hind wings that do not project from beneath the forewings at rest.

shrub A woody plant, typically with several major stems.

sound spectrogram A plot of sound frequency against time (also known as a sonogram or audiospectrogram).

sp. (*pl*, spp.) Abbreviation for species.

speciation The evolution of new species.

species Group of actually or potentially interbreeding populations that is reproductively isolated from other such groups.

spine Elongate, immovable pointed structure; in spurthroated grasshoppers and lubbers, a large spine called the prosternal spine is present between, or slightly in front of, the front legs, but it is sometimes reduced to a bluntly rounded elevation. Small but sharply pointed spines occur in rows along the tibiae in many groups.

spur A moveable spine-like structure found at the tip of the hind tibiae.

stridulation Sound production by rubbing two hard surfaces together; for example, the inner surface of the hind femur on the edges of the forewing in grasshoppers, or the file and scraper of the forewings in crickets and katydids.

stridulatory file A specialized portion of a vein on the forewing of crickets and katydids that bears a series of downward-projecting teeth that are engaged by the scraper during stridulation.

stridulatory vein The vein on the forewing that bears the stridulatory file.

stripe A narrow or broad line, of contrasting color, oriented along the length of the body.

subgenital plate A plate at the tip of the abdomen that covers the genital area from below.

suborder In taxonomy, a major division of an order; the suborders of the order Orthoptera are Caelifera and Ensifera.

superfamily A taxonomic category that is less inclusive than a suborder and more inclusive than a family. In this book, the superfamilies of the suborder Caelifera are Acridoidea (true grasshoppers), Tetrigoidea (pygmy grasshoppers), and Tridactyloidea (pygmy mole crickets); and the superfamilies of the suborder Ensifera are Tettigonioidea (katydids), Grylloidea (crickets), and Gryllacridoidea (camel crickets, Jerusalem crickets, and raspy crickets).

supra-anal plate A plate at the tip of the abdomen that covers the genital area from above.

sword-shaped antenna Antenna with flattened segments, expanding in width from the base and then decreasing in width toward the tip.

tarsus (*pl*, tarsi) The distal segment of the leg, the "foot."

taxon (*pl*, taxa) In taxonomy, a related group of organisms (for example, all the individuals belonging to an order, family, or subfamily).

tegmen (*pl*, tegmina) The thickened forewings.

tegminal Pertaining to the forewings or tegmina.

Tettigonioidea The superfamily of Orthoptera, in the suborder Ensifera, that contains the katydids.

thorax The second, or middle, of the three major body divisions of insects.

tibia (*pl*, tibiae) The long, thin fourth segment of the leg, between the femur and the tarsus.

toothstrike The impact of the scraper on a tooth of the stridulatory file.

tympanal organ An insect auditory organ that includes an eardrum-like tympanum.

tympanum (*pl*, tympana) A tightly stretched membrane located on the side of the abdomen (in grasshoppers) or on the front tibiae (katydids and crickets) that vibrates in response to airborne sounds.

ultrasonic Pertaining to sound frequencies too high to be heard by young, healthy human ears—that is, above 16–20 kHz.

univoltine Having one generation (complete life cycle) per year.

ventral Pertaining to the underside or below.

vertical Oriented up and down.

waveform A plot of sound amplitude against time; equivalent to an oscilloscopic display.

wing pads The partly developed wings, located at the juncture of the thorax and abdomen.

xeric Dry.

Additional Reading

(*) indicates that keys are included

Alexander R. D., A. E. Pace, and D. Otte. 1972. The singing insects of Michigan. Great Lakes Entomologist 5: 33–69. (*)

Bailey, W. J., and D. C. F. Rentz, eds. 1990. The Tettigoniidae: biology, systematics, and evolution. Springer-Verlag, Berlin. 395 pp.

Ball, E. D., E. R. Tinkham, R. Flock, and C. T. Vorhies. 1942. The grasshoppers and other Orthoptera of Arizona. University of Arizona Agricultural Experiment Station Technical Bulletin 93: 257–373. (*)

Bland, R. G. 2003. The Orthoptera of Michigan—biology, keys, and description of grasshoppers, katydids, and crickets, Michigan State University Extension Bulletin E-2815. 220 pp. (*)

Blatchley, W. S. 1920. Orthoptera of north eastern North America. Nature Publishing Co., Indianapolis, Indiana. (*)

Borror, D. J., C. A. Triplehorn, and N. F. Johnson. 1989. Introduction to the study of insects, 6th ed. Saunders, Philadelphia. 875 pp.

Capinera, J. L., ed. 1987. Integrated pest management on rangeland. A shortgrass prairie perspective. Westview Press, Boulder, Colorado. 426 pp.

Capinera, J. L., and T. S. Sechrist. 1982. Grasshoppers (Acrididae) of Colorado: Identification, biology, and management. Colorado State University Experiment Station Bulletin 584S. 161 pp. (*)

Capinera, J. L., C. W. Scherer, and J. M. Squitier. 1999. Grasshoppers of Florida (http://www.ifas.ufl.edu/~entweb/ghopper/ghopper.html). 70 pp. (*)

Capinera, J. L., C. W. Scherer, and J. M. Squitier. 2001. Grasshoppers of Florida. University Press of Florida. Gainesville. 143 pp. (*)

Chapman, R. F., and A. Joern, eds. 1990. Biology of grasshoppers. John Wiley and Sons, New York. 563 pp.

Coppock, S. 1962. Grasshoppers of Oklahoma. Oklahoma State University Agricultural Experiment Station Processed Series P-399. (*)

Dakin, M. E., Jr., and K. L. Hays. 1970. A synopsis of Orthoptera (*sensu lato*) of Alabama. Auburn University Agricultural Experiment Station Bulletin 404. 118 pp. (*)

Dethier, V. G. 1992. Crickets and katydids, concerts and solos. Harvard University Press, Cambridge. 140 pp.

Field, L. H., ed. 2001. The biology of wetas, king crickets and their allies. CABI, New York. 540 pp.

Froeschner, R. C. 1954. The grasshoppers and other Orthoptera of Iowa. Iowa State College Journal of Science 29: 163–354. (*)

Gangwere, S. K. 1961. A monograph on food selection in Orthoptera. Transactions of the American Entomological Society 87: 67–230.

Gangwere, S. K., M. C. Muralirangan, and M. Muralirangan, eds. 1997. The bionomics of grasshoppers, kaytdids, and their kin. CAB International, New York. 529 pp.

Goettel, M. S., and D. L. Johnson, eds. 1997. Microbial control of grasshoppers and locusts. Memoirs of the Entomological Society of Canada 171: 1–400.

Gwynne, D. T. 2001. Katydids and bush-crickets: reproductive behavior and evolution of the Tettigoniidae. Cornell University Press, Ithaca, New York. 317 pp.

Hanford, R. H. 1946. Identification of nymphs of the genus *Melanoplus* of Manitoba and adjacent areas. Scientific Agriculture 26: 147–180. (*)

Hebard, M. 1925. The Orthoptera of South Dakota. Proceedings of the Academy of Natural Sciences of Philadelphia 77: 33–155.

Hebard, M. 1928. The Orthoptera of Montana. Proceedings of the Academy of Natural Sciences of Philadelphia 80: 211–306.

Hebard, M. 1929. The Orthoptera of Colorado. Proceedings of the Academy of Natural Sciences of Philadelphia 81: 303–425.

Hebard, M. 1931. The Orthoptera of Kansas. Proceedings of the Academy of Natural Sciences of Philadelphia 83: 119–227.

Hebard, M. 1932. The Orthoptera of Minnesota. Minnesota Agricultural Experiment Station Technical Bulletin 85. 61 pp. (*)

Hebard, M. 1934. The Dermaptera and Orthoptera of Illinois. Bulletin Illinois Natural History Survey 20:125–279. (*)

Hebard, M. 1936. The Orthoptera of North Dakota. North Dakota Agricultural Experiment Station Technical Bulletin 284. 69 pp. (*)

Helfer, J. R. 1972. The grasshoppers, cockroaches, and their allies. W. C. Brown Company, Dubuque, Iowa. Repr., Dover Publications, 1987. 359 pp. (*)

Hubbell, T. H. 1936. A monographic revision of the genus *Ceuthophilus* (Orthoptera, Gryllacrididae, Rhaphidorphorinae). University of Florida Publications in Biological Science Vol. 2, No. 1. 551 pp.

Hubbell, T. H. 1960. The sibling species of the Alutacea group of the bird-locust genus *Schistocerca* (Orthoptera, Acrididae, Cyrtacanthacridinae). Museum of Zoology, University of Michigan, Miscellaneous Publication 116.

Huber, F., T. E. Moore, and W. Loher, eds. 1989. Cricket behavior and neurobiology. Cornell University Press, Ithaca, New York. 565 pp.

Lomer, C. J., and C. Prior, eds. 1992. Biological control of locusts and grasshoppers. CAB International, Wallingford, United Kingdom. 394 pp.

McDaniel, B. 1987. Grasshoppers of South Dakota. South Dakota State University Agricultural Experiment Station TB 89. 163 pp. (*)

Miller, C. K., R. L. Knight, L. C. McEwen, and T. L. George. 1994. Responses of nesting savannah sparrows to fluctuation in grasshopper densities in interior Alaska. Auk 111: 962–969.

Mulkern, G. B., K. P. Pruess, H. Knutson, H. F. Hagen, J. B. Campbell, and J. D. Lambley. 1969. Food habits and preferences of grassland grasshopper species of the north central Great Plains. North Dakota State University Agricultural Experiment Station Bulletin 481. 31 pp.

Otte, D. 1981. The North American grasshoppers. Vol. 1, Acrididae: Gomphocerinae and Acridinae. Harvard University Press, Cambridge. 275 pp. (*)

Otte, D. 1984. The North American grasshoppers. Vol. 2, Acrididae: Oedipodinae. Harvard University Press, Cambridge. 366 pp. (*)

Otte, D. 1994–2000. Orthoptera species file, Vols. 1–8. Orthopterists Society. Philadelphia.

Otte, D., D. C. Eades, and P. Naskrecki. 2004. Orthoptera species file online (version 2) (http://osf2.orthoptera.org/basic/HomePage.asp).

Pfadt, R. 2002. Field guide to common western grasshoppers. Wyoming Agricultural Experiment Station Bulletin 912.

Rees, N. E. 1973. Arthropod and nematode parasites, parasitoids, and predators of Acrididae in America north of Mexico. United States Department of Agriculture Technical Bulletin 1460. 288 pp.

Rehn, J. A. G., and H. J. Grant. 1961. A monograph of the Orthoptera of North America (north of Mexico). Vol. 1. Academy of Natural Sciences of Philadelphia Monograph 12. 255 pp. (*)

Richman, D. G., D. C. Lightfoot, C. A. Sutherland, and D. J. Ferguson. 1993. A manual of the grasshoppers of New Mexico. Orthoptera: Acrididae and Romaleidae. New Mexico State University Cooperative Extension Service Handbook 7. 112 pp. (*)

Strohecker, H. F., W. W. Middlekauff, and D. C. Rentz. 1968. The grasshoppers of California. California Insect Survey Bulletin 10. 177 pp. (*)

Uvarov, B. 1966. Grasshoppers and locusts. A handbook of general acridology. Vol. 1. Cambridge, United Kingdom. 481 pp.

Uvarov, B. 1977. Grasshoppers and locusts. A handbook of general acridology. Vol. 2. Cambridge, United Kingdom. 613 pp.

Vickery, V. R., and D. K. McE. Kevan. 1985. The grasshoppers, crickets, and related insects of Canada and adjacent regions. The insects and arachnids of Canada, Part 14. Agriculture Canada. 918 pp. (*)

Walker, T. J., and T. E. Moore. 2004. Singing insects of North America (http://buzz.ifas.ufl.edu/). (*)

Index

Species are listed by scientific name (as genus species and as species, genus) and by common name.